The Intext Series in ADVANCED MATHEMATICS
under the consulting editorship of

RICHARD D. ANDERSON
Louisiana State University

ALEX ROSENBERG
Cornell University

Introduction to
  Real Variable Theory

Introduction to Real Variable Theory

S. C. SAXENA
The University of Akron

S. M. SHAH
University of Kentucky

INTEXT EDUCATIONAL PUBLISHERS
College Division of Intext
Scranton   San Francisco   Toronto   London

ISBN 0-7002-2393-2

Copyright © 1972, International Textbook Company

Library of Congress Catalog Card Number: 72-185820

Preface

Those topics which generally are included under the title of Real Variables seem to form the foundation of mathematical analysis. This book is designed for use in advanced undergraduate courses in Real Variables or in that part of Advanced Calculus which deals with introductory Real Variable theory. Deep work in nearly every branch of mathematics requires a good knowledge of these topics. This book is intended to be a readable, but rigorous introduction to this subject. While there is emphasis on rigor and careful argument, there are also many examples and a great deal of discussion which is designed to build the reader's understanding and insight. It is the authors' hope that once the student has worked through the text, he (or she) need not only have been exposed to a beautiful part of mathematics, but will also have begun to develop that very important faculty—mathematical intuition.

The book begins with an introduction to sets, relations, and mappings. These topics may be covered quickly depending upon the background of the students, but topics on linear order relation should not be ignored.

Chapter 2 starts with the development of real numbers through Dedekind cuts. Here an instructor may find it expedient to skip the details and accept Cantor Dedekind theorem (Theorem 2.2.1) as a postulate. However, the time spent on Dedekind cuts is worthwhile. This chapter also introduces bounds and the greatest lower bound property. It ends with an introduction to complex numbers and to R^n.

In Chapter 3 the ideas of infinite sets and cardinal numbers are introduced. The emphasis would depend on instructor's taste and the needs of the students. The discussion of ordinal numbers (Section 3.2) is optional.

Chapter 4 introduces sequences and includes the concepts of limit superior and limit inferior. The chapter has a section on sequences in R^n.

Chapter 5 is probably the most important chapter in the book. The ideas of limit points, open sets, closed sets and metric spaces are discussed. Also, Bolzano-Weierstrass theorem, Heine-Borel theorem, and Cantor intersection theorem are proved. The concept of connectedness of a metric space is introduced. The chapter ends with the extension of these topics in R^n.

One may pass on to Chapter 7 from Chapter 5 or may discuss some topics of Chapter 6 which include perfect sets (including the Cantor set), Borel sets and Baire Category theorem. Chapter 7 contains the important topics of continuity, uniform continuity, and upper and lower semicontinuity. Continuity and connectedness in R^n are also discussed in this chapter.

Infinite series are considered in Chapters 8 and 11. One may come to

Chapter 8 right after Chapter 4. Some knowledge of integration (Chapter 9) is needed for Chapter 11. Chapter 9 deals with derivatives (including Dini derivatives) and Riemann integration.

In Chapter 10 we prove some inequalities of frequent occurrence in analysis and applications, and give a brief introduction to convex functions and improper integrals. The notations capital order O, small order o and asymptotic equivalence $\sim$ (due to E. Landau) are also given.

Chapter 11 includes sections on uniform convergence, Cesàro and Abel summability methods and a brief account of elementary transcendental functions (section 11.7 and 11.8). In earlier chapters we have assumed some properties of these functions to illustrate the theory there; but our account in sections 11.7 and 11.8 is independent of these properties. It may be noticed that the series given in the beginning of section 11.8 converge for all complex numbers and may be utilized to define the hyperbolic and circular functions of a complex variable. A similar remark applies to the series in (11.20) for the exponential function.

Chapters 12 and 13 deal with elementary measure theory and Lebesgue integration. Some knowledge of Chapter 6 is required. The idea of Lebesgue measure is introduced as a natural extension of the length of an interval. Measurability is defined in terms of external (outer) measure and internal (inner) measure of a set. We find it convenient to develop the basic properties of measure by restricting ourselves to bounded sets and then generalizing the same properties to unbounded sets.

Chapter 14 deals with Fourier series. Weierstrass theorems on approximating a continuous function by a trigonometric polynomial or an algebraic polynomial are proved.

Each chapter contains examples and exercises, some with hints. A reference such as Exercise 10.2 (7) indicates problem 7 in exercise 10.2.

In expressing our thanks the first person that comes to our mind and who deserves our sincere gratitude is Mr. Lance K. Parks. A former student—Mr. Parks not only made numerous stylistic changes, but also offered several mathematical suggestions. Another former student, Dr. Amy King, deserves our thanks for going through a good part of the manuscript and for several corrections. Thanks are also due to the reviewers of Intext Educational Publishers for their suggestions and generous reviews with special mention of Professor Richard D. Anderson who made several significant comments. A long list of typists deserves many thanks with special mention of (Mrs.) Wanda Jones and Sue Linn. Finally, we must thank Intext Educational Publishers and their Mathematics Editor, Mr. Charles J. Updegraph, for extreme patience and great cooperation.

S. C. Saxena
S. M. Shah

March 1972

Interdependence of Chapters

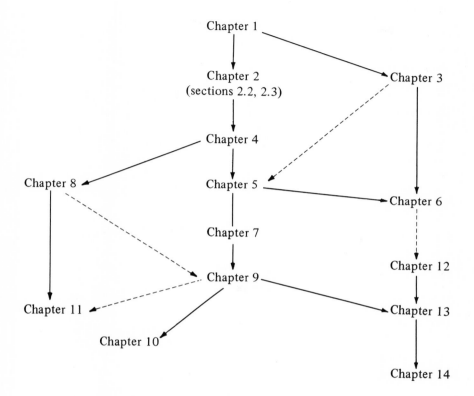

Dotted arrows indicate slight dependence on the chapter where the arrow starts, and the solid arrows imply heavy dependence.

Contents

Introduction to
  Real Variable Theory

1

Sets, Mappings, Relations and Linear Orders

1.1. Sets

Most courses and textbooks in Mathematics in post-sputnik era begin with some discussion of set theory. This book is no exception, for an adequate knowledge of rudimentary set theory is indispensable to the study of "Modern Analysis."

We start with two primitive (or undefined) terms—"element" and "set"; and an undefined relationship between them. We shall say an element "is a member of" a set or equivalently an element "belongs to" a set. The terms "belongs to" or "is a member of" are synonymous, and the words set, aggregate, class, and collection, are synonymous; as are "element," "object," "point," "member."

Most of the time in this book an arbitrary set will be denoted by a capital or upper case Roman letter, and an element by a small* or lowercase Roman letter (occasionally by a small Greek letter). The relationship "x is a member of S" is denoted by "$x \in S$," and the negation of this relationship by "$x \notin S$." It is assumed here that for any element x and for any set S either $x \in S$ or $x \notin S$, and, of course, the two possibilities are mutually exclusive.

EXAMPLE 1.1.1. The following are some examples of sets.
1. The set $\mathfrak{N}$ of all natural numbers (or positive integers). Here $1 \in \mathfrak{N}$, $2 \in \mathfrak{N}$etc., but $-1 \notin \mathfrak{N}$, $0 \notin \mathfrak{N}$, $1/2 \notin \mathfrak{N}$.
2. The set $\mathcal{J}$ of all integers (positive, negative or zero). In this case $1 \in \mathcal{J}$, $0 \in \mathcal{J}$, $-1 \in \mathcal{J}$, but $1/2 \notin \mathcal{J}$.
3. The set $\mathcal{Q}$ of all rational numbers. Here $1 \in \mathcal{Q}$, $-1 \in \mathcal{Q}$, $2/5 \in \mathcal{Q}$, but $\sqrt{2} \notin \mathcal{Q}$.

---

*In the discussion of R^n (Secs. 2.4, 4.4, 5.6, 6.6, 7.5, and 7.6) the capital Roman letters will be used to denote the elements of R^n.

4. The set R of all real numbers.
5. The set $S = \{1,2,3,4\}$.

The last example differs from the first four in as much as we can list all the elements of the set. This is always true for a finite set.*

We can write a finite set in two ways, either by listing all the elements of the set or by using a variable x and stating a rule applied to that variable. For instance, in (5) of Example 1.1.1, $S = \{x : x \in \mathfrak{N} \text{ and } x < 5\}$. This is read as S equals the set of all x such that x is an element of $\mathfrak{N}$ and x is less than 5. We shall come across the examples of the first four sets quite often, and we shall use the symbols $\mathfrak{N}$, g, $\mathcal{Q}$, R respectively, for these four sets, unless otherwise specified.

Now a few definitions.

Definition 1.1.1. *Let A and B be two sets. If every element of A is a member of B, then we say "A is contained in B" and write $A \subset B$ or $B \supset A$ (the last notation reads "B contains A").*

Symbolically, we may write $A \subset B$ iff $x \in A \Longrightarrow x \in B$. If $A \subset B$, we sometimes say that A is a subset of B. In the Example 1.1.1 we notice $S \subset \mathfrak{N} \subset \mathit{g} \subset \mathcal{Q} \subset R$. (Here and later, we use "iff" to mean "if and only if.")

Definition 1.1.2. *We define $A = B$ iff $A \subset B$ and $B \subset A$. In other words two sets are equal if every element of one is a member of the other and vice versa.*

Definition 1.1.3. *If $A \subset B$ and $A \neq B$, then we say that A is a proper subset of B.*

A is a proper subset of B iff $x \in A \Longrightarrow x \in B$, and $\exists$ at least one $y \in B$ such that $y \notin A$.

Definition 1.1.4. *The union of A and B (written as $A \cup B$) is defined as the set of all elements which are members of at least one of the two sets. $A \cup B = \{x : x \in A \text{ or } x \in B\}$.*

We may also write $x \in A \cup B \Longleftrightarrow x \in A$ or $x \in B$.

Definition 1.1.5. *The intersection of A and B (written as $A \cap B$) is the set of all those objects which belong to both A and B.*
Thus

$$A \cap B = \{x : x \in A \text{ and } x \in B\}$$

or

$$x \in A \cap B \Longleftrightarrow x \in A \text{ and } x \in B$$

---

*The terms "finite set" and "infinite set" will be defined in Chapter 3, but we assume here that the reader knows the meaning of the terms intuitively.

Definition 1.1.6. *The difference of A and B (written as A - B) is defined to be the set of all elements in A which are not in B:*

Definition 1.1.7. *If $B \subset A$, then $A - B$ is called the complement of B with respect to A.*

In a particular discussion we usually assume that every set A is a subset of a fixed set E (which may be called the entire set or universal set in the discussion). The set E consists of all the elements under consideration. Now if we consider the complement of A with respect to E we would simply call $E - A$ "the complement of A." We then write cA for the complement of A. In most parts of this book we are dealing with real numbers only, so that $E = R$. In this case the complement of Q, for example, is the set of all irrational numbers.

Definition 1.1.8. *A set is called empty or void if it has no member.*

Clearly, all empty sets are equal, so we usually refer to *the* empty set, and denote it by the symbol $\emptyset$.

An illustration of this set is the set of all integers which are positive as well as negative.

Definition 1.1.9. *Two sets A and B are called disjoint if their intersection is empty,* i.e., $A \cap B = \emptyset$.

The following are the theorems for union, intersection, and complementation of sets.

Theorem 1.1.1.   (a) $A \cup B = B \cup A$

Commutative laws

   (b) $A \cap B = B \cap A$.

Theorem 1.1.2.   (a) $(A \cup B) \cup C = A \cup (B \cup C)$

Associative laws

   (b) $(A \cap B) \cap C = A \cap (B \cap C)$.

Theorem 1.1.3.   (a) $A \cup A = A$

Idempotent laws

   (b) $A \cap A = A$.

Theorem 1.1.4.   (a) $A \cap (B \cup C) = (A \cap B) \cup (A \cap C)$

Distributive laws

   (b) $A \cup (B \cap C) = (A \cup B) \cap (A \cup C)$.

Theorem 1.1.5.   (a) $A \cup E = E$, $A \cup \emptyset = A$
   (b) $A \cap \emptyset = \emptyset$, $A \cap E = A$.

Theorem 1.1.6.   (a) $cA \cap cB = c(A \cup B)$

De Morgan's laws

   (b) $cA \cup cB = c(A \cap B)$.

Theorem 1.1.7. (a) $c(cA) = A$

(b) $A \cup cA = E$

(c) $A \cap cA = \phi$.

These theorems can be easily proved in a manner similar to the following proof.

Proof of Theorem 1.1.6 (a). To prove the equality of two sets X and Y it is necessary and sufficient to prove that $X \subset Y$ and $Y \subset X$.

$$x \in c(A \cup B) \Longrightarrow x \notin A \cup B \Longrightarrow x \notin A \text{ and } x \notin B$$

$$\Longrightarrow x \in cA \text{ and } x \in cB \Longrightarrow x \in (cA \cap cB). \quad (1.1)$$

This means $c(A \cup B) \subset cA \cap cB$.

Now

$$x \in (cA \cap cB) \Longrightarrow x \in cA \text{ and } x \in cB \Longrightarrow x \notin A \text{ and } x \notin B$$

$$\Longrightarrow x \notin (A \cup B) \Longrightarrow x \in c(A \cup B). \quad (1.2)$$

Thus, $(cA \cap cB) \subset c(A \cup B)$.   Combining (1.1) and (1.2) it follows that $c(A \cup B) = cA \cap cB$. $\blacksquare$

It must be noticed that in the above proof the argument used to prove (1.1) is reversible and we could replace $\Longrightarrow$ by $\Longleftrightarrow$ to save time and space. However, beginners are warned here not to assume that every argument is reversible.

The associative law for union and intersection (Theorem 1.1.2) enables us to generalize the concept of union to any number of sets. For example, if A, B, C are three sets, then we can write $A \cup B \cup C$ without any parentheses. The definitions of union and intersection can now be extended to any number of sets in the following way.

Definition 1.1.10. *Suppose for each member α of a fixed set A we have a set X_α. Then the union of all the sets X_α is the set $\{x : x \in X_\alpha \text{ for some } \alpha \in A\}$.*

We denote this set by $\bigcup_{\alpha \in A} X_\alpha$ (or $\cup X_\alpha$ if the set A is understood). The set A is called the index set for this union. For example, if $A = \{1,2,3\}$, then $\bigcup_{\alpha \in A} X_\alpha = X_1 \cup X_2 \cup X_3$. We define $\cap X_\alpha$ analogously.

Definition 1.1.11. *Let $\{X_\alpha\}$ be a class of sets. These sets are called mutually disjoint or pairwise disjoint if every two distinct sets of the class $\{X_\alpha\}$ have an empty intersection.*

It must be noticed here that if there are more than two sets in a class they may have an empty intersection, but may not be pairwise disjoint. For example, if we consider the sets $X_1 = \{1,2,3\}$, $X_2 = \{3,4,5\}$, $X_3 = \{4,5,6,7\}$, then $X_1 \cap X_2 \cap X_3 = \phi$. However, X_1, X_2, and X_3 are not pairwise disjoint.

Definition 1.1.12. *The Cartesian product of two sets A and B is defined as the set of all ordered pairs (x,y) where $x \in A$ and $y \in B$ and it is denoted by $A \times B$.*

Thus $A \times B = \{(x,y) : x \in A \text{ and } y \in B\}$.

We can give a geometrical interpretation to the Cartesian product if the sets involved are not too "intricate." For example, let $A = \{1,3,5\}$ and $B = \{2,4\}$. If we represent the set A by three points on the x-axis of the Euclidean plane and the set B by two points on the y-axis, then the set $A \times B$ consists of the six points in the plane having coordinates (1,2), (3,2), (5,2), (1,4), (3,4) and (5,4) as shown in the diagram.

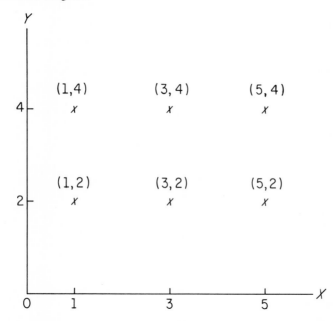

Had we taken $A = \{x : x \text{ is a real number and } 1 \leqslant x \leqslant 5\}$ and $B = \{y : y \text{ is a real number and } 2 \leqslant y \leqslant 4\}$, $A \times B$ would then consist of all the points inside and on the boundary of the rectangle with vertices (1,2), (5,2), (1,4) and (5,4).

The definition of Cartesian product can be extended to any number of sets in the same manner as we did for union and intersection. It may be shown that the Cartesian product is not associative nor commutative.

$$(A \times B) \times C \neq A \times (B \times C).$$

For

$$(A \times B) \times C = \{((a,b),c) : a \in A, b \in B, c \in C\}$$

and

$$A \times (B \times C) = \{(a,(b,c)) : a \in A, b \in B, c \in C\}.$$

$(A \times B) \times C$ is the set of all ordered pairs such that the first element is it-self an ordered pair, and the second element is in C, whereas in the case of $A \times (B \times C)$ the first element is in A and the second element is an ordered pair. In spite of this, $A \times B \times C$ is used to denote the set $\{(a,b,c) : a \in A, \ b \in B, c \in C\}$ without any confusion.

It is easy to show that $A \times B \neq B \times A$ unless $A = B$.

Exercise 1.1

1. Show that $A - B = A - (A \cap B) = (A \cup B) - B$
2. Show that $(A - B) - C = A - (B \cup C) = (A - B) \cap (A - C)$
3. The symmetric difference between two sets A and B, denoted by $A \triangle B$, is defined by $A \triangle B = (A - B) \cup (B - A)$
   (a) Show that $\triangle$ satisfies associative law and commutative law.
   (b) $A \cap (B \triangle C) = (A \cap B) \triangle (A \cap C)$
   (c) $A \triangle \emptyset = A, \ \ A \triangle A = \emptyset$
4. Show that
   (a) $\{x : |x - 2| < .5\} = \{x : x < 2 + .5\} \cap \{x : x > 2 - .5\}$
   (b) $\{x : |x - 2| \geqslant .5\} = \{x : x \geqslant 2.5\} \cup \{x : x \leqslant 1.75\}$
5. Show that
   (a) $\{x : |x| < 5\} \subset c \ \{x : |x| \geqslant 6\}$
   (b) $\{x : |x - 2| \geqslant 1\} \subset c \ \{x : |x - 2| < .5\}$
6. List all the elements of the following sets
   (a) $S = \{x : x < 7 \text{ and } x \in \mathfrak{N} \}$
   (b) $A = \{x : x \in \mathfrak{I} \text{ and } -3 < x < 7\}$.

1.2. Mappings

We now introduce the concept of mapping.

Definition 1.2.1. *A mapping f of a nonempty set X into a nonempty set Y is a set of ordered pairs $(x, f(x))$, with $x \in X$ and $f(x) \in Y$, such that for every $x \in X$ there is one and only one ordered pair $(x, f(x))$ with x as the first co-ordinate.*

For a beginner we may say that a mapping f of a nonempty set X into a nonempty set Y is a rule which associates with every member of X a unique member $f(x)$ of Y.

If f is a mapping of X into Y, we write $f : X \longrightarrow Y$. The element $f(x)$ of Y is called the image of x under the mapping f. The subset of Y which consists of all the images $f(x)$ for every $x \in X$ is called the *range* of the mapping. We de-note the range of f as $f(X) = \{f(x) : x \in X\}$. The set X is called the *domain* of the mapping f. The terms "mapping" and "function" are synonymous and are used interchangeably. Some authors restrict the use of "function" to those mappings whose range is a subset of reals, but we do not.

If X and Y are both subsets of reals, then the collection of all points (x,y) on the Euclidean plane such that $y = f(x)$ is called the *graph* of the function.

An element x in X is called an *inverse image* or a *preimage* of $f(x)$. It must be observed here that $f(x)$ may have more than one inverse image in X since more than one element in X may have the same image in Y. For example, if the mapping $f: X \longrightarrow Y$ is defined by $f(x) = |x|$, where $X = \mathcal{I}$, and $Y* = \mathfrak{N} \cup \{0\}$, then $f(1) = 1, f(-1) = 1, f(2) = 2, f(-2) = 2$, etc.

Definition 1.2.2. *If $f: X \longrightarrow Y$ is a mapping of X into Y and $f(X) = Y$ then the mapping f is called a surjective mapping or we say that f is a mapping of X "onto" Y.*

The terms "surjective" and "onto" are synonymous.

Definition 1.2.3. *If for any mapping $f: X \longrightarrow Y$ every image $f(x)$ in Y has a unique inverse image in X then the mapping is called an injective mapping or a one-one mapping.*

For a surjective mapping $f: X \longrightarrow Y$, if $y \in Y$, then $\exists \, x \in X$, such that $f(x) = y$.

For an injective mapping $f: X \longrightarrow Y$, if x_1 and $x_2 \in X$ then $x_1 \neq x_2 \Longrightarrow f(x_1) \neq f(x_2)$.

Definition 1.2.4. *A mapping is called a bijective mapping if it is both surjective and injective.*

Notice here that if f is a one-one mapping of X onto Y (which means f is a bijective mapping) then not only does $x \in X$ have a unique image in Y, but also every $y \in Y$ has a unique inverse image in X.

If there exists a bijective mapping of X into Y then X and Y are said to be in one-one correspondence.

When we speak of a mapping, unless additional information is given, we are referring to a many-one (read "many to one") mapping. Thus, this includes the possibility of the mapping being injective, surjective, bijective, or none of these.

EXAMPLE 1.2.1. Let $f(x) = x^2$. We are going to show that f can be injective, surjective or even bijective depending upon the choice of X and Y.

(i) Let X be the set of all real numbers and Y be the set of all real numbers; then $f: X \longrightarrow Y$, where $f(x) = x^2$, is a many-one mapping of X into Y. This mapping is neither injective (2 and -2 have the same image 4) nor surjective (no negative real number has an inverse image in X) and, consequently, not bijective.

---

*1, 2, 3 ... are the positive integers, whereas 0, 1, 2, ... are the nonnegative integers. Likewise 0, -1, -2, ... are the nonpositive integers and -1, -2, -3, ... are the negative integers.

(ii) Now let X denote the set of all nonnegative real numbers and Y be the set of all real numbers; then $f: X \longrightarrow Y$ is injective but not surjective. Note here we are excluding all negative numbers from X, thus making $f(x) = x^2$ a one-one mapping of X into Y.

(iii) Let X be the set of all real numbers and let Y be the set of all non-negative real numbers. Then $f: X \longrightarrow Y$ is a surjective mapping. Here, every element in Y has an inverse image in X. In fact every element (except zero) has two inverse images in X. The mapping is obviously not injective.

(iv) Finally, let each one of X and Y be the set of all nonnegative real numbers. Then $f: X \longrightarrow Y$ is a bijective mapping; i.e., a one-one mapping of the set of nonnegative real numbers onto itself.

Definition 1.2.5. *Let $f: X \longrightarrow Y$ be a bijective mapping. The inverse mapping of f, written as $f^{-1}: Y \longrightarrow X$ is defined as follows.*

$$f^{-1}(y) = x \iff f(x) = y \ \forall y \in Y, \text{ and } \forall x \in X.$$

Inverse mapping of f maps every element of the range of f into its preimage. In Example 1.2.1 (iv) the inverse function f^{-1} is such that $f^{-1}(x) = \sqrt{x}$.

It must be noticed here that if the mapping f were not one-one then $f^{-1}(y)$ (denoting inverse images of y) would consist of more than one element for at least one $y \in Y$. In this case we could not define the inverse mapping of f. It follows that for any mapping $f: X \longrightarrow Y$ the inverse mapping $f^{-1}: Y \longrightarrow X$ exists if and only if f is bijective. Furthermore, the mapping f^{-1} is always bijective. Even though the inverse mapping of a mapping f may not exist, we could still write $f^{-1}(y)$ to denote the set of preimages of y; that is, we write $f^{-1}(y) = \{x : f(x) = y\}$.

Theorem 1.2.1. Let $f: X \longrightarrow Y$ be a mapping and let $A \subset X, B \subset X$. We then have

(a) $f(\phi) = \phi$
(b) $A \subset B \Longrightarrow f(A) \subset f(B)$
(c) $f(A \cup B) = f(A) \cup f(B)$
(d) $f(A \cap B) \subset f(A) \cap f(B)$.

We shall prove property (c) listed above and the proofs of the remaining properties will be left as exercises.

Proof of property (c):

$$y \in f(A \cup B) \iff y \text{ is the image of an element in } A \cup B$$
$$\iff y \text{ is the image of some element in } A \text{ or}$$
$$\text{some element in } B$$
$$\iff y \in f(A) \text{ or } y \in f(B)$$
$$\iff y \in f(A) \cup f(B)$$
$$\iff f(A \cup B) = f(A) \cup f(B). \ \blacksquare$$

It must be remarked here that it would not be accurate to conclude from $y \in f(A \cup B)$ that $f^{-1}(y) \in A \cup B$, for $f^{-1}(y)$ may contain an element lying outside $A \cup B$. In the property (d) the equality of $f(A \cap B)$ and $[f(A) \cap f(B)]$ cannot be proved. The following example illustrates why we need $\subset$ rather than $=$ in property (d).

Let $f(x) = x^2$ and let X and Y both be the set of all real numbers [Example 1.2.1 (i)]. Let A be the set of all positive real numbers and let B be the set of all nonpositive real numbers. Then $A \cap B = \phi$ and $f(A \cap B) = \phi$, but $f(A) \cap f(B) \neq \phi$.

Theorem 1.2.2. If $f: X \longrightarrow Y$ is a mapping and if A_1 and A_2 are subsets of Y then we have the following properties:
  (i) $f^{-1}(\phi) = \phi$ and $f^{-1}(Y) = X$.
  (ii) $A_1 \subset A_2 \Longrightarrow f^{-1}(A_1) \subset f^{-1}(A_2)$.
  (iii) $f^{-1}(A_1 \cup A_2) = f^{-1}(A_1) \cup f^{-1}(A_2)$.
  (iv) $f^{-1}(A_1 \cap A_2) = f^{-1}(A_1) \cap f^{-1}(A_2)$.
  (v) $f^{-1}(cA) = cf^{-1}(A)$.
The proofs are left as exercises.

One may note here that the mapping f^{-1} (if it exists) is much better behaved than f. These results are true, however, whether or not f^{-1} exists.

It is important to observe there that in contrast to property (v) for f^{-1} nothing can be said about the relationship between $c[f(A)]$ and $f(cA)$. For example, if we consider the mapping $f(x) = x^2$ on the set of all real numbers into itself and if A is the set of all nonnegative real numbers then $f(A)$ is the set of all nonnegative real numbers and $c[f(A)]$ is the set of all negative real numbers. It is easy to see that $f(cA)$ is the set of all positive real numbers.

Definition 1.2.6. *On any set X the mapping $f(x) = x$ for every $x \in X$ is called the identity mapping on X and this mapping will be denoted by I_x.*
Such a mapping keeps every element of X fixed.

Definition 1.2.7. *Let $f: X \longrightarrow Y$ and $g: Y \longrightarrow Z$ be two mappings. The first one maps X into Y and the other maps Y into Z. The product* (or composition) of these mappings written as $gf: X \longrightarrow Z$ is the mapping which maps x into $g(f(x))$ for every $x \in X$. In other words, if f maps x into $f(x)$ in Y and if g maps $f(x)$ into $g(f(x))$ in Z, then gf is a mapping of X into Z such that the image of x under gf is $g(f(x))$.*

It must be noticed here that the product of the mapping f and g is written as gf (not as fg) though f is the first mapping and g is the second mapping. The reason for this convention is just a matter of convenience, for $gf(x) = g(f(x))$.

---

*The product gf should not be confused with the algebraic product; for example if $f = \log$ and $g = \sin$, then $gf(x) = \sin (\log x)$ whereas the algebraic product $g(x) f(x) = (\sin x)(\log x)$.

One should also observe that to define the product gf, the range of f must be contained in the domain of g.

It is an easy matter to show that if the inverse of a function $f: X \longrightarrow Y$ exists, then $f^{-1} f = I_x$.

EXAMPLE 1.2.2. Let $U = V = W = R$ (the set of all reals). Let $f: U \longrightarrow V$ be a mapping such that $f(x) = x^2 \; \forall \, x \in U$. Let $g: V \longrightarrow W$ be a mapping such that $g(y) = \sqrt{y} \; \forall \, y \in V$. In this case $gf: U \longrightarrow W$ is described by $gf(x) = \sqrt{x^2} = |x|$. It must be noticed here that g and f are not inverse functions of each other. In fact, the inverse of f does not even exist, since f is not bijective.

At this point we introduce the idea of restriction and extension of a mapping.

Definition 1.2.8. *Let* $f: X \longrightarrow Y$ *be a mapping and let* $D \subset X$. *Then* $f_D: D \longrightarrow Y$ *such that* $f_D(d) = f(d) \, \forall \, d \in D$, *is called the restriction of* f *on* D.

Restriction of f on D is also denoted by $f|D$. We may use the phrase "restrict f to D" to describe f_D. Notice, the only difference between a function and its restriction is in their domains.

In Example 1.2.2 if we restrict f to the set D of all nonnegative reals, then f_D and g are inverse of each other, for $gf_D(x) = |x| = x$ (x is a nonnegative real) and gf_D becomes an identity function.

Definition 1.2.9. *Let* $f: X \longrightarrow Y$ *be a mapping and let* $E \supset X$. *Then a mapping* $g: E \longrightarrow Y$ *is called an extension of* f *on* E *if* f *is the restriction of* g *on* X.

An extension of f on E may be denoted by $E|f$ or f^E. It is significant to note that if $e \in E$ and $e \notin X$, then $f^E(e)$ may be defined in any arbitrary way. Therefore, for a given function $f: X \longrightarrow Y$, there can be several extensions of f on E, and hence use of the article "an" preceding extension. On the other hand there is only one way to restrict f to a subset D of X.

Exercise 1.2.

In 1 and 2 $f: X \longrightarrow Y$ is a mapping, and $A \subset X$ and $B \subset Y$.

1. Prove that $A \subset f^{-1} f(A)$; moreover, $A = f^{-1} f(A)$ is true for every $A \Longleftrightarrow f$ is injective. Give an example of A, such that $A \neq f^{-1} f(A)$.

2. Prove that $ff^{-1}(B) \subset B$; moreover, $ff^{-1}(B) = B$ is true for every $B \Longleftrightarrow f$ is surjective. Given an example of B, such that $B \neq ff^{-1}(B)$.

3. We define a binary operation on a nonempty S as a mapping $O: S \times S \longrightarrow S$.
   (a) Show that addition on the set of rationals is a binary operation.
   (b) Show that the multiplication on the set of rational numbers is a binary operation.

4. Show that $f(A_1 \cap A_2) = f(A_1) \cap f(A_2)$ for all subsets A_1 and A_2 of X if and only if f is an injective mapping.

5. Show that $c[f(A)] \subset f(cA)$ is true for every subset A of $X \Longleftrightarrow f$ is surjective. Also show that $c[f(A)] = f(cA) \; \forall$ subset A of $X \Longleftrightarrow f$ is bijective.
6. Let $X = \{1,2,3\}$ and $Y = \{0,1\}$. List all the mappings $f: X \longrightarrow Y$. How many of these mappings are surjective? How many injective? Is there a bijective mapping $f: X \longrightarrow Y$? Give a restriction of f which is bijective.
7. Let $f: R \longrightarrow R$ be a mapping such that $f(x) = \cosh x = \frac{1}{2} (e^x + e^{-x})$. Give a restriction of f which is bijective.
8. Prove Theorems 1.2.2. and 1.2.1 (a), (c), and (d).
9. Let $f: X \longrightarrow Y$ be a bijective mapping, and let f^{-1} be its inverse. Show that $ff^{-1} = I_y$. How does I_y differ from I_x?

1.3. Relations

In mathematics as in everyday life we use the word "relation" to indicate some connection between elements of a set. By a "binary relation" on a set S we mean a symbol R (or a phrase) such that for any two points x and y in S either x is related to y or x is not related to y. We denote these latter two statements respectively as $x R y$ or $x \not R y$. For example, on the set of natural numbers we may say $2 < 4$. The symbol $<$ stands for a binary relation on N. Other examples of a binary relation are: "Is perpendicular to," "Is parallel to" on the set of all Euclidean lines on a plane, and "Is a multiple of" on the set of all natural numbers. In contrast to a binary relation there are ternary relations also, like "betweenness" on the set of points of a Euclidean line. However, in this book the term "relation" would always mean a "binary relation." Let us now give a formal definition of this term.

Definition 1.3.1. *A (binary) relation* R *on a set* S *is a subset of* $S \times S$. *We write* $x R y \Longleftrightarrow (x,y) \in R$, *and* $x \not R y \Longleftrightarrow (x,y) \notin R$.

To illustrate the relation $<$ on the set $S = \{1,2,3,4\}$ (Example 5, Sec. 1.1) we can say it is represented by the set $\{(1,2), (1,3), (1,4), (2,3), (2,4), (3,4)\}$, which is a subset of $S \times S$. It should be noticed that an element $(4,2)$ is not a member of this relation since 4 is not less than 2.

Definition 1.3.2. *A binary relation* R *defined on a set* S *is said to be*
(a) *reflexive if* $x R x$ *for every* $x \in S$.
(b) *symmetric if* $x R y \Longrightarrow y R x$
(c) *transitive if* $x R y$ *and* $y R z \Longrightarrow x R z$.

Definition 1.3.3. *A binary relation which satisfies all the three conditions is called an "equivalence relation."*

"Is equal to" is an example of an equivalence relation. "Is divisible by" is reflexive and transitive, but not symmetric. "Is perpendicular to" is symmetric, but neither reflexive nor transitive. The relation $<$ is transitive, but neither symmetric nor reflexive.

Definition 1.3.4. *A relation R on a set S is called determinate if for any two distinct points x and y ∈ S, either x R y or y R x.*

The relation $<$ defined on the set of all reals is determinate, but "Is perpendicular to" is not.

Definition 1.3.5. *A relation R on a set S is called asymmetric if x R y ⟹ y R̸ x for all x, y ∈ S.*

Again, the relation $<$ is asymmetric but "Is perpendicular to" is not.

Exercise 1.3

1. Give an example of a relation which is reflexive and symmetric but not transitive.
2. Give an example of a relation which is reflexive and transitive but not symmetric.
3. Give an example of a relation which is symmetric and transitive but not reflexive.
4. What is wrong with the following argument?
   "If a relation R is symmetric then, $aRb \Longrightarrow bRa$, and if it is transitive then, aRb and $bRc \Longrightarrow aRc$. Thus if a relation is symmetric and transitive then $aRb \Longrightarrow bRa$ and aRb and $bRa \Longrightarrow aRa$. Hence it must be reflexive."
5. The domain of a relation R defined on S is the subset D of $S = \{x : (x,y) \in R$ for some $y \in S\}$. Show that if $D = S$, then symmetry and transitivity of R on S would imply its reflexivity.
6. Let R be a relation on S with domain D. Show that R is a function $f : D \longrightarrow S$ if and only if $(x, y_1) \in R$ and $(x, y_2) \in R \Longleftrightarrow y_1 = y_2$.

1.4. Linear Orders

Definition 1.4. *A linear order is a relation $\leqslant$ on a nonempty set L such that the following conditions are satisfied.*
   (i) If $a, b \in L$ and if $a \neq b$ then either $a \leqslant b$ or $b \leqslant a$.
   (ii) If $a \leqslant b$ then $b \not\leqslant a$.
   (iii) If $a \leqslant b$ and $b \leqslant c$ then $a \leqslant c$.
If $a \leqslant b$, we may say "a is a precedent of b" or "a is less than b."

We can rephrase our definition as follows: "A linear order is a relation defined on a nonempty set such that it is determinate, asymmetric, and transitive."

The set L along with linear order relation is called a *linearly ordered set* (abbreviated los). If A is a nonempty subset of L, then elements of A will also satisfy the conditions (i), (ii), and (iii) of Def. 1.4.1 with respect to the relation $\leqslant$ of L. Hence A is also a linearly ordered set. The relation $\leqslant$ on A is said to be induced by L and is called a *suborder* of the original linear order.

EXAMPLE 1.4.1. The following are some examples of linearly ordered sets (los).

(i)   The set of all natural numbers $\mathfrak{N}$ with their usual order $<$.

(ii) The set $\mathscr{I}$ of all integers in their natural order.

(iii) The set $\mathfrak{Q}$ of rational numbers in the usual order.

(iv) The set R of all real numbers in their natural order.

(v)   The set $\{a_1 \bigotimes a_2 \bigotimes a_3 \bigotimes \ldots ; \bigotimes b_1 \bigotimes b_2 \bigotimes \ldots\}$, such that every a_i is a precedent of every b_i.

(vi) The set $\{1,2,3,4\}$ in its natural order.

It must be noticed that the linear order of the set in (vi) is a suborder of the linear order of $\mathfrak{N}$, which is a suborder of the linear order $\mathscr{I}$, which again is a suborder of the linear order of $\mathfrak{Q}$, which again is a suborder of linear order of R.

Definition 1.4.2. *An element a of a* los *L is called the first element of L if there does not exist any element x in L such that x $\bigotimes$ a.*

The first element of a los is necessarily unique.

Definition 1.4.3. *An element b of a* los *L is called the last element of L if there does not exist an element x of L such b $\bigotimes$ x.*

Again, the last element of a los must be unique.

The set of all natural numbers (in their usual order) has a first element but not a last element, whereas, the set of all negative integers (in their natural order) does not have a first element but has a last element. A finite los as in Example 1.4.1. (vi) has the first as well as the last element.

Definition 1.4.4. *A well-ordered set (denoted by* wos*) is a* los *such that every subset of this set has a first element.*

The relation of a wos is called a *well-ordering*.

Examples 1.4.1 (i), 1.4.1 (v) and 1.4.1 (vi) are the examples of well ordered sets.

Definition 1.4.5. *A* los *L is said to be unbounded in both directions if it does not have a first element nor a last element, i.e., for every element x of L, $\exists$ elements a and b of L $\ni$ a $< x <$ b.*

Examples 1.4.1 (ii), 1.4.1 (iii) and 1.4.1 (iv) are the examples of los which are unbounded in both directions whereas, those of (i) and (v) are not unbounded in both directions.

Definition 1.4.6. *A* los *L is called dense if for any two distinct points x and y of L $\exists$ an element $a \in L \ni x < a < y$.*

The linear orders of $\mathfrak{Q}$ and R are examples of dense linear orders.

It can be shown that in a dense linear order there are infinitely many points between any two distinct points and hence the name dense.

Definition 1.4.7. *A dense los L is called continuous if given any two non-empty disjoint subsets A and B of L such that (a) $A \cup B = L$ and (b) $a < b$ for all $a \in A$, $b \in B$, then either A has a last element or B has a first element.*

Of the examples mentioned earlier, (iv) is the only one with a continuous linear order.

The concept of continuity of a los will become clear in the next chapter.

The subdivision of a los L into sets A and B satisfying conditions (a) and (b) of Def. 1.4.7 is called a *cut*, and is denoted by $(A|B)$. In such a subdivision the following situations may arise:

1. A has a last element and B has a first element.
2. A has a last element but B does not have a first element.
3. A does not have a last element but B has a first element.
4. A does not have a last element and B does not have a first element.

In case 1 the cut $(A|B)$ is said to have a *jump*. If a linear order L is such that no cut has a jump then L is dense (prove).

In case 4 the cut $(A|B)$ is said to have a *gap*. If no cut of a dense linear order has a gap then by Def. 1.4.7 it is a continuous linear order.

It should not be difficult for the reader to construct cuts of a linear order with jumps or with gaps.

Definition 1.4.8. *Two linearly ordered sets L and L' with relations $<$ and $<'$ are said to be similar if there exists a bijective mapping $f:L \longrightarrow L'$ which preserves the linear order. That is, $a < b$ in $L \Longrightarrow f(a) <' f(b)$ in L'.*

For example the los $\{1 < 2 < 3 < 4\}$ is similar to $\{4 <' 3 <' 2 <' 1\}$.

Miscellaneous Exercises for Chapter 1

1. Show that $f(x) = \dfrac{1}{x}$ defined on the set of all nonzero real numbers is a symmetric relation. Is it transitive?
2. Let f be an injective function defined on a subset of real numbers, and let f^{-1} be its inverse. Show geometrically, that the graph of f^{-1} is the reflection of the graph of f through the line $y = x$.
3. Let $f:X \longrightarrow Y$ be a mapping. Show that f is injective if and only if there exists a mapping $g:Y \longrightarrow X$ such that $gf = I_x$.
4. Show that a mapping $f:X \longrightarrow Y$ is surjective $\Longleftrightarrow$ there exists a mapping $g:Y \longrightarrow X$ such that $fg = I_y$.
5. Let $f:X \longrightarrow Y$ and $g:Y \longrightarrow Z$ be two bijective mappings. Show that $(gf)^{-1} = f^{-1} g^{-1}$. What are the domain and range of this mapping?
6. Let $f: R \longrightarrow Y$ be a function defined by $f(x) = \sin x$, and $Y = \{y: -1 \leqslant y \leqslant 1, y \in R\}$. Using your knowledge of Calculus define a restriction of f such that f is bijective.

7. Let $f: \mathfrak{Q} \longrightarrow R$ be a function defined by $f(x) = x \; \forall x \in \mathfrak{Q}$. Describe an extension of f into R such that $f^R: R \longrightarrow R$ is neither injective nor surjective.

    Describe another extension of f into R which is bijective.

8. Show that the los $\{-\infty < 1 < 2 < 3 < \ldots\}$ is a well-ordered set and that it is similar to the well-ordered set of natural numbers.

9. Show that the set of all rational numbers is not a well-ordered set in their natural order.

2

Real Numbers

2.1. Dedekind Cuts

We shall start this chapter with the construction of the real number system, which may be accomplished in various ways. The method we are presenting here is due to Richard Dedekind (1831-1916) whose work entitled *Was Sind Und was Sollen die Zahlen?* ('What are and what should be numbers?') was a true masterpiece. An instructor may find it expedient to skip the Sec. 2.1 entirely, and accept Theorem 2.2.1 as a postulate.

Although some knowledge of rational numbers* is assumed here, we list the following properties:

1. *Field properties*. The set $\mathcal{Q}$ of all rationals forms a field, that is, there exist two binary operations $+$ and $\cdot$ such that the following conditions are satisfied.

   (i)   If a and $b \in \mathcal{Q}$, $a + b$ and $a \cdot b \in \mathcal{Q}$

   (ii) $(a + b) + c = a + (b + c)$ and $a \cdot (b \cdot c) = (a \cdot b) \cdot c$

   (iii) $a + b = b + a$ and $a \cdot b = b \cdot a$

   (iv)   There exist 0 and $1 \in \mathcal{Q}$ such that $a + 0 = 0 + a = a$ and $1 \cdot a = a \cdot 1 = a$

   (v)   For every $a \in \mathcal{Q}$, there exists $-a$ such that $a + (-a) = 0$, and if $a \neq 0$ there exists $a^{-1} \in \mathcal{Q}$ such that $a^{-1}a = 1$

   (vi) $(a + b)c = ac + bc, a(b + c) = ab + ac$.

2. *Ordering*. (a) Define $a < b$ if $b - a$ is positive. Then $<$ is a linear order on the set of all rational numbers.

   (b)   This los (linearly ordered set) of all rational numbers is unbounded, dense, and has gaps.

---

*Rational numbers may be defined as ordered pairs of integers, which in turn may be defined as ordered pairs of natural numbers. Natural numbers are obtained from Peano's postulates. For a discussion of these, one may consult E. G. H. Landau's *Foundations of Analysis*.

(c) If $a < b$, then $a + c < b + c$, $\forall c \in \mathfrak{Q}$, and $ad < bd$ if $d > 0$, $bd < ad$ if $d < 0$.

3. *Archimedean property.* For any two rational numbers a and b, there exists a natural number n such that $b < na$.

It will be shown in this chapter that the gaps created by rational numbers are filled by real numbers.

In this section lowercase letters $a, b, c, \ldots p, q, r, \ldots x, y, z$ will always denote rational numbers unless otherwise specified.

Definition 2.1.1. *The subdivision of the set $\mathfrak{Q}$ of all rational numbers into two nonempty disjoint sets L and U is called a cut (cf. Def. 1.4.7) if $L \cup U = \mathfrak{Q}$ and if*

(i) $p \in L$ and $q \in U \Longrightarrow p < q$

(ii) *L does not contain a largest rational number.**

We shall denote a cut by $(L \mid U)$. L will be called the *lower part* (abbreviated l.p.) of the cut and U the *upper part* (abbreviated u.p.).

For any cut $(L \mid U)$ and for any rational number p either $p \in L$ or $p \in U$. It follows that if $p \in L$ then any rational number $p' < p$ also belongs to L, and if $q \notin L$ then $q' > q$ is not in L. Furthermore, if $p \in L$ and $q \notin L$ then $p < q$.

Definition 2.1.2. *Two cuts $(L_1 \mid U_1)$ and $(L_2 \mid U_2)$ are said to be equal if the sets L_1 and L_2 are equal (which implies U_1 and U_2 are equal).*

Observe that a cut is completely determined by its lower part L with respect to $\mathfrak{Q}$, or by its upper part.

Definition 2.1.3. *$(L_1 \mid U_1) < (L_2 \mid U_2)$ if $p \in L_1 \Longrightarrow p \in L_2$ and $\exists$ at least one rational number $q \in L_2$ such that $q \notin L_1$. In other words L_2 properly contains L_1.*

If $(L_1 \mid U_1) < (L_2 \mid U_2)$ we may write $(L_2 \mid U_2) > (L_1 \mid U_1)$.

$(L_1 \mid U_1) \leqslant (L_2 \mid U_2) \Longrightarrow (L_1 \mid U_1) < (L_2 \mid U_2)$ or $(L_1 \mid U_1) = (L_2 \mid U_2)$.

$(L_1 \mid U_1) \geqslant (L_2 \mid U_2) \Longrightarrow (L_1 \mid U_1) > (L_2 \mid U_2)$ or $(L_1 \mid U_1) = (L_2 \mid U_2)$.

Theorem 2.1.1. If $(L_1 \mid U_1)$ and $(L_2 \mid U_2)$ are two cuts and if there is an element $a \in L_2$ such that $a \notin L_1$, then $(L_1 \mid U_1) < (L_2 \mid U_2)$.

Proof. Let $x \in L_1$. Either $x < a$ or $x = a$ or $x > a$. But $x \neq a$ by hypothesis. Furthermore $a \in U_1$, so $x > a \Longrightarrow x \in U_1 \Longrightarrow x \notin L_1$, a contradiction. Hence $x < a$ and since $a \in L_2$, this implies $x \in L_2$. Adding this to our hypothesis we conclude that L_2 properly contains L_1; i.e., $(L_1 \mid U_1) < (L_2 \mid U_2)$. ∎

Theorem 2.1.2. For any two cuts $(L_1 \mid U_1)$ and $(L_2 \mid U_2)$ either $(L_1 \mid U_1) = (L_2 \mid U_2)$ or $(L_1 \mid U_1) < (L_2 \mid U_2)$ or $(L_1 \mid U_1) > (L_2 \mid U_2)$.

---

*Equivalently, instead of (ii) we may suppose that U does not have a smallest member.

The proof of this theorem is an easy exercise, and is left for the reader.

The three possibilities mentioned in Theorem 2.1.2 are mutually exclusive for:

$$(L_1 | U_1) = (L_2 | U_2) \Longleftrightarrow L_1 = L_2$$
$$(L_1 | U_1) < (L_2 | U_2) \Longleftrightarrow L_1 \text{ is properly contained in } L_2$$
$$(L_1 | U_1) > (L_2 | U_2) \Longleftrightarrow L_1 \text{ properly contains } L_2.$$

Theorem 2.1.3. $(L_1 | U_1) < (L_2 | U_2)$ and $(L_2 | U_2) < (L_3 | U_3) \Longrightarrow (L_1 | U_1) < (L_3 | U_3)$.

The proof of this theorem follows immediately from Def. 2.1.3.

From these results it follows that the set of all cuts with the relation $<$ as defined in Def. 2.1.3 is a los. Later, we shall prove that this los is unbounded, dense, and continuous.

Theorem 2.1.4. Let r be a rational number and let $L = \{p : p < r, p \text{ is rational}\}$ and let $U = \mathcal{Q} - L$. Then $(L | U)$ is a cut and r is the smallest element of U.

Proof. It is obvious from the hypothesis that L and U are nonempty and $p \in L, q \in U \Longrightarrow p < q$.

It also follows that L does not have a largest number; for if a is any rational number in L, then $a < r$. Hence $a < \dfrac{a + r}{2} < r$, and it follows that $\dfrac{a + r}{2}$ is an element of L which is larger than a.

Therefore, L cannot contain a larger rational number. Furthermore, $r \not< r \Longrightarrow r \notin L \Longrightarrow r \in U$. Finally, $p < r \Longrightarrow p \in L \Longrightarrow p \notin U \Longrightarrow r$ is the smallest element of U. ∎

We shall denote the above cut by C_r; a cut so obtained is called a rational cut.

The cut C_0 (rational cut obtained by zero) will have negative rational numbers in its lower part, and its upper part will consist of all nonnegative rational numbers. This cut will also be denoted by $(M | P)$.

Definition 2.1.4. *A cut $(L | U)$ is called positive if L contains at least one positive rational number and is called negative if U contains at least one negative rational number.*

It is a simple exercise to prove that a positive cut $> (M | P) >$ a negative cut.

The next theorem enables us to define the addition of two cuts.

Theorem 2.1.5. Let $(L_1 | U_1)$ and $(L_2 | U_2)$ be two cuts, $L = \{x : x = a + b, a \in L_1 \text{ and } b \in L_2\}$ and $U = \mathcal{Q} - L$; then $(L | U)$ is a cut.

Proof. By hypothesis, L is nonvoid. If $p \in U_1$ and $q \in U_2$, then $p + q >$

$a + b$ for all $a \in L_1$ and $b \in L_2$ which implies $p + q \notin L \Longrightarrow p + q \in U$; i.e., U is nonvoid. Obviously, $L \cap U = \emptyset$.

Moreover, if $p \in L$ and $q \in U$, then $p < q$; for $q \leqslant p \Longrightarrow q \leqslant a + b$, $a \in L_1$, $b \in L_2$ where $a + b = p \Longrightarrow q - b \leqslant a \Longrightarrow q - b \in L_1 \Longrightarrow q - b = c$ where $c \in L_1 \Longrightarrow q = c + b$. Since $c \in L_1, b \in L_2, q \in L$, a contradiction.

Finally, L does not contain a largest member; for $d \in L \Longrightarrow d = a + b$, $a \in L_1, b \in L_2$; but $\exists\ a' \in L_1$ and $b' \in L_2$ such that $a < a'$ and $b < b'$ since neither L_1 nor L_2 contains a largest member. Letting $d' = a' + b'$ we obtain a rational number $d' \in L$ such that $d < d'$. Since $d \in L$ was arbitrary, it follows that L does not contain a largest member. Hence $(L|U)$ is a cut. ∎

Definition 2.1.5. *The cut constructed in the above theorem is called the sum of the cuts $(L_1|U_1)$ and $(L_2|U_2)$ and we write $(L|U) = (L_1|U_1) + (L_2|U_2)$.*

Theorem 2.1.6. Let $(L|U)$ be a cut and let ϵ be a positive rational number; then $\exists$ rational numbers $b \in L$ and $c \in U$ such that $c - b = \epsilon$.

Proof. Case I: Let $\epsilon\ (>0)$ be in L. If a is any rational number in U, then by Archimedean property there exists a positive integer m such that $m\epsilon > a$, and hence $m\epsilon$ is in U. Now consider the set $\{m\epsilon, (m - 1)\epsilon, \ldots, 2\epsilon, \epsilon\}$. Notice ϵ is in L and $m\epsilon$ is in U and some of them are in L and some in U. Choose the minimum $n\epsilon$ from this set such that $n\epsilon$ is in U, implying $(n - 1)\epsilon$ is in L.

If we now let $b = (n - 1)\epsilon$, and $c = n\epsilon$, the result follows.

Case II: Let ϵ be in U. If now 0 is in L then let $b = 0$ and $c = \epsilon$, and the result follows.

If 0 is in U, then by a process similar to that of Case I, we can show that there exists a negative integer $-n$, such that $-n\epsilon$ is in L, and $(-n + 1)\epsilon$ is in U.

Letting $b = -n\epsilon$, and $c = (-n + 1)\epsilon$, the result follows, and the proof is complete. ∎

Theorem 2.1.7. Let $(L_1|U_1)$, $(L_2|U_2)$ and $(L|U)$ be cuts such that $(L_1|U_1) < (L_2|U_2)$. Then $(L_1|U_1) + (L|U) < (L_2|U_2) + (L|U)$.

Proof. Let $L_1' = \{p + q : p \in L_1, q \in L\}$
$\qquad\quad L_2' = \{p + q : p \in L_2, q \in L\}$.

We must find an element $r \in L_2' \ni r \notin L_1'$. By hypothesis, $\exists$ an element $a \in L_2$ such that $a \notin L_1$. Since a is not the largest member of $L_2, \exists\ b \in L_2$, such that $a < b$. Letting $\epsilon = b - a$ in Theorem 2.1.6, we get rational numbers $c \in L$ and $d \notin L \ni d - c = \epsilon$. Then $d - c = b - a \Longrightarrow a + d = b + c$. Now $b \in L_2$ and $c \in L \Longrightarrow b + c \in L_2'$. Let $b + c = r\ (\epsilon L_2')$ But $a \notin L_1$ and $d \notin L \Longrightarrow a > p$ for every $p \in L_1$ and $d > q$ for every $q \in L_2 \Longrightarrow a + d > p + q$, for all $p + q \in L_1' \Longrightarrow a + d \notin L_1'$. But $a + d = b + c = r \in L_2'$, therefore, $L_1' \subset L_2'$; and the result follows. ∎

Theorem 2.1.8. The following results are valid for addition of two cuts:

(a) $(L_1|U_1) + (L_2|U_2) = (L_2|U_2) + (L_1|U_1)$ (commutative law)

(b) $[(L_1|U_1) + (L_2|U_2)] + (L_3|U_3) = (L_1|U_1) + [(L_2|U_2) + (L_3|U_3)]$

(associative law)

(c) $(L|U) + (M|P) = (L|U)$

(d) For any cut $(L|U)$ ∃ a cut $(L'|U')$ ∋ $(L|U) + (L'|U') = (M|P)$.

The proofs of (a) and (b) follow by making use of the commutative law and the associative law for addition of rational numbers. The proof of (c) is quite easy and is left for the reader. The proof of Theorem 2.1.8 (d) is as follows:

Given $(L|U)$, construct $(L'|U')$ as follows:

$L' = \{x: -x \in U$ and $-x$ is not the smallest member of $U\}$, and

$U' = \mathcal{Q} - L'$.

First we show that $(L'|U')$ is a cut. It is obvious that L' is nonempty. It can be easily shown that $U' = \{y: -y \in L$ or $-y$ is the smallest element of $U\}$. So U' is nonempty. Furthermore, $x \in L' \Longrightarrow -x \in U$ and $-x$ is not the smallest element of U. Now $y \in U' \Longrightarrow -y \in L$ or $-y$ is the smallest element of U. In either case $-y < -x$, so that $x < y$. Thus,

$$x \in L', y \in U' \Longrightarrow x < y$$

Also for any $x \in L$, $-x \in U$, but $-x$ is not the smallest element of U. Thus, there is a rational number $a < -x$, ∋ $a \in U$. Since the rational numbers are dense, ∃ a rational number $b \ni a < b < -x$. Since $b > a$, $b \in U$ and b is not the smallest element of U. Thus, $-b \in L'$ and since $b < -x$, $-b > x$. Therefore, x is not the largest element of L'; i.e., L' does not have a largest member.

Next we prove that $(L|U) + (L'|U') = (M|P)$. If $a \in$ l.p. of $(L|U) + (L'|U')$ then $a = b + c$ where $b \in L$ and $c \in L'$. Now $c \in L' \Longrightarrow -c \in U \Longrightarrow -c > b \Longrightarrow b + c < 0 \Longrightarrow a < 0 \Longrightarrow a \in M$. Conversely, if $a \in M$ then $a < 0$, so that $-a > 0$. Using Theorem 2.1.5 we can find rational numbers $x \in L$ and $y \in U$, such that y is not the smallest element of U and $y - x = -a$. Then $a = x - y$. Since $x \in L$ and $-y \in L'$, it follows that $a \in$ l.p. of $(L|U) + (L'|U')$.

$$\text{Hence } (L|U) + (L'|U') = (M|P)$$

The uniqueness of $(L'|U')$ follows immediately from Theorems 2.1.2 and 2.1.7, and that completes the proof. ∎

Definition 2.1.6. *The cut $(L'|U')$ constructed in Theorem 2.1.8 (d) is called the negative of $(L|U)$ and is written as* $-(L|U)$.

From Theorem 2.1.8 it follows that the set of all cuts with the addition defined in Def. 2.1.5 forms an Abelian group.*

---

*An Abelian group is a nonempty set Y with a binary operation 0 such that the operation is closed on the set, is associative and is commutative (the

Theorem 2.1.9. If $(L_1|U_1) < (L_2|U_2)$ then $\exists$ a rational number $r \ni$ $(L_1|U_1) < C_r < (L_2|U_2)$, that is, between any two distinct cuts there lies at least one rational cut.

Proof. By hypothesis, L_1 is properly contained in L_2; i.e., $\exists$ a rational number $a \in L_2 \ni a \notin L_1$.

Since a cannot be the largest rational number of $L_2 \ni$ a rational number $r \in L_2 \ni r > a$. Consider the cut $C_r = (L|U)$ with $L = \{x : x < r\}$. Now $r \notin L$ and $r \in L_2 \Longrightarrow (L|U) < (L_2|U_2)$. Also $a \in L$ $(a < r)$ and $a \notin L_1 \Longrightarrow (L_1|U_1) < (L|U)$. Thus, $(L_1|U_1) < C_r < (L_2|U_2)$, and that completes the proof. $\blacksquare$

Theorem 2.1.10. For any cut $(L|U)$, a rational number $r \in L$ if and only if $C_r < (L|U)$.

Theorem 2.1.11. If $(L_1|U_1)$ and $(L_2|U_2)$ are two cuts, then $\exists$ a unique cut $(L|U) , \ni (L_1|U_1) + (L|U) = (L_2|U_2)$.

The proofs of these theorems are simple and are left as exercises.

Definition 2.1.7. *The cut $(L|U)$ constructed in Theorem 2.1.11 is written as $(L_2|U_2) - (L_1|U_1)$.*

We shall now state a theorem which is not difficult to prove and will enable us to define the multiplication of two cuts.

Theorem 2.1.12. Let $(L_1|U_1)$ and $(L_2|U_2)$ be two positive cuts and let $L = M \cup \{xy : x \geqslant 0$ and $y \geqslant 0, x \in L_1$ and $y \in L_2\}$, M being the set of all negative rational numbers. Then $(L|U)$ is a cut $(U = \mathcal{Q} - L)$.

Definition 2.1.8. *The cut obtained in Theorem 2.1.12 is called the product of the two positive cuts $(L_1|U_1)$ and $(L_2|U_2)$ and we write $(L|U) = (L_1|U_1) \cdot (L_2|U_2)$.*

We shall now define the product of any two cuts by using the above definition.

Definition 2.1.9. *Let $(L_1|U_1)$ and $(L_2|U_2)$ be any two cuts. If either one of them is $(M|P)$ we define their product as $(M|P)$.*

If $(L_1|U_1) < 0$ ($\Longrightarrow -(L_1|U_1) > 0$) and $(L_2|U_2) > 0$, we define $(L_1|U_1) \cdot (L_2|U_2) = -\{[-(L_1|U_1)] \cdot (L_2|U_2)\}$.

If $(L_1|U_1) > 0$ and $(L_2|U_2) < 0$ ($\Longrightarrow -(L_2|U_2) > 0$) we define $(L_1|U_1) \cdot (L_2|U_2) = -(L_1|U_1) \cdot [-(L_2|U_2)]$.

---

commutative law is not required for non-Abelian groups). Also $\exists$ an identity element e in the group $\ni e \, 0 \, a = a \, 0 \, e = a$ for all $a \in Y$, and for every $a \in Y$, $\exists \, a^{-1} \in Y , \ni a^{-1} \, 0 \, a = a \, 0 \, a^{-1} = e$. For details, the reader may consult any book on abstract algebra.

If $(L_1 | U_1) < 0$ and $(L_2 | U_2) < 0$ we define $(L_1 | U_1) \cdot (L_2 | U_2) = [-(L_1 | U_1)] \cdot [-(L_2 | U_2)]$.

The above definition takes care of multiplication of any two cuts.

We now state the following results for multiplication without proofs.

Theorem 2.1.13.

(a) $[(L_1 | U_1) \cdot (L_2 | U_2)] (L_3 | U_3) = (L_1 | U_1) [(L_2 | U_2) \cdot (L_3 | U_3)]$

(associative law)

(b) $(L_1 | U_1) \cdot (L_2 | U_2) = (L_2 | U_2) \cdot (L_1 | U_1)$ (commutative law)

(c) $(L_1 | U_1) [(L_2 | U_2) + (L_3 | U_3)] = (L_1 | U_1) \cdot (L_2 | U_2) + (L_1 | U_1) \cdot (L_3 | U_3)$ (distributive law)

(d) If $(L_1 | U_1) \neq (M | P)$ then for every cut $(L_2 | U_2) \; \exists$ a unique cut $(L | U)$ such that $(L_1 | U_1) \cdot (L | U) = (L_2 | U_2)$.

From Theorem 2.1.8 and Theorem 2.1.13 it follows, that the set of cuts with addition and multiplication as defined forms a field.

It can be shown that the rational cuts satisfy all those properties which rational numbers do, and vice versa. Sometimes we express this by saying that there is an isomorphism between the set of all rational numbers and the set of all rational cuts. In this way a rational cut C_r can be identified with the rational number r.

Exercise 2.1

1. Prove Theorem 2.1.8 (a), Theorem 2.1.8 (b), and Theorem 2.1.8 (c).
2. Prove Theorem 2.1.11, Theorem 2.1.12, and Theorem 2.1.13.
3. Show that $-(M | P) = (M | P)$.
4. Prove that $(L | U)$ is positive if and only if $-(L | U)$ is negative.
5. Show that the system of cuts introduced in the last section is a dense linearly ordered set, and is unbounded in both directions.

2.2. Real Numbers

In the preceding section we constructed cuts of rational numbers and defined an order relation (which was a linear order relation) and two binary operations, addition and multiplication. We then proved and developed some results which are similar to the results of the number system in Algebra. Therefore, cuts can be treated as numbers.

One might also observe that these cuts satisfy the properties of an ordered field. The question now arises: "Are we sure of getting some cuts which are not rational?" The answer is in the affirmative. One example of such a cut is $(L | U)$ where $L = \{x : x^3 < 2 \text{ and } x \text{ is rational}\}$ and $U = \mathfrak{Q} - L$. It can be shown easily that $(L | U)$ is a cut and U does not have a smallest member; therefore, $(L | U)$

cannot be a rational cut, for in a rational cut, C_r, r is the smallest rational number of the upper part. This example shows that the set of all rational cuts is a proper subset of the set of all cuts, and illustrates the very significant fact that there are gaps (cf. Sec. 1.4) in the system of rational numbers. We shall show that these gaps are filled by the cuts.

Definition 2.2.1. *Cuts defined in Definition 2.1.1 will be called real numbers. A rational cut will be called a rational number and the cuts which are not rational will be called irrational numbers.*

From now on we will not restrict the use of the letters $a, b, c, \ldots, p, q, r, \ldots, x, y, z$, to rational numbers. Instead, we will use these symbols for all real numbers unless otherwise specified. In the remainder of this chapter a real number a will be identified by a cut $(L \mid U)$.

The next theorem will enable us to establish the fact that there are no gaps in the system of real numbers.

Theorem 2.2.1. (Cantor-Dedekind Theorem).
Let R be the set of all real numbers (cuts) and let R be decomposed into two nonempty disjoint sets A and B $(A \cup B = R)$ such that if $a \in A$ and $b \in B$ then $a < b$.* Then there exists one and only one real number c such that $a \leqslant c$ for all $a \in A$, and $c \leqslant b$ for all $b \in B$. Furthermore, c is either the largest element of A or the smallest element of B.

Proof. Either A has a largest element or A does not have a largest element. There is no third possibility.

Case I. If A has a largest element, say c, then $a \leqslant c$ for all $a \in A$ and since $a \in A$, it follows from the hypothesis that $c < b$ for all $b \in B$. Hence the result.

Case II. If A does not have a largest element, then we construct a cut $(L \mid U)$ of rational numbers in the following way:

Let $L = \{x : x$ is rational and $x \in A\}$ and $U = \mathcal{Q} - L$.

First we shall prove that $(L \mid U)$ is a cut. By hypothesis A is nonempty; hence, $\exists$ a real number $a \in A$. Let $a = (L_1 \mid U_1)$. There exists at least one rational number $r \in L_1$. From Theorem 2.1.10, $r < a \Longrightarrow r \in L \Longrightarrow L$ is nonempty. Since B is nonempty, $\exists$ at least one real number $b \in B$. Let $b = (L_2 \mid U_2)$. There is at least one rational number $p \in U_2$. Using Theorem 2.1.10, again we have $p \geqslant b \Longrightarrow p \in B \Longrightarrow p$ is a rational number not in $A \Longrightarrow p \in U$. Therefore U is nonempty. Furthermore, U consists of all rational numbers in B because L consists of all rational numbers in A. It follows immediately that $q \in L$, and $p \in U \Longrightarrow q < p$. Next, L does not have a largest member; for if r is a rational number in L then $r \in A$ and since A contains no largest real number (in this case) $\exists$ a real number $a \in A \ni r < a$. By Theorem 2.1.9, $\exists$ a rational number q such that $r < q < a \Longrightarrow q \in L$, and therefore r cannot be the largest rational number in L. This shows that $(L \mid U)$ is a cut.

---

*The linear order "$<$" as used here is of course that of Def. 2.1.3.

Let us write $c = (L|U)$. Let a be any real number of A. Since we have assumed that A does not have a largest real number, $\exists$ a real number $a' > a \ni a' \in A$. By Theorem 2.1.9, $\exists$ a rational number d such that $a < d < a'$; obviously, $d \in A \implies d \in L$. By Theorem 2.1.10, $d < c \implies a < c$.

Now if $b \in B$ and if r is any rational number such that $r < c$ then by Theorem 2.1.10, $r \in A \implies r \notin B \implies r < b$. What we have shown here is that for any $b \in B, r < c \implies r < b$. Thus, $c \leqslant b$. Furthermore, $c \in B$ because c could not be in A which proves that c is the smallest element of B. The proof of uniqueness of c is very simple. If there were two such numbers c and $c' \ni c \neq c'$, then either $c < c'$ or $c' < c$. In either case, $\exists$ a rational number r lying between c and c'. Suppose $c < r < c'$.

$$r > c \implies r \text{ cannot belong to } A$$

$$r < c' \implies r \text{ cannot belong to } B.$$

Therefore, we have a contradiction. Hence $c = c'$, which completes the proof. $\blacksquare$

This theorem enables us to illustrate a very important difference between the set of all rational numbers and the set of all real numbers. If, for example, we partition the set of all rational numbers into the two sets as follows: $A = \{x : x^3 < 2 \text{ and } x \text{ is rational}\}$ and $B = \mathcal{Q} - A$, then A and B satisfy the hypothesis of Dedekind's theorem but fail to satisfy the conclusion. For we cannot find any rational number $c, c \geqslant x$ for $x \in A$ and $c \leqslant y$ for $y \in B$. Of course, there is a real number, $\sqrt[3]{2}$, which would fill the gap in this example.

2.3. Bounds

Definition 2.3.1. *Let S be a nonempty set of real numbers. A number l is called a lower bound of S if $l \leqslant x$ for every $x \in S$.*

From the above definition it follows that if l is a lower bound of S then any number $l' \leqslant l$ is also a lower bound of S.

Similarly, an upper bound of a set S is a number $u \ni x \leqslant u$ for every $x \in S$.

If a set has a lower bound it is said to be bounded below, and if it has an upper bound it is said to be bounded above.

A set S is said to be bounded if and only if it is bounded above and bounded below; otherwise it is called unbounded. Notice here that an unbounded set may be unbounded above or unbounded below or both.

If a nonempty set S is bounded below then the set of all its lower bounds is bounded above, since for every lower bound l of $S, l \leqslant x$ for every $x \in S$.

Definition 2.3.2. *A real number g is called the greatest lower bound of a nonempty set S, if (i) g is a lower bound of S, and (ii) $g < x \implies x$ is not a lower bound of S.*

In other words, the greatest lower bound of a set is the largest of the lower bounds of S. We denote the greatest lower bound of S by glb (S) or inf (S) (stands for infimum of S).

A similar definition can be given for the least upper bound or supremum of a set S which is bounded above, and this is frequently written as lub (S) or sup (S) (supremum of S).

It is interesting to note that the glb of a set is the lub of the set of its lower bounds and the lub of a set is the glb of the set of its upper bounds. Also glb or lub of a set may or may not belong to the set.

EXAMPLES 2.3.1.

(a) The set $\mathfrak{N}$ of natural numbers is unbounded above, but is bounded below, and its glb is 1, which is in the set.

(b) The set $\left\{ 1, \dfrac{1}{2}, \dfrac{1}{3}, \dfrac{1}{n}, \ldots \right\}$ is bounded below (its glb is 0, which is not in the set) and bounded above (its lub is 1, which is in the set).

(c) A finite set is always bounded and contains both its glb and its lub.

(d) The set $\left\{ 1, 1 + \dfrac{1}{2}, 1 + \dfrac{1}{2} + \dfrac{1}{2^2}, \ldots, 1 + \dfrac{1}{2} + \cdots + \dfrac{1}{2^{n-1}}, \ldots \right\}$ is bounded above and below. The glb is 1 and the lub is 2. The lub is not in the set.

(e) $\left\{ \sin \dfrac{\pi}{6}, \sin \dfrac{2\pi}{6}, \sin \dfrac{3\pi}{6}, \ldots, \sin \dfrac{n\pi}{6}, \ldots \right\}$. The glb is -1 and the lub is 1 and both belong to the set.

When the glb of a set is in the set it is sometimes called the "minimum" of the set. Similarly we use the term "maximum" for the least upper bound of a set when it belongs to the set.

Theorem 2.3.1. If g is the glb of a set S, then for every $\epsilon > 0$, $\exists$ a number $x_0 \in S \ni x_0 < g + \epsilon$.

Proof. Suppose the theorem is false; i.e., there is no point of S which is less than $g + \epsilon$. Then $g + \epsilon \leqslant x$ for every $x \in S \Longrightarrow (g + \epsilon)$ is a lower bound of S. But $g + \epsilon > g$ and g is the glb of S, which is a contradiction; hence the result follows.

The following question may be raised here: If a set is bounded below, does the glb of the set always exist? The answer may not always be yes, for if we are dealing with the rational numbers only and if we consider the set $S = \{x : x > 0$ and $x^2 > 2\}$, then glb (S) is $\sqrt{2}$, which is not a rational number. In the system of rational numbers a set may be bounded below without having a glb in the system itself. This defect is overcome in the real number system as we observe in the following theorem.

Theorem 2.3.2. Let A be a nonempty set of real numbers. If A is bounded below then the glb of A always exists.

Proof. Let us construct a set S in the following manner: $x \in S$ if and only if $\exists$ an element $x_0 \in A \ni x_0 < x$. In other words each element of S is greater than at least one element of A. It follows that no lower bound of A is contained in S.

Now let $T = R - S$. If $x \in T$ then $\exists$ an element of A which is less than x; i.e., $x \in T \Longleftrightarrow x \leqslant y$ for every $y \in A$. Consequently, T consists of all the lower bounds of A. We shall now prove that T has the largest element.

T is nonempty, for A is bounded below. Also, since A is nonempty $\exists x_0 \in A$; but $\exists$ at least one real number $x > x_0$. Therefore, S is nonempty. Moreover, $a \in T$ and $b \in S \Longrightarrow a$ is a lower bound of A, whereas since b is not a lower bound we have $a < b$. We now see that T and S satisfy the hypothesis of Dedekind's theorem, and we conclude that $\exists$ an element $g \ni a \leqslant g$ for $a \in T$, $g \leqslant b$ for $b \in S$ and g is either the largest element of T or the smallest element of S.

If $g \in S$ then $\exists$ an element $x_0 \in A \ni x_0 < g$; but $\exists$ a rational number $p \ni x_0 < p < g \Longrightarrow p \in S$ and $p < g \Longrightarrow g$ cannot be the smallest element of S. Therefore g is the largest element of T, i.e., the glb of A. ∎

The existence of the glb of every set which is bounded below is sometimes called the "greatest lower bound property." Similarly we may define the "least upper bound property." It can be proved that the lub property is equivalent to the glb property. Each one of these properties is in turn equivalent to the property of completeness which we shall discuss in Chapter 4.

Definition 2.3.3. *Let a and b be two real numbers and let $a < b$. We now define intervals as follows*:
Closed interval $[a,b]$ = $\{x : x \text{ is real and } a \leqslant x \leqslant b\}$.
Open interval (a,b) = $\{x : x \text{ is real and } a < x < b\}$.
Semiopen interval $[a,b)$ = $\{x : x \text{ is real and } a \leqslant x < b\}$.
Semiopen interval $(a,b]$ = $\{x : x \text{ is real and } a < x \leqslant b\}$.
The points a and b are called the end-points of the intervals, a being the left end-point and b being the right end-point.

Each interval is a bounded set and a (the left end-point) is the glb, while b (the right end-point) is the lub. Furthermore, it is easy to show that a set of real numbers is bounded if and only if it can be enclosed in an interval.

To los R of real numbers we add two symbols $-\infty$ and ∞ such that for every $x \in R$, $-\infty < x < \infty$. We denote the set $R \cup \{-\infty, \infty\}$ by R_e, and call it the "Extended Real Number System." R_e is a los with $-\infty$ as the first element and ∞ as the last element. However, there is no second element or next to last element in R_e. It would be convenient to use $(-\infty, a)$, (a, ∞), $(-\infty, a]$, $[a, \infty)$. and $(-\infty, \infty)$ (which are subsets of R), as intervals of the extended real number system R_e. In the usual sense they should not be considered intervals of R, however.

Exercise 2.3

1. Prove that if a set S of real numbers is bounded above then the lub of S exists.

2. Let A and B be bounded sets and let $A \subset B$. Prove that $\sup A \leqslant \sup B$ and $\inf A \geqslant \inf B$.
3. Prove that if l is the lub of a nonempty set S, then for $\epsilon > 0$, there exists $x_0 \in S$ such that $x_0 > l - \epsilon$.
4. Prove that a set S is bounded $\iff$ there is a real number $M > 0$ such that $|x| < M$ for $x \in S$.

2.4. *n*-Dimensional Euclidean Space

In this section we give a brief introduction to the n-dimensional Cartesian product of the set R of all real numbers by itself. Readers not interested in this topic may skip this section (and similar sections in Chapters 4, 5, 6, and 7) without any loss of continuity.

Definition 2.4.1. *The set of all ordered n-tuples $\langle x_1, x_2, \ldots, x_n \rangle$ of real numbers (n being a natural number) is called an n-dimensional Euclidean space and is denoted by R^n or E^n.*

x_i is called the ith *coordinate* of the point $\langle x_1, x_2, \ldots, x_n \rangle$. Two points $\langle x_1, x_2, \ldots, x_n \rangle$ and $\langle y_1, y_2, \ldots, y_n \rangle$ are said to be equal if and only if $x_1 = y_1$, $x_2 = y_2, \ldots, x_n = y_n$.

The point $\langle 0, 0, \ldots, 0 \rangle$ is called the *origin* of R^n.

Definition 2.4.2. *For any point X $\langle x_1, x_2, \ldots, x_n \rangle$ of R^n, the point $\langle 0, \ldots, 0, x_i, 0, \ldots, 0 \rangle$ is called the ith projection of X.*

For a fixed i, it is easy to see that there is a 1-1 correspondence between all ith projections of points of R^n and the set R.

Definition 2.4.3. *If S is a subset of R^n, then the set of all ith coordinates of the points of S is called the ith coordinate of S and is denoted by S_i. That is, S_i is a subset of R and $S_i = \{x_i : \langle x_1, \ldots, x_i, \ldots, x_n \rangle \in S\}$. It is also called the ith coordinate set.*

It may be cautioned that a set S is not necessarily the Cartesian product of its coordinates. For example, the set $\{\langle 0,1 \rangle, \langle 2,3 \rangle\}$ of R^2 has coordinate sets $S_1 = \{0,2\}$ and $S_2 = \{1,3\}$; but is *not equal to* $S_1 \times S_2$. In fact, different sets may have the same coordinates.

It must also be observed that every nonempty set has nonempty coordinates.

Besides having a collection of points, there should be a structure in R^n which retains the basic properties of R. In other words, from any property of R^n we could obtain the corresponding one for R just by letting $n = 1$.

We now introduce the following operations in R^n.

Definition 2.4.4. *Let $X = \langle x_1, x_2, \ldots, x_n \rangle$ and $Y = \langle y_1, y_2, \ldots, y_n \rangle$ be two points of R^n. Then*

$$X + Y = \langle x_1 + y_1, x_2 + y_2, \ldots, x_n + y_n \rangle$$

$$-X = \langle -x_1, -x_2, \ldots, -x_n \rangle$$

$$cX = \langle cx_1, cx_2, \ldots, cx_n \rangle$$

$$X \cdot Y = x_1 y_1 + x_2 y_2 + \cdots + x_n y_n.$$

It must be noticed that $X \cdot Y$ is a real number and not a member of R^n except when $n = 1$, whereas $X + Y, -X, cX$ are points of R^n.

It is an easy matter to see that if $X_1, X_2, \ldots, X_n$ are the projections of a point X, then $X = X_1 + X_2 + \cdots + X_n$.

The readers familiar with vectors may easily note that the points of R^n can be treated as elements of a vector space, $X + Y$ being the vector sum of X and Y, and $X \cdot Y$ the dot product.

Definition 2.4.5.   *The norm of a point $X = \langle x_1, x_2, \ldots, x_n \rangle$ of R^n which is denoted by $\|x\|$ is defined as*

$$\sqrt{X \cdot X} = \sqrt{\sum_{k=1}^{n} x_k^2}.$$

Next we give the definition of distance between two points.

Definition 2.4.6.   *Let $X = \langle x_1, x_2, \ldots, x_n \rangle$ and $Y = \langle y_1, y_2, \ldots, y_n \rangle$ be two points of R^n. Then the distance between X and Y is defined as*

$$\|Y - X\| = \sqrt{(Y - X) \cdot (Y - X)} = \sqrt{\sum_{k=1}^{n} (y_k - x_k)^2}.$$

This distance is obviously a nonnegative real number and satisfies certain properties of a "metric space." Such spaces will be discussed in Chapter 5.

In the special case of the system of real numbers (when $n = 1$), the expression for the distance simply reduces to $|X - Y|$.

From the definition it follows that $\|X\|$ represents the distance of X from the origin of R^n, and the distance between X and Y is the norm of $Y - X$. Thus one can define distance in terms of "norm" and vice versa.

Now we discuss some properties of a norm.

Theorem 2.4.1.   For any point X in R^n
(a) $\|X\| \geqslant 0$
(b) $\|X\| = 0$ if and only if $X = \langle 0, 0, \ldots, 0 \rangle$
(c) $\|cX\| = |c| \, \|X\|$
(d) $\|X + Y\| \leqslant \|X\| + \|Y\|.$

Proof. The proofs of (a), (b), and (c) are simple and are left as exercises. We prove (d) only. By definition of a norm, we have

$$\|X + Y\|^2 = (X + Y) \cdot (X + Y)$$
$$= X \cdot X + 2X \cdot Y + Y \cdot Y$$
$$\|X - Y\|^2 = X \cdot X - 2X \cdot Y + Y \cdot Y.$$

Since $\|X + Y\|^2 \geqslant 0$ and so is $\|X - Y\|^2$, we have $\|X\|^2 + \|Y\|^2 \geqslant 2X \cdot Y$ and $\|X\|^2 + \|Y\|^2 \geqslant -2X \cdot Y$. Thus

$$|X \cdot Y| \leqslant \frac{\|X\|^2 + \|Y\|^2}{2}. \tag{2.1}$$

Now let $X' = \dfrac{X}{\|X\|}$, $Y' = \dfrac{Y}{\|Y\|}$, which means $\|X'\| = 1 = \|Y'\|$ and $X' \cdot Y' = \dfrac{X \cdot Y}{\|X\| \|Y\|}$. Substituting X' for X and Y' for Y in (2.1) we get

$$\frac{|X \cdot Y|}{\|X\| \|Y\|} \leqslant 1$$

which implies $|X \cdot Y| \leqslant \|X\| \|Y\|$. \hfill (2.2)

Thus

$$\|X + Y\|^2 = \|X\|^2 + \|Y\|^2 + 2X \cdot Y$$
$$\leqslant \|X\|^2 + \|Y\|^2 + 2\|X\| \|Y\|$$
$$= (\|X\| + \|Y\|)^2.$$

Hence $\|X + Y\| \leqslant \|X\| + \|Y\|$, and the proof is complete.

This result [Theorem 2.4.1(d)] is called the Minkowski's inequality. Minkowski (1864-1909) made contributions in geometry and convex sets. His work also had significant effect on the theory of relativity. The inequality $|X \cdot Y| \leqslant \|X\| \|Y\|$ which we obtained in the course of the last proof is called the Cauchy-Bunyakowskii-Schwarz inequality or sometimes just the Schwarz inequality. Victor Bunyakowskii (1804-89) made generalization of Cauchy inequality for integrals. H. Schwarz (1843-1921) who was a student of Karl T. Weierstrass (to be mentioned later) made some important contributions in complex analysis.

Next we seek a simple generalization of open intervals and closed intervals as follows.

Definition 2.4.7. (a) *An open box* ℬ *in* R^n *is the Cartesian product of n open intervals.* We write

$$\mathcal{B} = \{\langle x_1, x_2, \ldots, x_n \rangle : a_i < x_i < b_i, \ i = 1, 2, \ldots, n\}.$$

(b) *Similarly, a closed box* ℭ *in* R^n *is the Cartesian product of n closed intervals.* We write

$$\mathcal{C} = \{\langle x_1, x_2, \ldots, x_n \rangle : a_i \leqslant x \leqslant b_i, \ i = 1, 2, \ldots, n\}.$$

It is obvious from these definitions that the *i*th coordinate of an open box is an open interval and that of a closed box is a closed interval.

There is yet another way of generalizing open intervals.

Definition 2.4.8. *An open ball around a point P in R^n of radius ϵ is the set* $\{X : \|X - P\| < \epsilon\}$. *P is called the center of the ball.*

A closed ball may be defined in a similar way. In order to show that these definitions are natural generalizations of intervals, we prove the following result.

Theorem 2.4.2. Every coordinate set of an open ball is an open interval.

Proof. Let $P = \langle p_1, p_2, \ldots, p_n \rangle$ be the center of an open ball $\mathcal{B}$ and ϵ be its radius. Then

$$\mathcal{B} = \{X : \|X - P\| < \epsilon\} \quad \text{where } X = \langle x_1, x_2, \ldots, x_n \rangle.$$

The *i*th coordinate of X is x_i and that of P is p_i. And since

$$|x_i - p_i| \leqslant \sqrt{\sum_{i=1}^{n} (x_i - p_i)^2} < \epsilon, \quad \text{therefore } x_i \in (p_i - \epsilon, p_i + \epsilon).$$

Furthermore, every x_i in the open interval $(p_i - \epsilon, p_i + \epsilon)$ is the coordinate of at least one (in fact, infinitely many) points of the open ball $\mathcal{B}$, and that completes the proof.

Next, we introduce the important concept of boundedness.

Definition 2.4.9. *A set S of R^n is said to be bounded if there exists a real number $M > 0$ such that $\|P\| \leqslant M$ for every $P \in S$.*

Geometrically, we may say that a set S is bounded if and only if it can be enclosed in a ball of radius M and centered at the origin. Of course, a bounded set could also be enclosed in a ball whose center may not be at the origin. For a bounded set, we can define its diameter as follows:

Definition 2.4.10. *Let S be a set of R^n. Then if the set $\{\|X - Y\| : X, Y \in S\}$ of nonnegative real numbers is bounded above, then its least upper bound is called the diameter of S and the set S is said to have a finite diameter. We denote it by* diam (S).

If $\{\|X - Y\| : X, Y \in S\}$ is not bounded above, then we write diam $(S) = \infty$.

For example, the set $\{X : \|X - P\| < \epsilon\}$ has a finite diameter, namely ϵ. On the other hand, the set of all points in R^n with rational coordinates has an infinite diameter. One might guess that a bounded set must have a finite diameter and vice versa; this is confirmed by the next theorem.

Theorem 2.4.3. A nonempty set S of R^n is bounded if and only if it has a finite diameter.

Proof. If S is bounded, then there exists a real number $M > 0$ such that $\|P\| \leqslant M$ for every $P \in S$.

Now $\|X - Y\| \leqslant \|X\| + \|Y\| \leqslant M + M = 2M$ for $X, Y \in S$. Hence the set $\{\|X - Y\| : X, Y \in S\}$ is bounded and as such S has a finite diameter.

Conversely, if S has a finite diameter say d, then $d = \ell$ub $\{\|X - Y\|, X, Y \in S\}$. Now if P is a fixed point of S, then $\|X\| = \|X - P + P\| \leqslant \|X - P\| + \|P\| \leqslant d + \|P\|$, which means that S is bounded and that completes the proof.

Next we prove the following theorem.

Theorem 2.4.4. A set S of R^n is bounded if and only if its every coordinate set is bounded.

Proof. Let $S_1, S_2, \ldots, S_n$ be the coordinates of S. Since S is bounded there is a positive real number M such that $\|X\| \leqslant M$ for every $X \in S$.

Now if S_i is the ith coordinate of S and x_i be an arbitrary point of S_i, then there is a point $X \in S$ such that the ith coordinate of X is x_i. Now since $\|X\| \leqslant M$, we have $|x_i| \leqslant \|X\| \leqslant M$ and this is true for every $x_i \in S_i$; $i = 1, 2, \ldots, n$.

Conversely, if every S_i is bounded, then there are positive reals M_i; $i = 1, 2, \ldots, n$; such that $|x_i| \leqslant M_i$ for $x_i \in S_i$.

Let $M = M_1 + M_2 + \cdots + M_n$. Now if $X \in S$ and $X = \langle x_1, x_2, \ldots, x_n \rangle$, then from the corollary of Theorem 2.4.1 we have

$$\|X\| \leqslant |x_1| + |x_2| + \cdots + |x_n| \leqslant M_1 + M_2 + \cdots + M_n = M$$

which completes the proof.

Next we introduce functions whose domains are subsets of R^n and their ranges are in R^m. In general, n and m may not be equal. If $m = 1$, that is the range is a subset of real numbers, then we say the function is a real-valued function. For example, $f: R^2 \longrightarrow R$ where $f(x, y) = x^2 + y^2$ is a real-valued function. In elementary calculus we refer to such a function as a real-valued function of two variables.

For a function $f: D^n \longrightarrow R^m$, where D^n is an n-dimensional subset of R^n, we may write $f(X) = \langle y_1, y_2, \ldots, y_m \rangle$ where $X = \langle x_1, x_2, \ldots, x_n \rangle$. This function consists of m real-valued functions $f_i : D^n \longrightarrow R$; $i = 1, 2, \ldots, m$ such that

$$f_1 \langle x_1, x_2, \ldots, x_n \rangle = y_1$$
$$f_2 \langle x_1, x_2, \ldots, x_n \rangle = y_2$$
$$\vdots$$
$$f_m \langle x_1, x_2, \ldots, x_n \rangle = y_m.$$

Definition 2.4.11. $f_1, f_2, \ldots, f_m$ *described above are called components or coordinate functions of f.*

To illustrate this, let us consider the following examples.

EXAMPLE 2.4.1. Let $f: R^2 \longrightarrow R^3$ be a function such that

$$f\langle x_1, x_2 \rangle = \left\langle \frac{4x_1}{4 + x_1^2 + x_2^2}, \frac{4x_2}{4 + x_1^2 + x_2^2}, \frac{4 - x_1^2 - x_2^2}{4 + x_1^2 + x_2^2} \right\rangle.$$

In that case

$$f_1 \langle x_1, x_2 \rangle = \frac{4x_1}{4 + x_1^2 + x_2^2}$$

$$f_2 \langle x_1, x_2 \rangle = \frac{4x_2}{4 + x_1^2 + x_2^2}$$

$$f_3 \langle x_1, x_2 \rangle = \frac{4 - x_1^2 - x_2^2}{4 + x_1^2 + x_2^2}.$$

The real-valued functions f_1, f_2, f_3 are the coordinate functions of f.

The function $f: R^2 \longrightarrow R^3$ maps the entire plane (R^2) onto the unit sphere with center at the origin but without the point $\langle 0,0,-1 \rangle$. This missing point may be regarded as the South Pole of the sphere. In geometry the three equations given above (describing the coordinate functions) are called parametric equations of a sphere. x_1 and x_2 are called the parameters.

EXAMPLE 2.4.2. Consider the function $f: I \longrightarrow R^2$ where $I = [0, 2\pi)$ and $f(x) = \langle a \cos x, b \sin x \rangle$. This function maps the semiopen interval $[0, 2\pi)$ onto an ellipse of an Euclidean plane. Its coordinate functions are

$$f_1(x) = a \cos x$$

$$f_2(x) = b \sin x$$

which are, of course, the parametric equations of the ellipse.

We conclude this section with the introduction of a very special type of function in R^n.

Definition 2.4.12. *A function $f: R^n \longrightarrow R^m$ is called linear if $f(aX + bY) = af(X) + bf(Y)$; $X, Y \longrightarrow R^n$ and $a, b \in R$.*

The following theorem describes the form of a linear function.

Theorem 2.4.5. A real-valued function $f: R^n \longrightarrow R$ is linear if and only if $f(X) = c_1 x_1 + c_2 x_2 + \cdots + c_n x_n$; $X = \langle x_1, \ldots, x_n \rangle, c_1, \ldots, c_n \in R$.

Proof. Write $X = \langle x_1, \ldots, x_n \rangle = x_1 \langle 1,0, \ldots, 0 \rangle + x_2 \langle 0,1,0, \ldots 0 \rangle + x_n \langle 0, \ldots, 0,1 \rangle$ and let

$$f(\langle 1,0, \ldots, 0 \rangle) = c_1, f(\langle 0,1,0, \ldots, 0 \rangle) = c_2, \ldots, f(\langle 0, \ldots, 0, 1 \rangle) = c_n.$$

Now if $f: R^n \longrightarrow R$ is linear, then

$$f(X) = f(x_1\langle 1, \ldots, 0\rangle + x_2\langle 0, 1, \ldots, 0\rangle + \cdots + x_n \langle 0, \ldots, 1\rangle)$$
$$= x_1 f(\langle 1, 0, \ldots, 0\rangle) + x_2 f(\langle 0, 1, \ldots, 0\rangle) + x_n f(\langle 0, \ldots, 1\rangle)$$
$$= c_1 x_1 + c_2 x_2 + \cdots + c_n x_n.$$

Conversely, if $f(X) = c_1 x_1 + \cdots + c_n x_n$; then

$$f(aX + bY) = f(\langle ax_1 + by_1, ax_2 + by_2, \ldots, ax_n + by_n\rangle)$$
$$= c_1 (ax_1 + by_2) + c_2 (ax_2 + by_2) + \cdots + c_n (ax_n + by_n)$$
$$= a(c_1 x_1 + \cdots + c_n x_n) + b(c_1 y_1 + \cdots + c_n y_n)$$
$$= af(X) + bf(Y).$$

Hence f is linear and the proof is complete. ∎

Exercise 2.4

1. Prove Theorem 2.4.1 (a), (b), (c).
2. If $P, Q \in R^n$, then show that $(\|P + Q\|)^2 = \|P\|^2 + \|Q\|^2$ is true if and only if $P \cdot Q = 0$.
3. Prove that for any two points X, Y in R^n, we have $\|X - Y\| = 0$ if and only if $X = Y$, and $\|X - Y\| + \|Y - Z\| \geqslant \|X - Z\|$.
4. Show that $\|X - Y\| \geqslant \|X\| - \|Y\|$ for any two points X, Y in R^n.
5. Let S be the set of all those points of $R^n (n \geqslant 2)$ which have rational coordinates. Show that every coordinate set of the complement of S is R. (*Hint:* The complement of S does not consist of only those points whose coordinates are irrational.)
6. Show that a function $f: R^n \longrightarrow R^m$ is linear if and only if $y_i = f_i\langle x_1, x_2, \ldots, x_n\rangle = a_{i_1} x_1 + a_{i_2} x_2 + \cdots + a_{i_n} x_n$ where f_i is the ith coordinate function, and $a_{i_1}, a_{i_2}, \ldots, a_{i_n}$ are real numbers.

2.5. Complex Numbers

In this section we give a very brief introduction to "complex numbers." Here we have a special case of R^n when $n = 2$.

Definition 2.5.1. *The points* $\{\langle x,y\rangle : x, y \in R^2\}$ *are called complex numbers.* x *is called the* real part *of the complex number* $\langle x,y\rangle$ *and* y *is called the* imaginary part.

In this section we shall write z for the complex number $\langle x, y\rangle$.

Definition 2.5.2. *The norm of* z *written as* $\|z\|$ $(= \sqrt{x^2 + y^2})$ *is called the* *modulus or absolute value of z and is sometimes denoted by* mod *z or even by* $|z|$.

The addition of complex numbers is defined in exactly the same way as in R^n—that is,

$$\langle x,y \rangle + \langle u,v \rangle = \langle x+u, y+v \rangle.$$

Definition 2.5.3. *If* $x \neq 0$, *then* $\tan^{-1}(y/x)$ *is called the argument of* $z = \langle x,y \rangle$ *and is written as* arg *z. If* $x = 0$, $y > 0$ *then the argument of z is* $\pi/2$.

Next we introduce another operation: multiplication of complex numbers (*not to be confused* with the dot product).

Definition 2.5.4. *Let* $z = \langle x,y \rangle$ *and* $w = \langle u,v \rangle$ *be two complex numbers. Then* $zw = \langle xu - yv, xv + yu \rangle$.

One reason we are paying special attention to R^2 is because that it is a field. Moreover, it is impossible to define an operation of multiplication in R^n (for $n \geqslant 3$) which will make it (R^n) a field; though, in R^4 one can define a multiplication which is *not commutative*, but satisfies all other postulates of a field. Such a structure is called a skewfield, and points of R^4 having this structure are called *quaternions* which were studied by Sir William Rowan Hamilton (1805–1865). Going back to R^2, the addition and multiplication of complex numbers satisfy all the postulates of a field which were given in Sec. 2.1. We state this in the form of a theorem.

Theorem 2.5.1. The set R^2 of all complex numbers with the addition and multiplication as described above is a field.

The proof of this theorem is left as an exercise.

The field of complex numbers is not an ordered field like those of rational numbers and real numbers, but it has an additional property and that is of being "algebraically closed," which implies that every polynomial with complex coefficients has a root in R^2. We shall not elaborate it here. Readers interested in this topic may consult Birkhoff and Maclane, *A Survey of Modern Algebra.*

One may observe here that we have introduced complex numbers without using the so-called imaginary quantity $i = \sqrt{-1}$. In fact, we can identify i by the complex number $\langle 0,1 \rangle$. In that case, we can write any complex number $\langle x,y \rangle$ as follows:

$$\langle x,y \rangle = \langle x,0 \rangle + \langle 0,1 \rangle \langle y,0 \rangle$$

and if we identify $\langle x,0 \rangle$ by x, then

$$\langle x,y \rangle = x + iy.$$

In particular $\langle 0,y \rangle = iy$. In view of this it is convenient to call complex numbers $\{\langle x,0 \rangle : x \in R\}$ as "real" and $\{\langle 0,y \rangle : y \in R\}$ as "imaginary."

Definition 2.5.5. *If* $z = \langle x, y \rangle$ *is a complex number, then the complex number* $\langle x, -y \rangle$ *is called the conjugate of* z *and is denoted by* $\bar{z}$.

Some properties of a conjugate of a complex number are given in the next result.

Theorem 2.5.2. If $\bar{z}$ and $\bar{w}$ are conjugates of complex numbers z and w, respectively, then:

    (a) z is the conjugate of $\bar{z}$

    (b) $z + \bar{z}$ is real

    (c) $z\,\bar{z} = \langle \|z\|^2, 0 \rangle$

    (d) $\overline{z + w} = \bar{z} + \bar{w}$

    (e) $\overline{zw} = \bar{z}\,\bar{w}$.

Proof. We prove only (e). Let $z = \langle x, y \rangle$ and $w = \langle u, v \rangle$. Then $\bar{z} = \langle x, -y \rangle$ and $\bar{w} = \langle u, -v \rangle$. Using Def. 2.5.4, $zw = \langle xu - yv, xv + yu \rangle$, and $\bar{z}\,\bar{w} = \langle xu - yv, -xv - yu \rangle$ which mean $\overline{zw} = \bar{z}\,\bar{w}$ and the proof is complete.

One of the most important characteristics of a field is that the "division" by a nonzero element is possible. It simply amounts to multiplication by the inverse of the element. It can be easily shown that if

$$z = \langle x, y \rangle \text{ and } w = \langle u, v \rangle \neq \langle 0, 0 \rangle$$

then,

$$\frac{z}{w} = \left\langle \frac{xu + yv}{u^2 + v^2}, \frac{xu - yv}{u^2 + v^2} \right\rangle .$$

Exercise 2.5

1. Show that if $zw = 1$ and $z \neq \langle 0, 0 \rangle$ then $w = \dfrac{1}{z^2}$.

2. Show that if $z + w$ and zw are both real, then either $w = \bar{z}$ or z and w are both real.

3. Identify points in R^2 for which $\|z - 1\| + \|z + 1\| = 4$.

4. Let $z = \langle x, y \rangle \neq \langle 0, 0 \rangle$. If $z + \dfrac{1}{z} = \langle u, 0 \rangle$, then $y = 0$ or $x^2 + y^2 = 1$.

Miscellaneous Exercises for Chapter 2

1. A nonempty set S of reals is said to be *path-connected* if for every two distinct points a, b $(a < b)$ in S, every point x between a and b $(a < x < b)$ must also belong to S. Prove the following assertions.

(i) S is path-connected $\Longleftrightarrow S$ has any one of the following forms: $(-\infty,\infty)$, $(-\infty,a), (-\infty,a]$, $[a,\infty)$, (a,∞), (a,b), $[a,b]$, $[a,b)$, or $(a,b]$.

(ii) S is path-connected and has no lower bound nor an upper bound $\Longleftrightarrow S = (-\infty,\infty)\ (= R)$.

(iii) S is path-connected and has a lower bound but not an upper bound $\Longleftrightarrow S = [a,\infty)$ or (a,∞) depending upon whether the glb (S) belongs or does not belong to S.

(iv) S is path-connected and has an upper bound, but not a lower bound $\Longleftrightarrow S = (-\infty,a]$ or $(-\infty,a)$ depending upon whether the lub (S) belongs to or does not belong to S.

(v) S is path-connected and is bounded $\Longleftrightarrow S = [a,b]$ or $[a,b)$ or $(a,b]$ or (a,b). Specify the conditions for each one of these cases.

2. Let S be a nonempty bounded set. Let $L = \{l : l$ being a lower bound of $S\}$ and $U = \{u : u$ being an upper bound of $S\}$. Show that glb (S) = lub (L), and lub (S) = glb (U).

3. Show that for any real number x, there exists an integer $m(\epsilon\ \mathcal{I})$ for which $m < x \leqslant m + 1$.

4. Show that for every two real numbers a and b $(a < b)$, there exists a rational number r (in $\mathcal{Q}$) and an *irrational number* i such that $a < r < b$ and $a < i < b$.

5. Let $A \subset B$.
   (a) If $\exists\, x \in A \ni x \leqslant y\ \forall\, y \in B$, then show that *glb* (A) = *glb* (B) = x
   (b) If $\exists\, a \in A \ni b \leqslant a\ \forall\, b \in B$, then show that *lub* (A) = *lub* (B) = a.

3

Set Theory

3.1. Cardinal Numbers

In the post-sputnik era we hear the term "modern mathematics" used even by a layman who hardly knows much about the subject. If a single Mathematician deserved the title of "Father of Modern Mathematics" it should go to George Cantor (1845-1918). A student of Karl T. Weierstrass (1815-97), Cantor brought a revolution in the foundations of mathematics—a revolution which is still continuing (fortunately). We are going to discuss some of his work briefly in this chapter. Cantor's set theory has introduced many paradoxes (Sec. 3.2). Expressions like "the set of all sets," and "the set of all elements" always lead us into contradictions. To avoid inconsistencies we shall assume that there does not exist any set of *all sets*, and also there does not exist any set of *all elements*. Formally, it is enough to state the following axiom.

Axiom: Given a set S, there always exists an element x such that $x \notin S$.

The reader may skip Sec. 3.2 if he so chooses without any break of continuity. We start with a couple of definitions.

Definition 3.1.1. *A set A is said to be equivalent to a set B if $\exists$ a bijective mapping of A onto B. We denote this relation by $A \sim B$.*

It can easily be shown that the relation "$\sim$" is an equivalence relation.

Definition 3.1.2. *A set S is called finite if it is empty or if there is a natural number n such that S is equivalent to the set $\{1,2,3,\ldots,n\}$. A set is called infinite if it is not finite.*

In this chapter we are going to discuss one of the most important and powerful theorems in "set theory"—the Schroeder-Bernstein theorem. This theorem is of a general nature, and we therefore present it before formalizing the notion of cardinal number. The proof which we offer is based on a lemma of Banach.

Banach's Lemma. Let A and B be any two sets. If $f: A \longrightarrow B$ and $g: B \longrightarrow A$ are injective mappings, then $\exists$ a subset S of A such that $g[B - f(S)] = A - S$.

Proof. Let $h: A \longrightarrow A$ be the composition of f and g; i.e., $h = gf$. Let $T = A - g(B)$. If we consider

$$S = T \cup h(T) \cup h(h(T)) \cup h(h(h(T))) \cup \ldots \qquad (3.1)$$

then

$$h(S) = h(T) \cup h(h(T)) \cup h(h(h(T))) \cup \ldots \qquad (3.2)$$

and finally, upon substitution of (3.2) into (3.1), we have

$$S = T \cup h(S).$$

Since f and g are injective, it follows that

$$y \notin f(S) \Longleftrightarrow g(y) \notin gf(S) = h(S) \Longleftrightarrow g(y) \notin T \cup h(S) = S.$$

Thus $g(y) \in A - S \Longleftrightarrow y \in B - f(S)$.

$$\text{Therefore,} \ g[B - f(S)] = A - S$$

which proves the Lemma. $\blacksquare$

Theorem 3.1.1 (Schroeder-Bernstein Theorem). Let A and B be two sets. If there exists an injective mapping of A into B and an injective mapping of B into A, then $A \sim B$.

Proof. Let $f: A \longrightarrow B$ and $g: B \longrightarrow A$ be the existing maps. From Banach's Lemma $\exists S \subset A$ such that

$$g[B - f(S)] = A - S.$$

Moreover, $A - S \subset g(B)$.

We define a map $\phi: A \longrightarrow B$ by

$$\phi(x) = f(x) \qquad \text{if } x \in S,$$
$$\phi(x) = g^{-1}(x) \qquad \text{if } x \in A - S.$$

Evidently ϕ is injective. Since

$$\phi(A) = \phi(S \cup [A - S]) = \phi(S) \cup \phi(A - S)$$
$$= f(S) \cup g^{-1}(A - S)$$
$$= f(S) \cup [B - f(S)]$$
$$= B,$$

ϕ is indeed bijective. The existence of the map ϕ establishes the equivalence of A and B. ▮

The following corollary of this theorem is quite useful.

Corollary: If $A \supset A_1 \supset A_2$ and if $A \sim A_2$, then $A \sim A_1$.

We shall encounter numerous applications of the Schroeder-Bernstein theorem in this chapter.

Definition 3.1.3. *A set S which is equivalent to $\mathfrak{N}$ (the set of all natural numbers) is called denumerable or enumerable.*

Any denumerable set can be written in the form $\{a_1, a_2, a_3, \ldots, a_n, \ldots\}$, since it can be put in 1-1 correspondence with the set of all natural numbers. The process of writing a denumerable set in this form is called *enumeration*. On occasion, we use phrases like "denumerably many times" and "infinitely many times."

We are now going to introduce the concept of cardinal numbers—a concept which is easy to understand but sometimes difficult to define.

Definition 3.1.4. *All sets which are equivalent to each other have a common characteristic which we designate as the cardinal number (or the power) of each one of those sets.*

The cardinal number of the empty set is defined to be 0 (zero). We designate 1 as the cardinal number of all singletons; and we would naturally say that the power of all sets which are equivalent to the set $\{a_1, a_2\}$ $(a_2 \neq a_1)$ is 2, and so on. It seems easy to assign a cardinal number for a finite set.

We shall denote the cardinal number of the class of all denumerable sets by $\aleph_0$ (read as aleph nought, $\aleph$ being the first letter in the Hebrew alphabet). This notation was used by George Cantor, and is still in common use.

At this point we would like to make the following two statements which have far reaching consequences in mathematics.

Axiom of Choice. **If a nonempty set is divided into a class of nonempty pairwise disjoint subsets then there exists at least one set which contains exactly one element of each one of these subsets. This set may be called a "cross section" of the class.**

Zermelo's Well-Ordering Principle. Every set can be well-ordered.

In 1904 Ernest Zermelo proved the statement of this principle which shocked the mathematical world. Many mathematicians refused to accept it. One of them, Emile Borel (1871–1956), trying to find some fault with the proof of this theorem discovered that Zermelo had tacitly assumed the Axiom of Choice. Borel subsequently showed that the axiom of choice is logically equivalent to Zermelo's well-ordering principle. For the proof of the equivalence, see R. L. Wilder, *Foundations of Mathematics*.

There are many results of set theory which make use of the Axiom of

Choice, though some mathematicians still refuse to assume this axiom. We shall indicate whenever we make use of this axiom.

By definition of cardinal numbers, it is obvious that two cardinal numbers a and b are equal, if there exist two sets A and B with cardinal numbers a and b, respectively, such that $A \sim B$.

We now define the "inequality" of cardinal numbers.

Definition 3.1.5. *Let a and b be two cardinal numbers. Let A and B be two sets having cardinal numbers a and b respectively. We say $a < b$ if there exists a 1-1 mapping of A into B but there is no 1-1 mapping of B into A.*

If $a < b$ we sometimes write $b > a; a \leqslant b \iff a < b$ or $a = b$.

The definition given above and other definitions and results involving cardinal numbers are independent of the specific choice of the sets A and B. In other words, if we select two other sets A' and B' such that $A' \sim A$ and $B' \sim B$ then A' and B' serve just as well in describing the relationship between a and b.

For any two sets A and B with cardinal numbers a and b respectively we have the following mutually exclusive possibilities:

1. There exists a 1-1 mapping of A into B but there does not exist a 1-1 mapping of B into A.
2. There does not exist a 1-1 mapping of A into B but there exists a 1-1 mapping of B into A.
3. There exists a 1-1 mapping of A into B and there exists a 1-1 mapping of B into A.
4. There exists neither a 1-1 mapping of A into B nor a 1-1 mapping of B into A.

From Def. 3.1.5, the case 1 implies $a < b$, and the case 2 suggests $b < a$. If the case 3 holds, then $a = b$ by the Schroeder-Bernstein theorem.

If we assume Zermelo's well-ordering principle then we can show that case 4 is an impossibility. The proof will be given later. (Cf. Theorem 3.2.4).

Therefore, accepting the axiom of choice we claim that any two cardinal numbers are comparable; i.e., the relation $<$ as defined above (Definition 3.1.5) is a determinate relation. We shall show that this relation is a linear order relation.

Theorem 3.1.2. *If a, b, and c are cardinal numbers and if $a < b$ and $b < c$, then $a < c$.*

The proof of this theorem is simple and follows from the definition of $a < b$.

Since the asymmetry of this relation is obvious from the definition, we can now conclude that this relation is a linear order.

Definition 3.1.6. *Let a and b be two cardinal numbers and let A and B be two disjoint sets with cardinal numbers a and b respectively. We then define $(a + b)$ as the cardinal number of $A \cup B$.*

Definition 3.1.7. *Let a and b be two cardinal numbers and let A and B be any two sets having cardinal numbers a and b respectively. Then the product $a \cdot b$ is defined to be the cardinal number of $A \times B$.*

The cardinal number of an infinite set is sometimes called a *transfinite cardinal number.* We observe that $\aleph_0$ is a transfinite cardinal number. The question immediately arises: "Is $\aleph_0$ the smallest transfinite cardinal number?" The answer is in the affirmative as a consequence of the following theorem.

Theorem 3.1.3. Every infinite set has a denumerable subset.

Proof. Let S be an infinite set. Since S is nonempty ∃ an element $a_1 \in S$, and because S is not finite ∃ an element $a_2 (\neq a_1) \in S$. In fact for any natural number n there are n distinct elements, $a_1, a_2, \ldots, a_n$, of S; again, since S is not finite we can find an element a_{n+1} in S distinct from these n elements. In this way we obtain a well-ordered set $\{a_1, a_2, \ldots, a_n, a_{n+1}, \ldots\}$ which is a subset of S and which can be put in 1-1 correspondence with the set of all natural numbers. ∎

In the proof of the above theorem we have indirectly made use of the Axiom of choice.

Theorem 3.1.4. Every infinite set is equivalent to one of its proper subsets.

Proof. Let S be an infinite set. From the last theorem S has a denumerable subset T. Let $T = \{a_1, a_2, \ldots, a_n, \ldots\}$. Let $T_1 = \{a_2, a_3, \ldots, a_n, \ldots\} = T - \{a_1\}$, and let $S_1 = (S - T) \cup T_1$. Obviously S_1 is a proper subset of S ($a_1 \in S$ and $a_1 \notin S_1$). Consider the mapping $f: S \longrightarrow S_1$ defined as follows:

$$f(x) = x \text{ for every } x \in (S - T)$$

and

$$f(a_j) = a_{j+1} \quad \text{for every } a_j \in T \quad (j = 1, 2, \ldots).$$

This mapping is a bijective mapping of S onto S_1; therefore, $S \sim S_1$. This completes the proof. ∎

The above theorem was taken as the definition of an infinite set by Dedekind. It is easy to show that if a set can be put in 1-1 correspondence with one of its proper subsets then it cannot be a finite set and hence must be an infinite set according to our definition. However, to establish Theorem 3.1.4 we had to make use of Theorem 3.1.3, which depends upon the axiom of choice. For this reason many authors prefer to use Dedekind's definition of an infinite set. Since we have accepted the axiom of choice we are justified in using this approach.

Theorem 3.1.5. The set of all integers is denumerable.

Proof. Consider the following mapping on the set of all integers $\mathcal{J}$.

$$\left.\begin{array}{l} f(0) = 1 \\ f(n) = 2n \\ f(-n) = 2n + 1. \end{array}\right\} \quad n \text{ being a natural number}$$

The function $f\colon \mathcal{J} \longrightarrow \mathfrak{N}$ is a bijective mapping of $\mathcal{J}$ onto $\mathfrak{N}$; therefore, the result follows.

Theorem 3.1.5 may be generalized as follows.

Theorem 3.1.5(a). The union of any two denumerable sets is denumerable. The proof is left as an exercise.

Theorem 3.1.6. The set of all rational numbers is denumerable.

Proof. First we shall prove the result for the set of all positive rational numbers. We can represent a positive rational number by an ordered pair (m,n) of natural numbers (corresponding to m/n), where m and n are relatively prime. If P is the set of all positive rational numbers we can write $P = \{(m,n)\colon m,n \in \mathfrak{N}, m \text{ and } n \text{ relatively prime}\}$. Let $S = \{(m,n)\colon m \in \mathfrak{N} \text{ and } n \in \mathfrak{N}\}$. Obviously P is a proper subset of S. Consider the mapping $f\colon S \longrightarrow \mathfrak{N}$ defined as follows:

$$f[(m,n)] = \frac{1}{2}(m + n - 1)(m + n - 2) + n.$$

This is a one-one mapping of S onto $\mathfrak{N}$; therefore, $S \sim \mathfrak{N}$. Now $\mathfrak{N}$ can be identified with the set $M = \{(m,1)\colon m \in \mathfrak{N}\}$.

We then have $S \sim \mathfrak{N} \sim M \Longrightarrow S \sim M$. Since $M \subset P \subset S$, we conclude that $P \sim \mathfrak{N}$.

Since the positive rationals and negative rationals are in 1-1 correspondence, it follows from Theorem 3.1.5(a) that the union of these sets and $\{0\}$, that is the set of rational numbers, is denumerable.* The proof is now complete.

---

*Many books prove this result by diagonal method first used by George Cantor. The set S is written in the following manner.

(m,n) is in the mth row and nth column.

Let $a_1 = (1,1)$, $a_2 = (2,1)$, $a_3 = (1,2)$, $a_4 = (3,1)$, ... It is not difficult to note that the mapping f in the proof of Theorem 3.1.6 puts the elements of S in exactly the same order as indicated in the diagram; for example, $f(1,1) = 1$, $f(2,1) = 2$, ...

Theorem 3.1.6 establishes that the cardinal number of the set of all rational numbers is also $\aleph_0$. Having proved the denumerability of the set of all rational numbers we may raise the following question: "Is the set of all real numbers denumerable?" The answer to this question is provided in the next theorem. First some terminology:

Definition 3.1.8. *A set is said to be countable if it is equivalent to a subset of* $\mathfrak{N}$.
It follows that a countable set is either finite or denumerable.

Definition 3.1.9. *A set which is not countable is called nondenumerable. In other words a nondenumerable set is an infinite set which is not denumerable.*

Theorem 3.1.7. The set of all real numbers is nondenumerable.

Proof. It will be enough to prove that the set of all real numbers in $(0,1)$ is nondenumerable.

Any real number in $(0,1)$ can be written as an infinite decimal of the form $.d_1 d_2 d_3 \ldots$, where $d_1, d_2, \ldots$ are digital numbers $0, 1, 2, \ldots, 9$. To prove our theorem we use the indirect method. Since the set $(0,1)$ is infinite, it is either denumerable or nondenumerable.

Suppose $(0,1)$ is denumerable. In other words we can enumerate the set of all real numbers in $(0,1)$ in the following way:

$$\{a_1, a_2, \ldots, a_n, \ldots\}, \text{where}$$

$$a_1 = .d_{11} d_{12} d_{13} \quad \cdots \cdots \cdots$$
$$a_2 = .d_{21} d_{22} d_{23} \quad \cdots \cdots \cdots$$
$$a_3 = .d_{31} d_{32} d_{33} \quad \cdots \cdots \cdots$$
$$\cdots \cdots \cdots \cdots \cdots \cdots \cdots \cdots \cdots$$
$$\cdots \cdots \cdots \cdots \cdots \cdots \cdots \cdots \cdots$$
$$a_n = .d_{n1} d_{n2} d_{n3} \quad \cdots \cdots \cdots$$
$$\cdots \cdots \cdots \cdots \cdots \cdots \cdots \cdots \cdots$$

every d_{ij} being a digital number.
Now construct a real number a in the following manner*: $a = .d_1 d_2 d_3 \ldots$,

---

*In constructing a we should avoid 0 or 9 for d_i's since many rational numbers in $(0,1)$ may have two decimal expansions, one of them having 0 recurring and the other 9 recurring; for example:

$$\frac{1}{2} = .5000 \ldots = .4999 \ldots$$

$$\frac{1}{4} = .2500 \ldots = .2499 \ldots.$$

One could, however, choose any other digital number.

where $d_i = 1$ if $d_{ii} \neq 1$ and $d_i = 2$ if $d_{ii} = 1$; $i = 1, 2, 3, \ldots$. Clearly $a \in (0,1)$ and a is different from $a_1, a_2, \ldots, a_n, \ldots$, which gives us a contradiction. Therefore it is impossible to enumerate the set of all real numbers in $(0,1)$. $\blacksquare$

This theorem establishes a very important fact. There are infinite sets which are not denumerable.

In the next few theorems we shall show that any two intervals (regardless of their lengths) are equivalent, and that any interval is equivalent to the set of all real numbers.

Theorem 3.1.8. An open interval (a,b) is equivalent to any other open interval (c,d).

Proof. Let $x \in (a,b)$. Let us consider the following mapping on (a,b):

$$f(x) = c + \left(\frac{d-c}{b-a}\right)(x-a)$$

It is easy to see that $x \in (a,b) \Longleftrightarrow f(x) \in (c,d)$, and that f is a 1-1 mapping of (a,b) onto (c,d). Hence the result. $\blacksquare$

Using the same mapping we can prove that any two closed intervals are equivalent, but the remaining cases are not so straightforward.

Theorem 3.1.9. Any two intervals are equivalent.

Proof. We shall prove the result for the case when one of the intervals is open and the other closed. A similar argument will hold if either of them is a semiopen interval.

Let $[a,b]$ be a closed interval and let (c,d) be an open interval.

Since $a < b$, $\exists$ two real numbers a_1 and b_1 such that $a \leqslant a_1 \leqslant b_1 \leqslant b$.

Similarly, $\exists$ two real numbers c_1 and $d_1 \ni c < c_1 < d_1 < d$.

Construct the open interval $(a_1, b_1) \subset [a,b]$ and $[c_1, d_1] \subset (c,d)$. By the last theorem,

$$(a_1, b_1) \sim (c,d)$$

and

$$[a,b] \sim [c_1, d_1].$$

By the Schroeder-Bernstein theorem $[a,b]$ is equivalent to (c,d). $\blacksquare$

Theorem 3.1.10. Any interval is equivalent to the set R of all real numbers.

Proof. Let $x \in \left(-\frac{\pi}{2}, \frac{\pi}{2}\right)$ The function $f(x) = \tan x$ is a bijective mapping of $\left(-\dfrac{\pi}{2},\dfrac{\pi}{2}\right)$ onto R which establishes equivalence between $\left(-\dfrac{\pi}{2},\dfrac{\pi}{2}\right)$ and R. Since any interval is equivalent to $\left(-\dfrac{\pi}{2},\dfrac{\pi}{2}\right)$, by the last theorem, the result follows.

The cardinal number of the set R is denoted by c, the first letter of the word continuum.

Theorem 3.1.11.
(a) $c + \aleph_0 = c$
(b) $c + c + c + \cdots = c$
(c) $c \cdot c = c$.

Proof. We shall leave the proof of (a) to the reader, but we shall prove the much stronger result in (b).

Consider the intervals $[0,\frac{1}{2}), [\frac{1}{2},\frac{2}{3}), [\frac{2}{3},\frac{3}{4}), \ldots$ Each one of these intervals has cardinal number c, and these intervals are pairwise disjoint. Also their union is $[0,1)$, the cardinal number of which is c. Hence, $c + c + c + \cdots = c$.

Proof of (c). Let $A = [0,1)$. Let $B = [0,1) \times [0,1) = \left\{(x,y): \begin{matrix} 0 \leqslant x < 1 \\ 0 \leqslant y < 1 \end{matrix}\right\}$.
We have to prove that the cardinal number of B is the same as that of A. It is very easy to see that $\exists$ a 1-1 mapping of A into B defined by $f(x) = (x,0)$ for every $x \in A$. It suffices to exhibit a 1-1 mapping of B into A. For every $(x,y) \in B$, $x \in [0,1)$ and $y \in [0,1)$, so we can write

$$x = .a_1\, a_2\, a_3\, \ldots, \qquad y = .b_1\, b_2\, \ldots,$$

where

$$a_1, a_2, \ldots, b_1, b_2, \ldots$$

are all digital numbers. Let us assume that none of these fractions has recurring 9's.

Now consider the mapping $g: b \longrightarrow A$, where

$$g(x,y) = (.a_1 a_2\, \ldots, .b_1 b_2\, \ldots) = .a_1 b_1 a_2 b_2\, \ldots *$$

This is a 1-1 mapping of B into A. The result now follows from the Schroeder-Bernstein theorem. ∎

We now introduce exponentiation of cardinal numbers.

---

*It is interesting to observe that $g(0,.1) = .0100.\ldots$ If, however, we write .1 as $.09999\ldots$ then $g(0,.1) = .000909090\ldots \neq .0100.\ldots$ That is why we assume then that none of these fractions has recurring 9's. It is also important to note that the mapping g is not surjective, even though it is injective; e.g., there is no point of B with image $.191919\ldots$ under g.

Definition 3.1.11. *Let a ($\neq 0$) and b ($\neq 0$) be two cardinal numbers. Let A and B be two nonempty sets having cardinal numbers a and b respectively. We define a^b as the cardinal number of the set of all mappings on B with ranges in A.*

To illustrate this exponentiation we take $A = \{a_1, a_2\}$ and $B = \{b_1, b_2, b_3\}$. The following mappings of B into A can be defined:

$f_1 : B \longrightarrow A$, where $f_1(b) = a_1$, $\forall b \in B$.
$f_2 : B \longrightarrow A$, where $f_2(b_1) = a_1$, $f_2(b_2) = a_2, f_2(b_3) = a_1$.
$f_3 : B \longrightarrow A$, where $f_3(b_1) = a_1$, $f_3(b_2) = a_1, f_3(b_3) = a_2$.
$f_4 : B \longrightarrow A$, where $f_4(b_1) = a_1$, $f_4(b_2) = a_2, f_4(b_3) = a_2$.
$f_5 : B \longrightarrow A$, where $f_5(b_1) = a_2$, $f_5(b_2) = a_1, f_5(b_3) = a_1$.
$f_6 : B \longrightarrow A$, where $f_6(b_1) = a_2$, $f_6(b_2) = a_2, f_6(b_3) = a_1$.
$f_7 : B \longrightarrow A$, where $f_7(b_1) = a_2$, $f_7(b_2) = a_1, f_7(b_3) = a_2$.
$f_8 : B \longrightarrow A$, where $f_8(b) = a_2$, $\forall b \in B$.

We notice that there are 8 mappings of B into A. This can, however, be shown without writing all the above mentioned mappings, as follows: For any mapping of B into A every element in B can be mapped in 2 (cardinal number of A) ways. Therefore in this case there are $2 \cdot 2 \cdot 2 = 2^3$ mappings of B into A.

Theorem 3.1.12. For any set A with cardinal number a, the cardinal number of the set of all subsets of A is 2^a.

Proof. Let F be the set of all mappings on A into the set $\{0,1\}$, and let S be the set of all subsets of A. We shall show that there is a 1-1 correspondence between F and S. For any mapping $f_a (\in F)$ on A into $\{0,1\}$ associate a subset $X \in S$ of A as follows: For any element $x \in A$, $x \in X$ if $f_a(x) = 1$ and $x \notin X$ if $f_a(x) = 0$. For example, the mapping which maps every element of A into 0 will correspond to $\emptyset$ (a subset of A), and the mapping which maps every element of A into 1 will correspond to A itself; the mapping which maps only a particular element x_0 of A into 1 will correspond to the subset $\{x_0\}$, and so on. This correspondence is clearly one-to-one, and therefore the cardinal number of S is the same as that of F, i.e., 2^a. ∎

The set of all subsets of A is sometimes denoted by 2^A because of this theorem.

We have established the existence of two transfinite cardinal numbers $\aleph_0$ and c, and we know that $\aleph_0 < c$. The question now arises: "Is there any cardinal number greater than c?" The answer is "yes" as a consequence of the following theorem.

Theorem 3.1.13. For any cardinal number a, $2^a > a$.

Proof. Let A be a set with cardinal number a and let $B = 2^A$. There exists a 1-1 mapping of A into B, for the mapping $f(x) = \{x\}$ for every $x \in A$ is such a mapping ($\{x\}$ being a subset of A is an element of B). Using the indirect method we shall show that there does not exist a 1-1 mapping of B into A.

Suppose there is a 1-1 mapping g of B into A. Let $S = \{x : x \in A$, and $x \notin g^{-1}(x)\}$.* S is a subset of A and therefore $S \in B$. Let $g(S) = x_0$. Since g is a 1-1 mapping of B into $A, g^{-1}(x_0)$ must be unique $\Longrightarrow g^{-1}(x_0) = S$.

Now if $x_0 \in S$, then $x_0 \in g^{-1}(x_0) \Longrightarrow x_0 \notin S$, which is a contradiction.

If $x_0 \notin S$, then $x_0 \notin g^{-1}(x_0) \Longrightarrow x_0 \in S$ which is again, a contradiction. Therefore, it is not possible to have a 1-1 mapping of B into A. Hence, the result. ∎

For any cardinal number a there corresponds a set A with that cardinal number, but then 2^A has cardinal number $2^a > a$. Therefore there is no largest cardinal number.

Theorem 3.1.14. $2^{\aleph_0} = c$.

Proof. We shall prove that the set $2^{\mathfrak{N}}$ (the set of all subsets of natural numbers) is equivalent to $[0,1)$.

First construct a 1-1 mapping on the set of all subsets of natural numbers $2^{\mathfrak{N}}$ in the following manner. If A is any subset of natural numbers $(A \in 2^{\mathfrak{N}})$ then

$$f(A) = .d_1 \, d_2 \, d_3 \ldots$$

where $d_n = 1$ if $n \in A$, and $d_n = 2$ if $n \notin A$. $f(A)$ is in $[0,1)$ and has decimal expansion in terms of 1 and 2 only. For example, if $A = \{1,3,5,7\}$ then

$$f(A) = .1 \, 2 \, 1 \, 2 \, 1 \, 2 \, 1 \, 2 \, 2 \, 2 \ldots$$

This mapping is clearly injective, but not surjective. Therefore f is a 1-1 mapping of $2^{\mathfrak{N}}$ into $[0,1)$.

Now we shall show that $\exists$ a 1-1 mapping of $[0,1)$ into $2^{\mathfrak{N}}$. To prove this, we first express every real number in $[0,1)$ in its binary expansion. If $x \in [0,1)$ then $x = .b_1 b_2 b_3 \ldots$, where $b_n = 0$ or 1; and 1 is not recurring (to avoid duplicate expansions of a number). Now we define

$$g(x) = A, \quad \text{where } A = \{n : b_n = 1\}.$$

As an example, we consider the binary expansion of $1/4$, which is $.0100 \ldots$ In this case, $b_1 = 0$, $b_2 = 1$, and $b_n = 0$ for $n \geqslant 3$, so that $g(1/4) = \{2\}$. g is clearly a 1-1 mapping of $[0,1)$ into $2^{\mathfrak{N}}$. The result now follows from the Schroeder-Bernstein theorem. ∎

Having established that $2^{\aleph_0} = c$ and $2^{\aleph_0} > \aleph_0$, the question may arise whether there is any cardinal number greater than $\aleph_0$ and less than c (surely

---

*Since g is 1-1, $g^{-1}(x)$ consists of a single subset C of A, and we can ask whether x is in C or not.

for any finite cardinal number $n > 1$ there is always a cardinal number between n and 2^n). This problem has never been solved and is called the "continuum problem." However, some important results in mathematics have been proved by assuming that there is no cardinal number between $\aleph_0$ and c. This assumption is called the "continuum hypothesis." According to this hypothesis c is the second transfinite cardinal number, the only transfinite cardinal number less than c being $\aleph_0$. This hypothesis can be further generalized as follows:

Generalized Continuum Hypothesis. Let $\aleph_1 = 2^{\aleph_0}$, $\aleph_2 = 2^{\aleph_1}$, $\aleph_3 = 2^{\aleph_2}$, and so on. We assume that for any nonnegative integer k, there does not exist a cardinal number a such that $\aleph_k < a < \aleph_{k+1}$.

No one has produced a counterexample which would contradict the "generalized continuum hypothesis." However, the independence of this hypothesis from the axioms of "set theory" was established in 1963 by a brilliant young mathematician, Paul J. Cohen of Stanford University.

Exercise 3.1

1. If a, b, and d are three cardinal numbers then prove the following results.
   (i) $a^b \cdot a^d = a^{b+d}$ (ii) $(a^b)^d = a^{b \cdot d}$
   (iii) $a < b \implies a^d \leqslant b^d$ (iv) $a < b \implies d^a \leqslant d^b$
2. Show that (i) $\aleph_0^{\aleph_0} = c$ (ii) $\aleph_0^c = 2^c$ (iii) $c^{\aleph_0} = c$ (iv) $c^c = 2^c$
3. Prove that for any transfinite cardinal number t, $t + t = t$ and $t \cdot t = t$. (Use the axiom of choice).
4. Let a and b be two transfinite cardinal numbers. Show that $a + b = a \cdot b = \max \{a,b\}$.
5. Let the cardinal number of the class of all real valued functions of a real variable be denoted by f. Show that $f = c^c = 2^c = \aleph_2$.
6. Prove Theorem 3.1.2.
7. Prove Theorem 3.1.5(a).
8. Prove that the union of a denumerable number of denumerable sets is denumerable; i.e., that $S = \bigcup_{n=1}^{\infty} S_n$ is denumerable if each S_n is denumerable.
9. Prove that
$$\aleph_0 + n = \aleph_0$$
$$\aleph_0 + \aleph_0 = \aleph_0$$
$$\aleph_0 + \aleph_0 + \aleph_0 + \cdots = \aleph_0$$
$$\aleph_0 \cdot \aleph_0 = \aleph_0.$$
10. Show that the set of all points of R^n (cf. Sec. 2.4) with rational coordinates is denumerable.
11. If S is a subset of R^n and $S_1, S_2, \ldots, S_n$ are its coordinate sets then prove that the cardinal number of $S \geqslant$ the cardinal number of every S_i; and the

cardinal number of $S \leqslant$ the cardinal number of $S_1 \times S_2 \times \cdots \times S_n$. Give examples where strict inequality holds in each of these cases.

3.2. Order Types and Ordinal Numbers

In this section we discuss ordinal numbers and some of their fascinating properties. The reader may skip this section entirely, without any loss of continuity.

Definition 3.2.1. *All linearly ordered sets which are similar to each other are said to have the same order type.*

Here again an order type is defined as a common characteristic of a class of linear orders.

Definition 3.2.2. *The order type of a well-ordered set is called the ordinal number of that set.*

The ordinal number of the empty set is 0 and of the well ordered set $\{a\}$ is 1. The ordinal number of $\{a_1 \lessdot a_2\}$ is 2, and so on.

The ordinal number of the set of all natural numbers in their usual order is denoted by ω. It is important to note here that the order type of a linear order may be changed by the rearrangement of the elements. For example, if we arrange the natural numbers in the following manner:

$$\{1 \lessdot 3 \lessdot 5 \lessdot, \ldots; \lessdot 2 \lessdot 4 \lessdot \ldots\}$$

the linear order so obtained is not similar to the set of all natural numbers.

Definition 3.2.3. *Let A be a well-ordered set. If $a \in A$, then $\{x : x \lessdot a\}$ is called a left section (or an initial segment) of A determined by a, and is denoted by A_a.*

Definition 3.2.4. *Let α and β be two ordinal numbers. Let A and B be two well-ordered sets with ordinal numbers α and β, respectively. We say $\alpha < \beta$ if A is similar to a left section of B (and B is not similar to a left section of A).*

The statement within parenthesis is redundant as a consequence of a theorem which will be proved later (cf. Theorem 3.2.2).

Theorem 3.2.1. A well-ordered set is not similar to any of its left sections.

Proof. By way of contradiction we assume that a well ordered set A is similar to one of its left sections say $A_a (a \in A)$. Let $f : A \longrightarrow A_a$ be the bijective mapping which describes the similarity between A and A_a. Then

$$f(a) \in A_a \Longrightarrow f(a) \lessdot a \Longrightarrow f(f(a)) \lessdot f(a) \Longrightarrow f(f(f(a))) \lessdot f(f(a))$$

and so on. It is also obvious that $f(a), f(f(a))$, etc. are all distinct elements of A_a. Let A_1 be the set of all such elements. A_1 then must have a first element. Let a_1 be the first element of A_1. But again $f(a_1) \bigcirc a_1$, and thus $f(a_1) \in A_1$. Therefore we get the contradiction. ▌

Corollary. If A_a and A_b are two left sections of a wos (well-ordered set) A, and if A_a and A_b are similar, then $a = b$.

The proof of this corollary is simple and is left for the reader.

Theorem 3.2.2. If A is similar to a left section of B, then B cannot be similar to a left section of A.

Proof. Let A be similar to B_b, a left section of B. Let $f: A \longrightarrow B_b$ be the mapping which describes this similarity. Now if B is similar to a left section of A, say A_a, then there exists an order preserving bijective mapping $g: B \longrightarrow A_a$. The mapping $gf: A \longrightarrow A$ shows that A is similar to one of its left sections $g(B_b)$, which is a contradiction. ▌

We now state the following theorem without proof.

Theorem 3.2.3. If A and B are two well-ordered sets, then precisely one of the following statements is true:

    (i) A is similar to a left section of B

    (ii) B is similar to a left section of A

    (iii) A is similar to B

The proof of this theorem makes use of the principle of transfinite induction and can be found in R. L. Wilder's *Foundations of Mathematics*. This theorem is not based on the axiom of choice, however. Using Zermelo's well-ordering principle, we are now in a position to prove the comparability of any two cardinal numbers.

Theorem 3.2.4. Let A and B be any two sets. Then either A is equivalent to a subset of B or B is equivalent to a subset of A.

Proof. Assuming the well-ordering principle, both A and B can be well-ordered. The theorem now follows by the application of Theorem 3.2.3.

Definition 3.2.5. *Let A and B be two well-ordered disjoint sets with ordinal numbers α and β respectively. $\alpha + \beta$ is defined to be the ordinal number of the well-ordered set $C = A \cup B$, such that in constructing the linear order on C the order relations of A and B are preserved, and every element in A precedes every element in B (i.e., if $a \in A$ and $b \in B$, then $a \bigcirc b$.).*

It is very surprising to note that addition of ordinal numbers is not commutative. For example, let $A = \{a_1 \bigcirc a_2\}$ and let B be the set of all natural numbers in their usual order. By Def. 3.2.6, $(2 + \omega)$ is the ordinal number of $\{a_1 \bigcirc a_2 \bigcirc 1 \bigcirc 2 \bigcirc \ldots\}$ which is similar to the set of all natural numbers.

Hence $2 + \omega = \omega$,

Now $(\omega + 2)$ is the ordinal number of $\{1 \lessdot 2 \lessdot 3 \lessdot \ldots ; \lessdot a_1 \lessdot a_2\}$, which is not similar to the set of all natural numbers in their usual order.

$$\text{So } \omega + 2 \neq \omega,$$

$$\omega + 2 \neq 2 + \omega.$$

In general it can be shown that $n + \omega = \omega$, but $\omega + n \neq \omega$. Also it is easy to observe that $\omega < \omega + n$.

Definition 3.2.6. *Let α and β be two ordinal numbers and let A and B be two well ordered sets with ordinal numbers α and β respectively. We define $\alpha \cdot \beta$ to be the ordinal number of the well ordered set $A \times B$ with the following linear relation:*

$$(a,b) < (a',b') \Longleftrightarrow \text{either} \quad b < b' \text{ or } \quad b = b' \text{ and } a < a',$$

where $a, a' \in A$ and $b, b' \in B$.

To illustrate the above definition we consider the following example.

Let A be the well-ordered set of all natural numbers in their usual order. Let $B = \{b_1 < b_2 < b_3\}$. We define the following linear order relation on $A \times B$:

$$\{(1,b_1) < (2,b_1) < (3,b_1) < \ldots ; < (1,b_2) < (2,b_2) < (3,b_2) < \ldots ;$$

$$< (1,b_3) < (2,b_3) < (3,b_3) < \ldots \}.$$

Observe that $(3,b_1) < (2,b_2) < (1,b_3)$ since $b_1 < b_2 < b_3$.

In this example the ordinal number of the well-ordered set so constructed is $\omega \cdot 3$. It is easy to observe that $\omega \cdot 3 \neq \omega$.

To find $3 \cdot \omega$ we construct the well-ordered set $(B \times A)$ as follows:

$$\{(b_1,1) \lessdot (b_2,1) \lessdot (b_3,1) \lessdot (b_1,2) \lessdot (b_2,2) \lessdot (b_3,2) \lessdot (b_1,3)$$

$$\lessdot (b_2,3) \lessdot (b_3,3) \lessdot \ldots \}.$$

This wos is similar to the well-ordered set A, for the mapping

$$f[3(n - 1) + 1] = (b_1,n)$$

$$f[3(n - 1) + 2] = (b_2,n)$$

$$f[3(n - 1) + 3] = (b_3,n)$$

is a bijective, order preserving mapping. Therefore, $3 \cdot \omega = \omega$, which also proves that $3 \cdot \omega \neq \omega \cdot 3$.

Like addition, the multiplication of ordinal numbers is not commutative, for we can easily show that $n \cdot \omega = \omega$ and $\omega < \omega \cdot n$.

We shall state the following fundamental properties of addition and multiplication of ordinal numbers without proof.

(a) $\alpha + (\beta + \gamma) = (\alpha + \beta) + \gamma$

(b) $\alpha(\beta\gamma) = (\alpha\beta)\gamma$

(c) $\alpha(\beta + \gamma) = \alpha\beta + \alpha\gamma$.

It is important to note here that multiplication is distributive over addition only from the left. The other form of the distributive law, that is, $(\beta + \gamma)\alpha = \beta\alpha + \gamma\alpha$ does not always hold, as illustrated by the following:

$$(\omega + 3) \cdot 2 = (\omega + 3) + (\omega + 3) = \omega + (3 + \omega) + 3$$

$$= \omega + \omega + 3 = \omega \cdot 2 + 3.$$

$$(\omega + 3) \cdot 2 \neq \omega \cdot 2 + 3 \cdot 2.$$

$\omega \cdot \omega$ is sometimes written as ω^2.

Definition 3.2.8. *Let α and β be two ordinal numbers such that $\alpha < \beta$. If there does not exist any ordinal number γ such that $\alpha < \gamma < \beta$, then we say α is the immediate predecessor of β and β is the immediate successor of α.*

Ordinal numbers which have immediate predecessors are called *ordinal numbers of first kind.* For example, 3, $\omega + 1$, $\omega^2 + \omega \cdot 2 + 1$ are ordinal numbers of the first kind.

Ordinal numbers which do not have immediate predecessors are called *ordinal numbers of the second kind*; for example, ω, ω^2, $\omega \cdot 2$, $\omega^2 + 3\omega$.

Theorem 3.2.5. If α is an ordinal number and if $S = \{\beta : \beta < \alpha, \beta$ is an ordinal number$\}$, then S is well-ordered and its ordinal number is α.

Proof. $\alpha = 0 \Longrightarrow S$ is empty and in this case, the theorem is trivially true.

We can now assume that $\alpha > 0$ and that S is nonempty. Let A be a well-ordered set with ordinal number α. Now for every $x \in A$, A_x is a left section of A whose ordinal number is less than α, say β. Obviously $\beta \in S$. Defining $f(x) = \beta$ gives a mapping $f : A \longrightarrow S$ which is injective because of the corollary of Theorem 3.2.1. f is also bijective, for if $\beta \in S$, then $\beta < \alpha$; but then β is the ordinal number of some left section A_x of A, implying that $f(x) = \beta$. Also, the mapping is order preserving, since $x_1 < x_2 \Longrightarrow A_{x_1}$ is a left section of $A_{x_2} \Longrightarrow f(x_1) < f(x_2)$. Hence S is similar to A, and we get the result. |

It must be observed here that the proof fails if we do not include 0 in our ordinal number system.

Theorem 3.2.6. Every set of ordinal numbers is well-ordered.

Proof. Let S be a set of ordinal numbers. S is certainly a linear order. Let A be a nonempty subset of S and consider $\alpha \in A$. If α is the first element of A the theorem is proved. If α is not the first element of A, consider $A_1 = \{\beta : \beta < \alpha\}$. Clearly, $A_1 \cap A \neq \emptyset$. Let $A_2 = A_1 \cap A$. By Theorem 3.2.5, A_1 is well ordered and A_2, being a nonempty subset of A_1, must have a first element, say γ; moreover, $\gamma < \alpha$. If γ is not the first element of A then $\exists$ a δ in A such that $\delta < \gamma \Longrightarrow \delta < \alpha \Longrightarrow \delta \in A_1 \Longrightarrow \delta \in A_2$, which contradicts the fact that γ is the first element of A. This completes the proof of the theorem. |

From Theorem 3.2.5, it follows that if we write all ordinal numbers less than a given ordinal number in their natural order, then the well ordered set so obtained has ordinal number which is greater than every ordinal number in the set but is less than or equal to every other ordinal number. For example, the set of all finite ordinal numbers has ordinal number ω which is the smallest transfinite ordinal number.

The class of denumerable well-ordered sets has many transfinite ordinal numbers which are expressible in terms of polynomials in ω. If we write all ordinal numbers of the class of countable sets in their natural order, then the well ordered set so obtained has an ordinal number which is denoted by Ω. It is now obvious that Ω cannot be the ordinal number of a countable set. Therefore, Ω may be called the first nondenumerable (or uncountable) ordinal number.

We conclude this chapter with a well-known paradox.

Burali-Forti Paradox. Let S be the set of *all* ordinal numbers. Then S is well ordered and has an ordinal number which is greater than any of the ordinal numbers in S, and therefore cannot belong to S.

The term "set of all ordinal numbers" has become self-contradictory because of this paradox.

Exercise 3.2

1. Prove that $\omega \cdot n = \omega + \omega + \ldots$ (n times).
2. Show that $(\omega \cdot n) + m < \omega^2$.
3. Prove that if α, β, γ are ordinal numbers, then
   (a) $\alpha + (\beta + \gamma) = (\alpha + \beta) + \gamma$
   (b) $\alpha (\beta\gamma) = (\alpha\beta)\gamma$
   (c) $\alpha (\beta + \gamma) = \alpha\beta + \alpha\gamma$.

4

Sequences and Limits

4.1. Convergence and Limit

This chapter deals with "sequences and their limits." The foundation of this subject was firmly laid by A. L. Cauchy (1789–1857) who may be regarded as the father of modern analysis. His work on sequences, series, and complex variables made a great impact on analysis.

Definition 4.1.1. *A sequence is a "well-ordering" which is similar to the well-ordering of all natural numbers. The elements of the well-ordered set of the sequence are called* terms *of the sequence.*

From the definition it follows that we can write a sequence as $\{a_1, a_2, \ldots, a_n, \ldots\}$. Here a_1 is called the first term of the sequence, a_2 the second term, and so on; a_n is called the nth term or the general term of the sequence. Such a sequence is often written as $\{a_n\}$; for example, $\left\{\dfrac{1}{n}\right\}$ denotes the sequence $1, \dfrac{1}{2}, \dfrac{1}{3}, \dfrac{1}{4}, \ldots, \dfrac{1}{n}, \ldots$ and $\{(-1)^n\}$ denotes the sequence $\{-1, 1, -1, 1, \ldots, (-1)^n, \ldots\}$. In the latter sequence there are only two distinct elements yet there are infinitely many terms of the sequence, and this sequence should not be confused with the set $\{-1, 1\}$. Such a set may be called the *associated set* of the sequence.

A sequence may also be regarded as a function (mapping) with the set of natural numbers as the domain and the associated set as the range.

In Secs. 4.1, 4.2, 4.3 we shall consider only the sequences of real numbers, where the associated sets are subsets of R.

Definition 4.1.2. *A sequence $\{a_n\}$ of real numbers is said to be a Cauchy sequence if for all $\epsilon > 0$, $\exists\, n_0 \in \mathfrak{N}$, $\ni |a_n - a_m| < \epsilon$ for $n, m \geqslant n_0$.*

The above definition is called the *Cauchy's criterion.* Intuitively, this criterion implies that for sufficiently large values of n, the terms of the sequence are getting closer and closer to each other.

In the real number system there is an important concept of "limit," which we are now going to define.

Definition 4.1.3. *A sequence $\{a_n\}$ is said to converge to a limit L if for all $\epsilon > 0$, $\exists n_0 \in \mathfrak{N}$, $\ni |a_n - L| < \epsilon$ for $n \geqslant n_0$.* We write $\lim_{n \to \infty} a_n = L$. *A sequence having a limit is said to be convergent.*

By a *neighborhood* of a point x_0 we mean an open interval containing the point x_0. An ϵ-*neighborhood* of the point x_0 is the open interval $(x_0 - \epsilon, x_0 + \epsilon)$. For examples, $\left(-\dfrac{1}{3}, \dfrac{1}{2}\right)$ is a neighborhood of 0, and $(-\epsilon, \epsilon)$ is an ϵ-neighborhood of 0.

The limit of a sequence in terms of neighborhoods can then be defined as follows.

Definition 4.1.3(a). *A sequence $\{a_n\}$ is said to have a limit L if every neighborhood of L contains all but a finite number of terms of the sequence.*

The proof of the equivalence of Defs. 4.1.3 and 4.1.3(a) is as follows:

If every neighborhood of L contains all but a finite number of terms of a sequence $\{a_n\}$, then for $\epsilon > 0$ the ϵ-neighborhood of L contains all but a finite number (say n_0) of terms of the sequence. In other words $a_n \in (L - \epsilon, L + \epsilon)$ for $n \geqslant n_0$, that is $|a_n - L| < \epsilon$ for $n \geqslant n_0$. Therefore, Def. 4.1.3(a) $\Longrightarrow$ Def. 4.1.3.

Conversely, if L is the limit of the sequence $\{a_n\}$ according to Def. 4.1.3 and if (a,b) is an arbitrary neighborhood of L, let $\epsilon = \min \{(L - a), (b - L)\}$. For this $\epsilon > 0$, $\exists n_0 \in \mathfrak{N}$, $\ni |a_n - L| < \epsilon$ for $n \geqslant n_0$; i.e., $a_n \in (L - \epsilon, L + \epsilon)$, for $n \geqslant n_0$. Therefore $(L - \epsilon, L + \epsilon)$ contains all except at most n_0 terms of the sequence. But $(a,b) \supset (L - \epsilon, L + \epsilon)$. Consequently, the neighborhood (a,b) of L contains all but a finite number of terms of the sequence, Def. 4.1.3(a) is satisfied, and the proof is complete.▐

Theorem 4.1.1. If a sequence has a limit then it is a Cauchy sequence.

Proof. Let L be the limit of $\{a_n\}$. Then for all $\epsilon > 0$ and $n_0 \in \mathfrak{N}$ we have $|a_n - L| < \dfrac{\epsilon}{2}$ for $n \geqslant n_0$. For $m, n \geqslant n_0$, using the triangle inequality we get $|a_n - a_m| = |a_n - L + L - a_m| \leqslant |a_n - L| + |a_m - L| < \dfrac{\epsilon}{2} + \dfrac{\epsilon}{2} = \epsilon$, which is precisely the Cauchy criterion.▐

Later (Theorem 4.2.7) we shall prove that if a sequence of real numbers is a Cauchy sequence then it must have a limit. This property is called the *property of completeness*, and is formally defined as follows.

Definition 4.1.4. *A system S is called complete if every Cauchy sequence in S has a limit $L, L \in S$.*

We may remark here that the system of rational numbers is not complete since $\{2, 2.7, 2.71, 2.718, \ldots\}$ is a sequence of rational numbers and its limit is e (the base for natural logarithms) which is not a rational number. The sequence, of course, is a Cauchy sequence. The system of real numbers, however, is complete. We have the following logical equivalence.

Theorem 4.1.2. Greatest lower bound property (or equivalently lub property) $\Longleftrightarrow$ property of completeness.

The proof of this theorem is omitted.

The limit of a sequence, if it exists, is always unique as is shown by the following theorem.

Theorem 4.1.3. If a sequence has a limit, then the limit must be unique.

Proof. Let $\{a_n\}$ have limits L and L', and suppose $L \neq L'$. Let $\epsilon = \frac{1}{2} |L - L'|$ (notice $\epsilon > 0$). Now the ϵ-neighborhoods of L and L' are disjoint and both of them contain all but a finite number of terms of the sequence, which is impossible. Hence $L = L'$, which proves the theorem.∎

Theorem 4.1.4. Every Cauchy sequence is bounded.

Proof. Let $\{a_n\}$ be a Cauchy sequence. For $\epsilon = p$ (p being a fixed positive number) $\exists n_0 \in N \ni |a_n - a_m| < p$ for $n, m \geqslant n_0$. In particular,

$$|a_n - a_{n_0}| < p, \text{ for } n \geqslant n_0$$

Now

$$|a_n| = |a_n - a_{n_0} + a_{n_0}| \leqslant |a_n - a_{n_0}| + |a_{n_0}| < p + |a_{n_0}|, \text{ for } n \geqslant n_0$$

Let

$$M = \max \{|a_1|, |a_2|, \ldots, |a_{n_0 -1}|, |a_{n_0}| + p\}$$

Then $|a_k| < M$, for $k = 1, 2, 3, \ldots$ which proves the theorem.∎

Corollary. Every convergent sequence is bounded.

The converse of this theorem is not necessarily true. However, there are some special types of sequences (monotonic sequences) which, if bounded, are convergent and satisfy Cauchy criterion. We shall prove this result in Theorem 4.1.9.

EXAMPLE 4.1.1.

(a) Consider the sequence $\{3.1, 3.14, 3.141, \ldots, 3.d_1 d_2 \ldots d_n, \ldots\}$ where d_n is the nth decimal figure in the expansion of π. This is a sequence of rational numbers and is convergent according to Cauchy's criterion. Its limit is π which is not a rational number.

(b) Consider the sequence $\left\{1, 1, \frac{1}{2}, 2, \frac{1}{3}, 3, \frac{1}{4}, 4, \ldots\right\}$. Here $a_{2n} = n$, and $a_{2n-1} = \frac{1}{n}$. This sequence is not convergent. It is interesting to note, however, that every neighborhood of 0 contains infinitely many terms of the sequence, and at the same time it excludes infinitely many terms also.

We are now going to discuss some properties of convergent sequences.

Theorem 4.1.5. If $\{a_n\}$ and $\{b_n\}$ are two convergent sequences, then the sequence $\{a_n + b_n\}$ is convergent, and if the $\lim_{n \to \infty} a_n = L_1$ and $\lim_{n \to \infty} b_n = L_2$, then $\lim_{n \to \infty} (a_n + b_n) = L_1 + L_2$.

Proof. For arbitrary $\epsilon > 0$, the hypothesis guarantees that $\exists\ n_1 \in \mathfrak{N}$, $\exists\ |a_n - L_1| < \frac{\epsilon}{2}$ for $n \geqslant n_1$, and that $\exists\ n_2 \in \mathfrak{N}$, $\exists\ |b_n - L_2| < \frac{\epsilon}{2}$ for $n \geqslant n_2$. Letting $n = \max \{n_1, n_2\}$, it follows that $|(a_n + b_n) - (L_1 + L_2)| = |(a_n - L_1) + (b_n - L_2)| \leqslant |a_n - L_1| + |b_n - L_2| < \frac{\epsilon}{2} + \frac{\epsilon}{2} = \epsilon$ for $n \geqslant n_0$, establishing the result.

Theorem 4.1.6. If $\{a_n\}$ converges to zero and the sequence $\{b_n\}$ is bounded; then the sequence $\{a_n \cdot b_n\}$ converges to zero.

Proof. Since $\{b_n\}$ is bounded, $\exists\ M > 0 \ni |b_n| < M$ for every n. Given $\epsilon > 0$, $\exists\ n_0 \in \mathfrak{N} \ni |a_n| < \frac{\epsilon}{M}$ for $n \geqslant n_0$, since $\{a_n\}$ converges to zero.

For $n \geqslant n_0$, $|a_n \cdot b_n| = |a_n| \cdot |b_n| < \frac{\epsilon}{M} \cdot M = \epsilon$. This means that $\{a_n \cdot b_n\}$ converges to zero, and that completes the proof.

Note here that the conclusion of this theorem is true if $\{b_n\}$ is a convergent sequence. For by Theorem 4.1.4, a convergent sequence is bounded.

Theorem 4.1.7. If the sequences $\{a_n\}$ and $\{b_n\}$ converge to L and L' respectively, then $\{a_n \cdot b_n\}$ converges to $L \cdot L'$.

Proof. If either L or L' is zero then the result follows from the last theorem. Therefore we may assume that $L \neq 0$ and $L' \neq 0$. By Theorem 4.1.4, $\{a_n\}$ is bounded. Thus there exists $M > 0 \ni |a_n| < M$ for every n. Using the convergence of $\{a_n\}$ and $\{b_n\}$ we know that for every $\epsilon > 0$ there is $n_1 \in \mathfrak{N} \ni |a_n - L| < \dfrac{\epsilon}{2|L'|}$ for $n \geqslant n_1$.

Also $\exists\, n_2 \in \mathfrak{N} \ni |b_n - L'| < \dfrac{\epsilon}{2M}$ for $n \geqslant n_2$. Let $n_0 = \max\ \{n_1, n_2\}$. Then
$$|a_n b_n - LL'| = |a_n b_n - a_n L' + a_n L' - LL'| = |a_n(b_n - L') + L'(a_n - L)| \leqslant |a_n| \cdot |b_n - L'| + |L'|\,|a_n - L|.$$

Consequently, for $n \geqslant n_0$, $|a_n b_n - LL'| < M \cdot \dfrac{\epsilon}{2M} + |L'| \cdot \dfrac{\epsilon}{2|L'|} = \epsilon.$ Hence the result.

Theorem 4.1.7(a). If $\lim\limits_{n \to \infty} a_n = L$ and $\lim\limits_{n \to \infty} b_n = L'$, and if $b_n \neq 0$ for every n and $L' \neq 0$, then $\lim\limits_{n \to \infty} \dfrac{a_n}{b_n} = \dfrac{L}{L'}$.

The proof is left as an exercise.

Definition 4.1.5. *Two sequences $\{a_n\}$ and $\{b_n\}$ are said to be equivalent if for $\epsilon > 0\ \exists\, n_0 \in \mathfrak{N} \ni |a_n - b_n| < \epsilon$ for $n \geqslant n_0$. In other words the sequences $\{a_n\}$ and $\{b_n\}$ are equivalent if $\lim\limits_{n \to \infty} (a_n - b_n)$ exists and is zero. It is not necessary here for $\lim a_n$ and $\lim b_n$ to exist, however, as will be shown by an example following the next theorem.*

Theorem 4.1.8. If $\{a_n\}$ and $\{b_n\}$ are any two equivalent sequences then either both of them converge to the same limit or neither of them converges.

Proof. If neither $\{a_n\}$ converges nor $\{b_n\}$, then the theorem is satisfied. If, however, one of them converges (to a limit) we shall show that the other will also converge to the same limit.

Suppose $\lim\limits_{n \to \infty} a_n = L$. Now $\lim\limits_{n \to \infty} (a_n - b_n) = 0$. From Theorem 4.1.5, $\lim\limits_{n \to \infty} b_n = \lim\limits_{n \to \infty} [(b_n - a_n) + a_n]$ exists and $\lim\limits_{n \to \infty} b_n = \lim\limits_{n \to \infty} (b_n - a_n) + \lim\limits_{n \to \infty} a_n = \lim\limits_{n \to \infty} a_n$, which proves our result.

EXAMPLE 4.1.2:

(a)   The sequences $\left\{\dfrac{n^3 + 1}{n^2}\right\}$ and $\left\{\dfrac{n^3 + n + 1}{n^2}\right\}$ are equivalent for
$$\left|\frac{n^3 + 1}{n^2} - \frac{n^3 + n + 1}{n^2}\right| = \frac{1}{n} < \epsilon,$$
for $n > \dfrac{1}{\epsilon}$, but neither of these sequences converges. Observe here that $\lim\limits_{n \to \infty} (a_n - b_n)$ exists and is equal to zero, though $\lim\limits_{n \to \infty} a_n$ and $\lim\limits_{n \to \infty} b_n$ do not exist.

(b) The sequences $\left\{\dfrac{n}{n+1}\right\}$ and $\left\{\dfrac{n+1}{n}\right\}$ are equivalent and both have the limit 1.

Definition 4.1.6. *A sequence $\{a_n\}$ is said to be monotonic if either $a_n \leqslant a_{n+1}$ for every n, or $a_n \geqslant a_{n+1}$ for every n. In the first case, the sequence is called monotonically nondecreasing (montonically increasing if $a_n < a_{n+1}$ for every n), and in the latter case the sequence is called monotonically nonincreasing (monotonically decreasing if $a_n > a_{n+1}$ for every n.)*

For the sake of convenience we shall drop the term "monotonically" from "monotonically nonincreasing" and "monotonically nondecreasing."

Theorem 4.1.9. If a nonincreasing sequence is bounded below then it is convergent and its limit is its glb.

Proof. Let $\{a_n\}$ be a nonincreasing sequence which is bounded below. By the greatest lower bound property it has a greatest lower bound, say L. For $\epsilon > 0 \; \exists \, n_0 \in \mathfrak{N} \ni a_{n_0} < L + \epsilon$. Since $\{a_n\}$ is nonincreasing, $a_n \leqslant a_{n_0}$ for $n \geqslant n_0$,

$$a_n < L + \epsilon \quad \text{for} \quad n \geqslant n_0 . \qquad (4.1)$$

Also since L is a lower bound of a_n, we have $a_n \geqslant L$ for every n,

$$\Longrightarrow a_n > L - \epsilon \qquad \text{for every } n . \qquad (4.2)$$

Combining (4.1) and (4.2), we get

$$|a_n - L| < \epsilon \qquad \text{for } n \geqslant n_0$$

i.e., $\{a_n\}$ converges to L.

Exercise 4.1

1. Prove that if a nondecreasing sequence is bounded above then it is convergent. What is the limit of such a sequence?
2. Determine which of the following sequences are convergent and find the limit of each convergent sequence.

   (a) $\left\{1, -1, \dfrac{1}{2}, -\dfrac{1}{2}, \dfrac{1}{3}, -\dfrac{1}{3}, \ldots \ldots\right\}$. $a_n = ?$

   (b) $\left\{\dfrac{1}{2}, -\dfrac{1}{2}, \dfrac{2}{3}, -\dfrac{2}{3}, \dfrac{3}{4}, -\dfrac{3}{4}, \ldots \ldots\right\}$. $a_n = ?$

   (c) $\{1, 1.7, 1.73, \ldots\}$, $a_n = 1.d_1 d_2 d_3 \ldots d_n$, d_k being the kth decimal figure in $\sqrt{3}$.

   (d) $\left\{\sin \dfrac{\pi}{3}, \sin \dfrac{2\pi}{3}, \ldots, \sin \dfrac{n\pi}{3}, \ldots\right\}$.

3. (a) If $\lim\limits_{n \to \infty} a_n = L$, prove that $\lim\limits_{n \to \infty} (ka_n) = kL$, k being any real number.

    (b) If $\lim\limits_{n \to \infty} a_n = L$ and $\lim\limits_{n \to \infty} b_n = L'$, show that

$$\lim_{n \to \infty} (a_n - b_n) = L - L'.$$

4. Prove Theorem 4.1.7(a).

5. Show that if $\{a_n\}$ converges to l then the sequence $\left\{ \dfrac{a_1 + a_2 + \cdots + a_n}{n} \right\}$ also

    converges to l.

    [*Hint*: Let $a_n - l = \alpha_n$. Then $\{\alpha_n\}$ converges to 0, and $\dfrac{a_1 + a_2 + \cdots a_n}{n} - l =$

    $\dfrac{\alpha_1 + \alpha_2 + \ldots + \alpha_n}{n}$. Now establish that $\left\{ \dfrac{\alpha_1 + \alpha_2 + \ldots + \alpha_n}{n} \right\}$ converges to

    zero.]

6. Let $a_n = 1 + \dfrac{1}{1!} + \dfrac{1}{2!} + \dfrac{1}{3!} + \cdots + \dfrac{1}{n!}$. Show that $\{a_n\}$ is increasing and

    bounded above, and hence convergent.

$$\left(Hint: a_n < 1 + \left[1 + \frac{1}{2} + \frac{1}{2^2} + \cdots + \frac{1}{2^n} \right]. \right)$$

7. Without using Theorem 4.1.7(a), prove that if $\lim\limits_{n \to \infty} a_n = L \neq 0$, where $a_n \neq 0$

    for any n, then $\lim\limits_{n \to \infty} \dfrac{1}{a_n} = \dfrac{1}{L}$.

4.2. Limits Superior (upper limits) and Limits Inferior (lower limits)

There are many bounded sequences which are not convergent. For such sequences the idea of limit superior and limit inferior is useful. First, we introduce the idea of a subsequence.

Definition 4.2.1. *Let $\{a_n\}$ be a given sequence. For every increasing sequence $\{n_1, n_2, \ldots, n_k, \ldots\}$ of natural numbers, the sequence $\{a_{n_k}\}$ ($k = 1, 2, \ldots$) is called a subsequence of $\{a_n\}$.*

For example, $\{a_1, a_4, a_7, a_{10}, \ldots\}$ is a subsequence of $\{a_1, a_2, a_3, \ldots\}$. However, $\{a_{10}, a_4, a_7, a_1, a_{13}, \ldots\}$ is not a subsequence of $\{a_1, a_2, a_3, \ldots\}$. In other words, the linear order of a subsequence of $\{a_n\}$ should be the same as that of the original sequence.

Theorem 4.2.1. A sequence $\{a_n\}$ converges to $l \Longleftrightarrow$ every subsequence of $\{a_n\}$ converges to l.

Proof. If $\{a_n\}$ converges to l then for $\epsilon > 0$ there exists a natural number m such that $|a_n - l| < \epsilon$ for $n \geq m$. Now if $\{a_{n_k}\}$ is a subsequence of $\{a_n\}$, then $n_m \geq m$, which would imply $|a_{n_k} - l| < \epsilon$, for $n_k \geq n_m$; hence the subsequence $\{a_{n_k}\}$ also converges to l.∎

The converse follows trivially from the fact that $\{a_n\}$ is a subsequence of itself.∎

In Sec. 4.1 we proved that every convergent sequence is bounded, but the converse of this statement is not necessarily true. For example $\{(-1)^n\}$ is bounded but not convergent. We do have the following result.

Theorem 4.2.2. Every bounded sequence of real numbers has a convergent subsequence.

The proof of this theorem depends upon Bolzano-Weierstrass theorem, and we shall give it in Chapter 5. At present we accept it without proof.

Let $\{a_n\}$ be a bounded sequence of real numbers. As we said earlier, it must have at least one convergent subsequence. Let $\{L_\alpha\}$ be the set of all subsequential limits of $\{a_n\}$; that is, each L_α is the limit of a convergent subsequence of $\{a_n\}$. It can be easily shown that the set $\{L_\alpha\}$ is bounded (in fact, the glb of $\{a_n\}$ is a lower bound of $\{L_\alpha\}$ cf. Exercise 4.2(5)).

Definition 4.2.2. *The glb of $\{L_\alpha\}$ is called the limit inferior or the lower limit of $\{a_n\}$. It is denoted by* $\liminf_{n \to \infty} a_n$ *or* $\underline{\lim}_{n \to \infty} a_n$ *or simply by* $\underline{\lim} a_n$.

A similar definition is given for the limit superior or the upper limit of $\{a_n\}$, which is denoted by $\limsup_{n \to \infty} a_n$ *or by* $\overline{\lim} a_n$ *or simply by* $\overline{\lim} a_n$.

It must be cautioned here that lower limit of a sequence is not necessarily the glb of that sequence, nor is the upper limit the lub. In fact the lower limit may not be even a lower bound [Example 4.2.1(c)].

EXAMPLE 4.2.1.
(a) The sequence $\{-1, 1, -1, 1, \ldots\}$ has lower limit -1 and upper limit 1.
(b) The sequence $\left\{-\frac{1}{2}, 0, \frac{1}{2}, 0, -\frac{2}{3}, 0, \frac{2}{3}, 0, -\frac{3}{4}, 0, \frac{3}{4}, 0, \ldots\right\}$ has lower limit -1,
and upper limit 1. Notice here that 0 is also a subsequential limit.
(c) The sequence $\left\{-2, 2, -\frac{3}{2}, \frac{3}{2}, -\frac{4}{3}, \frac{4}{3}, \ldots\right\}$ has lower limit -1, and upper
limit 1. It is interesting to note, however, that none of the terms of the sequence lies in $(-1, 1)$. -1 is not a lower bound, nor 1 an upper bound.

The following result is obvious from the definition.

Theorem 4.2.3. $\underline{\lim} a_n \leq \overline{\lim} a_n$.
The proof is left for the reader.

Also we have the following result:

Theorem 4.2.4.   A sequence $\{a_n\}$ converges to $l \Longleftrightarrow \overline{\lim} \, a_n = \underline{\lim} \, a_n = l$.

The proof follows easily from the Def. 4.2.2 and Theorem 4.2.1.

Lower and upper limits of a sequence may be defined in various other ways. The following theorem gives some alternate definitions for the lower limit.

Theorem 4.2.5.   Let $\{a_n\}$ be a bounded sequence, then the following statements are logically equivalent.

(a) $L = \lim_{n \to \infty} \inf a_n$ (according to Def. 4.2.2).

(b)   For $\epsilon > 0$ there is natural number n_0 such that $a_n > L - \epsilon$ for $n \geqslant n_0$, and $a_n < L + \epsilon$ for infinitely many terms of the sequence.

(c)   Let $\alpha_n = \text{glb} \, \{a_n : n \geqslant m\}$, then $L = \text{lub} \, \{\alpha_m : m \geqslant 1\}$.

The proof of this theorem is left as a challenging exercise.

A similar theorem for the upper limit may be stated as follows.

Theorem 4.2.6. $U = \lim_{n \to \infty} \sup a_n \Longleftrightarrow$ for $\epsilon > 0$ there is a natural number n_0 such that $a_n < U + \epsilon$ for $n \geqslant n_0$, and $a_n > U - \epsilon$ for infinitely many terms of $\{a_n\} \Longleftrightarrow U = \text{glb} \, [\beta_m : m \geqslant 1]$ where $\beta_m = \text{lub} \, \{a_n : n \geqslant m\}$.

The proof is left as an exercise.

We now generalize the idea of upper and lower limits to include the unbounded sequences.

Definition 4.2.2.   (a)   *If a sequence* $\{a_n\}$ *is unbounded above then we define* $\lim \sup a_n = +\infty$.

(b)   *If the sequence* $\{a_n\}$ *is unbounded below then we define* $\lim \inf a_n = -\infty$.

(c)   *If the sequence* $\{a_n\}$ *is unbounded below and there is no subsequential limit then we write* $\overline{\lim} \, a_n = \underline{\lim} \, a_n = \lim a_n = -\infty$.

(d)   *If the sequence* $\{a_n\}$ *is unbounded above and there is no subsequential limit then we say* $\overline{\lim} \, a_n = \underline{\lim} \, a_n = \lim a_n = +\infty$.

EXAMPLE 4.2.2.   (a)   The sequence $\{-1, 1, -2, 2, \ldots, (-1)^n n, \ldots\}$ has upper limit $+\infty$ and lower limit $-\infty$.

(b)   For the sequence $\{1, 2, 3, \ldots, n, \ldots\}$, $\overline{\lim} \, a_n = \underline{\lim} \, a_n = +\infty$.

(c)   For the sequence $\{-1, -2, -3, \ldots, (-n), \ldots\}$ $\overline{\lim} \, a_n = \underline{\lim} \, a_n = -\infty$.

(d)   For the sequence $\{1, 0, 2, 0, 3, 0, \ldots\}$ $\overline{\lim} \, a_n = +\infty$, $\underline{\lim} \, a_n = 0$.

(e)   For the sequence $\{-1, 0, -2, 0, -3, 0, \ldots\}$ $\overline{\lim} \, a_n = 0$, $\underline{\lim} \, a_n = -\infty$.

We conclude this section by proving the important theorem on completeness of the real-number system.

Theorem 4.2.7. Every Cauchy sequence of real numbers has a limit.

Proof. Let $\{a_n\}$ be a Cauchy sequence of real numbers. Then for $\epsilon > 0$, there is a natural number, n_i, such that $|a_n - a_m| < \dfrac{\epsilon}{2}$ for $n, m \geqslant n_i$. By Theorem 4.1.4, $\{a_n\}$ must be bounded and then, according to Theorem 4.2.2, it must have a convergent subsequence, say $\{a_{n_k}\}$.

Let $l = \lim_{n_k \to \infty} a_{n_k}$. Therefore, for $\epsilon > 0$, there is a natural number n_j such that $|a_{n_k} - l| < \dfrac{\epsilon}{2}$ for $n_k \geqslant n_j$.

Let $n_0 = \max \{n_i, n_j\}$. Then for $n \geqslant n_0$ and $n_k \geqslant n_0$, we have

$$|a_n - l| = |a_n - a_{n_k} + a_{n_k} - l|$$
$$\leqslant |a_n - a_{n_k}| + |a_{n_k} - l|$$
$$< \frac{\epsilon}{2} + \frac{\epsilon}{2}$$
$$= \epsilon.$$

Hence, $\lim_{n \to \infty} a_n = l$, and the proof is complete.

Exercise 4.2.

1. Find the lower and upper limits of the following sequences.
   (a) $\{[2 - (-1)^n]^n\}$.
   (b) $\left\{(-1)^n + \cos \dfrac{n\pi}{4}\right\}$
   (c) $\left\{\cos \dfrac{n\pi}{4} + \cos n\right\}$
   (d) $\left\{e^{2n} \cos \dfrac{n\pi}{2}\right\}$
   (e) $\left\{(-1)^n \left(2 + \dfrac{5}{n}\right)\right\}$.

2. Prove Theorem 4.2.5.

3. Prove Theorem 4.2.6.

4. If $\{a_n\}$ and $\{b_n\}$ are two sequences and there is $n_0 \in \mathfrak{N}$ such that $a_n \leqslant b_n$ for $n \geqslant n_0$, then show that $\underline{\lim} \, a_n \leqslant \underline{\lim} \, b_n$, and $\overline{\lim} \, a_n \leqslant \overline{\lim} \, b_n$.

5. Let $\{L_\alpha\}$ be the set of all subsequential limits of $\{a_n\}$, and let A be the associated set of $\{a_n\}$. Show that a lower bound of A is always a lower bound of $\{L_\alpha\}$, and an upper bound of A is always an upper bound of $\{L_\alpha\}$. (*Hint:* Prove by way of contradiction.) Show by examples that the converses of these statements are not necessarily true.

4.3. Another Approach to Real-Number System

In Chapter 2 we used Dedekind's cut to construct real numbers from rationals. In doing so we used the property of linear order relation. We can obtain real number system by using the idea of Cauchy sequences of rationals— that is, all those sequences whose associated sets are subsets of $\mathfrak{Q}$ (rational numbers) and satisfy Cauchy's criterion. We divide them into equivalence classes such that two sequences belong to the same class if they are equivalent to each other (according to Def. 4.1.5), keeping in mind that each one of them is a Cauchy sequence—and these are the sequences of rational numbers.

We can now identify each equivalence class by a real number—the real number which is the limit of each one of the sequences of the class. For example, 0 can be identified by the class of all sequences of rational numbers which are equivalent to $\left\{\dfrac{1}{n}\right\}$; and 1 can be identified by the class of all sequences which are equivalent to $\left\{\dfrac{n}{n+1}\right\}$, and so on.

Suppose now $\{a_n\}$ is a sequence of the class which represents the real number α, and $\{b_n\}$ is a sequence of the class which represents the real number β, then $\alpha + \beta$ is obtained from the class which contains the sequence $\{a_n + b_n\}$ (cf. Theorem 4.1.6), and $\alpha \cdot \beta$ from the class equivalent to $\{a_n \cdot b_n\}$ (cf. Theorem 4.1.7). This way, a development of real numbers can take place by using Cauchy sequences of rationals, but we shall not attempt to do it here.

In order to relate this representation of real numbers with that of Dedekind's cuts, we proceed as follows.

Let a real number r be identified by a cut $(L|U)$. Let p_1 be a rational number in L. Obviously, p_1 is not the largest rational in L. Now consider the set $\left\{p_1 + \dfrac{1}{n} : n \in \mathfrak{N}\right\}$. This set is bounded above by $p_1 + 1$. Choose the largest element of this set which is in L. This would, of course, correspond to the smallest value of n. Denote it by p_2. Now consider the set $\left\{p_2 + \dfrac{1}{n} : n \in \mathfrak{N}\right\}$, and repeat the same process. Continuing this indefinitely, we obtain an increasing sequence $\{p_n : p_1 < p_2 < \ldots\}$ of rationals. It can be shown that the limit of this sequence is r.

To obtain a Dedekind cut from a class of Cauchy sequences of rationals which identifies a real number r, we proceed as follows:

Let $L = \{\{p_n\} : p_n < r\}$, and $U = \{\{p_n\} : p_n \geqslant r\}$. $(L|U)$ gives the cut which identifies the real number r.

Finally, we observe that a decimal representation, say $i.d_1 d_2 \ldots d_n \ldots$ (where i is an integer and the d's are digital numbers), of a real number r can be identified with the Cauchy sequence $\{i.d_1, i.d_1 d_2, \ldots, i.d_1 d_2 \ldots d_n, \ldots\}$ of rational numbers. Obviously, the real number r is the limit of this sequence.

4.4. Sequences in R^n

A sequence of points in R^n is defined in the same way as in R. To avoid any confusion with n of R^n, we denote a sequence by $\{P_m\}$ instead of $\{P_n\}$.

In order to generalize the results of Secs. 4.1 and 4.2, we merely have to replace the absolute value sign $|\ |$ by the norm $\|\ \|$. As an illustration:

Definition 4.4.1. *A sequence of points $\{P_m\}$ of R^n converges to L if for $\epsilon > 0$ there is a natural number m_0 such that $\|P_m - L\| < \epsilon$ for $m \geqslant m_0$. We write $\lim\limits_{m \to \infty} P_m = L$.*

From this definition it follows that every open ball with center at L contains all but a finite number of terms of $\{P_m\}$.

To investigate sequences in R^n it is very convenient to consider their coordinate sequences of real numbers which we introduce in our next definition.

Definition 4.4.2. *Let $\{P_m\}$ be a sequence in R^n. If we write $P_m = \langle p_1^m, p_2^m, \ldots, p_n^m \rangle$ then the sequence $\{p_i^m\}$ of real numbers is called the ith coordinate sequence. It is stipulated that m is a natural number and $i = 1, 2, \ldots, n$.*

Now we prove the theorem which would facilitate the discussion of sequences in R^n.

Theorem 4.4.1. A sequence $\{\langle p_1^m, p_2^m, \ldots, p_n^m \rangle\}$ of points of R^n converges to $L = \langle l_1, l_2, \ldots, l_n \rangle$ if and only if every coordinate sequence $\{p_i^m\}$ converges to l_i.

Proof. Let $P_m = \langle p_1^m, p_2^m, \ldots, p_n^m \rangle$. If $\{P_m\}$ converges to L, then for $\epsilon > 0$ there is a natural number m_0 such that $\|P_m - L\| < \epsilon$ for $m \geqslant m_0$, which means

$$\sqrt{\sum_{k=1}^{n} (p_k^m - l_k)^2} < \epsilon \qquad \text{for } m \geqslant m_0.$$

Now

$$|p_i^m - l_i| = \sqrt{(p_i^m - l_i)^2} \leqslant \sqrt{\sum_{k=1}^{n} (p_k^m - l_k)^2} < \epsilon \text{ for } m \geqslant m_0.$$

Hence, $\{p_i^m\}$ converges to l_i.

Conversely, if $\{p_i^m\}$ converges to l_i for every i, then for $\epsilon > 0$ there is a natural number m_i such that

$$|p_i^m - l_i| < \frac{\epsilon}{n} \qquad \text{for } n \geqslant m_i; \quad i = 1, 2, \ldots, n.$$

Now let $m_0 = \max \{m_1, m_2, \ldots, m_n\}$. From the Minkowski inequality it follows that

$$\sqrt{\sum_{k=1}^{n} (p_k^m - l_k)^2} \leqslant |p_1^m - l_1| + |p_2^m - l_2| + \cdots + |p_n^m - l_n|$$

$$< \frac{\epsilon}{n} + \frac{\epsilon}{n} + \cdots + \frac{\epsilon}{n} = \epsilon \text{ for } m \geqslant m_0$$

which means that $\{P_m\}$ converges to L and that proves the result.

Next, we discuss the completeness property of R^n. First let us introduce "Cauchy sequences" in R^n.

Definition 4.4.2. *A sequence* $\{P_m\}$ *is called a Cauchy sequence if for* $\epsilon > 0$ *there exists a natural number* m_0 *such that* $\|P_m - P_k\| < \epsilon$ *for* $m, k \geqslant m_0$.

Theorem 4.4.2. A sequence $\{P_m\}$ in R^n is a Cauchy sequence if and only if every coordinate sequence of $\{P_m\}$ is a Cauchy sequence of real numbers.

The proof of this theorem is quite similar to that of Theorem 4.4.1 and is left as an exercise.

To assert the completeness property of R^n, we prove the following theorem.

Theorem 4.4.3. A sequence $\{P_m\}$ of R^n is a Cauchy sequence if and only if $\{P_m\}$ converges to a limit L.

Proof. Let $P_m = \langle p_1^m, p_2^m, \ldots, p_n^m \rangle$. If $\{P_m\}$ is a Cauchy sequence, then every $\{p_i^m\}$ is a Cauchy sequence of real numbers and from the property of completeness of real numbers it follows that $\{p_i^m\}$ would converge to a limit l_i. Let $L = \langle l_1, l_2, \ldots, l_n \rangle$. Now using Theorem 4.4.1, we infer that $\lim_{m \to \infty} P_m = L$.

Conversely, if $\lim_{m \to \infty} P_m = L$, where $L = \langle l_1, l_2, \ldots, l_n \rangle$; then by Theorem 4.4.1 it follows that $\lim_{m \to \infty} p_i^m = l_i$ which would imply that every $\{p_i^m\}$ is a Cauchy sequence (Theorem 4.1.1). Now using Theorem 4.4.2, we conclude that $\{P_m\}$ is a Cauchy sequence. ∎

The boundedness of a convergent sequence of R^n follows from the next theorem.

Theorem 4.4.4. Every convergent sequence of R^n is bounded.

The proof of this theorem is left as an exercise.

It may be fairly obvious to the reader that the converse of the above theorem is not necessarily true, but we do have a counterpart of Theorem 4.2.2 in R^n.

Theorem 4.4.5. Every bounded sequence in R^n has a convergent subsequence.

The proof depends upon Theorems 4.4.1 and 4.2.2 and is left as an exercise.

From the preceding theorems, it is quite obvious that the discussion of the sequences in R^n can easily be reduced to that of sequences in R. Then one only has to examine the coordinate sequences of the original one.

For example, the sequence $\left\{\left\langle \dfrac{1}{n}, \dfrac{1}{n} \right\rangle\right\}$ in R^2 converges to $\langle 0,0 \rangle$ whereas the sequence $\left\{\left\langle \dfrac{1}{n}, n \right\rangle\right\}$ does not converge.

Exercise 4.4

1. Prove Theorem 4.4.2.
2. Prove Theorem 4.4.4 (*Hint*: Use Theorems 4.4.1 and 4.1.4.)
3. Prove Theorem 4.4.5.
4. Test the convergence of the following sequences.

   (a) $\left\{\left\langle -\dfrac{m-1}{m}, (-1)^m \right\rangle\right\}$ in R^2.

   (b) $\left\{\left\langle \dfrac{1}{m}, \dfrac{m+1}{m}, \sin\dfrac{\pi}{m} \right\rangle\right\}$ in R^3.

   (c) $\left\{\left\langle \dfrac{1}{m+2}, \dfrac{2m+1}{3m+7}, \cos\dfrac{1}{m}, \sqrt{m} \right\rangle\right\}$ in R^4.

5. Define the limit superior and the limit inferior of a sequence in R^n in such a way that they are the natural generalizations of similar concepts in R.
6. Construct a sequence of points of R^n such that it has rational coordinates and is a Cauchy sequence, yet its limit does not have rational coordinates.

Miscellaneous Exercises for Chapter 4

1. Let $\{a_n\}$ be a sequence with the limit l, and let $a_n \geqslant K$ for every n. Show that $l \geqslant K$.
2. Prove that the convergence and limit of a sequence are not affected, if a finite number of terms of the sequence are altered.
3. Give an example of a sequence $\{a_n\}$ which does not converge, but $\{|a_n|\}$ converges.
4. Does convergence of $\{a_n\}$ necessarily imply convergence of $\{|a_n|\}$? Prove or disprove.
5. Prove that (i) $\lim\limits_{n\to\infty} \sup (a_n + b_n) \leqslant \lim\limits_{n\to\infty} \sup a_n + \lim\limits_{n\to\infty} \sup b_n$, and

(ii) $\lim_{n \to \infty} \inf (a_n + b_n) \geq \lim_{n \to \infty} \inf a_n + \lim_{n \to \infty} \inf b_n$. Give examples when equalities in (i) and (ii) do not hold.

6. Construct a sequence $\{a_n\}$ with four distinct subsequential limits such that $\lim_{n \to \infty} \sup a_n = -1$ and $\lim_{n \to \infty} \inf a_n = -2$.

7. Construct a sequence $\{a_n\}$ such that $\lim_{n \to \infty} \sup a_n = \infty$, and $\lim_{n \to \infty} \inf a_n = 0$.

8. Construct a sequence $\{a_n\}$ such that $\lim_{n \to \infty} \sup a_n = -1$ and $\lim_{n \to \infty} \inf a_n = -\infty$.

9. Using the relationship between Cauchy sequences of rationals and Dedekind cuts (as indicated in Sec. 4.3) establish an isomorphism between them —that is, a 1-1 correspondence such that additions and multiplications would be preserved.

10. Prove that every monotonic bounded sequence in R is convergent.

5

Point Set Theory

5.1. Limit Points

In this chapter we develop some of the basic set-theoretic concepts of the real-number system. We introduce the important notion of "limit point" and establish a number of results—among which the powerful Bolzano-Weierstrass theorem and the Heine-Borel theorem are paramount—having far-reaching consequences in all areas of analysis. We then generalize many results in R^n and introduce the concept of metric spaces.

The first result which we obtain is the theorem on nested intervals. To this end we let $l(I)$ denote the length, $b - a$, of the interval I, with end points a and b (naturally of the form $[a,b]$, (a,b), $[a,b)$, or $(a,b]$).

Definition 5.1.1. *A sequence $\{I_n\}$ of intervals is called a nest of intervals (or a set of nested intervals) if $I_1 \supset I_2 \supset \cdots I_n \supset \cdots$ and the sequence $\{l(I_n)\}$ converges to zero*; i.e., *the length of the interval approaches zero as $n \longrightarrow \infty$.*

We now have the following result.

Theorem 5.1.1. (Theorem on Nested Intervals). The intersection of a nest of closed intervals is a single point.

Proof. Let $\{I_n\}$ be a nest of closed intervals, $I_n = [a_n, b_n]$, $n = 1, 2, \ldots$. From the definition it follows that

$$[a_1, b_1] \supset [a_2, b_2] \supset \ldots [a_n, b_n] \supset \ldots,$$

and hence that

$$a_1 \leqslant a_2 \leqslant a_3 \leqslant \ldots, \quad \text{and} \quad b_1 \geqslant b_2 \geqslant b_3 \geqslant \ldots.$$

Indeed, we have a nonincreasing sequence $\{b_n\}$ which is bounded below (every a_n is a lower bound).

By Theorem 4.1.9 the sequence $\{b_n\}$ converges to its glb; call it b. It must be true, then that

$$a_n \leqslant b \leqslant b_n \quad \text{for every } n$$

$$b \in [a_n, b_n] \quad \text{for every } n$$

$$b \in \bigcap_{n=1}^{\infty} I_n .$$

We shall show that this is the only point of intersection of these intervals.

Suppose $c \neq b$ is another point such that $c \in \bigcap_{k=1}^{\infty} I_k$. This would imply

$a_k \leqslant c \leqslant b_k \ \forall \ k \in \mathfrak{N}.$

Now let $\epsilon = |b - c|$. Since $l\{I_n\} \longrightarrow 0$ as $n \longrightarrow \infty$, there is an interval $[a_m, b_m]$ of the nest such that $b_m - a_m < \epsilon$. But this is impossible since $a_m \leqslant b \leqslant b_m$ and $a_m \leqslant c \leqslant b_m$. The proof is now complete.* ∎

It may be remarked here that b is also the lub $\{a_n\}$ (Prove it).

The reader may verify that for a nest of closed intervals, if $l(I_n)$ does not approach zero[†] then the intersection contains more than one point. By way of example, consider the sequence of closed intervals $\left\{\left[-\dfrac{n+1}{n}, \dfrac{n+1}{n}\right]\right\}$; $\{l(I_n)\}$ does not converge to zero, and therefore the intersection is not a single point. (The intersection is, in fact, the interval $[-1,1]$).

The conclusion of this theorem does not necessarily hold if we consider a nest of open intervals; for example, the nest of intervals $\left\{\left(0, \dfrac{1}{n}\right)\right\}$ has empty intersection.

An example of a nest of open intervals whose intersection is a single point is $\left\{\left(-\dfrac{1}{n}, \dfrac{1}{n}\right)\right\}$. The intersection in this case is $\{0\}$.

Let (a,b) be a neighborhood of a point x_0. If we delete that point x_0 from (a,b), *the set $(a,b) - \{x_0\}$ may be called a "deleted neighborhood" of the point* x_0.

Definition 5.1.2. *Let S be a set. A point x_0 is called a limit point (or a point of accumulation or a cluster point) of S if every deleted neighborhood of x_0 contains at least one point of S.*

---

*It must be noticed that the proof of this theorem makes use of the glb property. In fact, it can be shown that the "glb property" is equivalent to the 'nested interval property.'

†Some authors define a nest without assuming that $l(I_n) \longrightarrow 0$ as $n \longrightarrow \infty$.

It must be remarked here that a limit point of a set S may or may not belong to S; for example, the end points of each interval (closed, open or semi-open) are limit points of that interval.

Although the definition requires only one point of the set S in each deleted neighborhood of a limit point, we can clearly formulate a much stronger result:

Theorem 5.1.2. x_0 is a limit point of a set $S \Longleftrightarrow$ every neighborhood of x_0 contains infinitely many points of S.

Proof. Let (α,β) be an arbitrary neighborhood of the limit point x_0. Then $(\alpha,\beta) - \{x_0\}$ contains a point x_1 of S, $x_1 \neq x_0$. Let

$$\epsilon_1 = \min \{|x_1 - x_0|, |x_0 - \alpha|, |\beta - x_0|\}.$$

The deleted neighborhood $(x_0 - \epsilon_1, x_0 + \epsilon_1) - \{x_0\}$ must contain a point x_2 of S, and this point is certainly distinct from x_1 or x_0. Next, let $\epsilon_2 = |x_2 - x_0|$. Then $\epsilon_2 < \epsilon_1$, and we conclude that the deleted neighborhood $(x_0 - \epsilon_2, x_0 + \epsilon_2) - \{x_0\}$ must contain a point x_3 of S, necessarily distinct from x_0, x_1, x_2. Proceeding inductively, we obtain a sequence

$$\{x_1, x_2, x_3, \ldots\}$$

of distinct points of S, each contained in (α,β). Thus, each neighborhood of x_0 contains infinitely many points of S.

The converse follows directly from the definition, and the theorem is established. |

Related to Theorem 5.1.2 is the following more powerful result, the proof of which is left to the reader.

Theorem 5.1.3. The point x_0 is a limit point of a set $S \Longleftrightarrow$ there is a sequence of distinct points of S which converges to x_0.

Definition 5.1.3. *A point* $x_0 \in S$ *is called an isolated point of S if it is not a limit point of S.*

The terminology of this definition is rather descriptive, since x_0 being an isolated point simply means that there is some deleted neighborhood of x_0 which contains no point of S. Notice, however $x_0 \in S$.

EXAMPLE 5.1.1.
(a) The set of all natural numbers consists of isolated points only.
(b) No point in an interval is an isolated point of that interval.
(c) Each set with only finitely many elements obviously consists entirely of isolated points.

Theorems 5.1.2 and 5.1.3 indicate that if a set has a limit point then that set must necessarily be infinite. The converse of this statement is not always true, however, as the Example 5.1.1(a) above indicates. A sufficient condition for a set to have a limit point is given in the following very important theorem, which is due to two mathematicians—Bernard Bolzano (1781–1848), whose work was not recognized until twenty years after his death, and Karl T. Weierstrass (1815–97), who was a strong exponent of rigor. Weierstrass made many significant contributions in Analysis, and is considered one of the three founders of Analytic function theory—the other two being Cauchy and Riemann (who will be discussed later). Incidentally, George Cantor was one of Weierstrass' students.

Theorem 5.1.4. (Bolzano-Weierstrass theorem): Every bounded infinite set of real numbers has at least one limit point.

Proof. Let S be a bounded infinite set. We divide the set of all real numbers into two sets A and B in the following manner: Let $x \in B$ if there are infinitely many points of S less than x; otherwise, $x \in A$. Now B is nonempty since S is infinite and bounded, and as such every upper bound of S belongs to B; moreover, every lower bound of S is in A, guaranteeing that A is nonempty.

Consider any two points $a \in A$ and $b \in B$. There are at most finite number of points of S less than a, yet there must be infinitely many points of S greater than a but less than b. Consequently, $a < b$; i.e., every point of A is less than every point of B. By the Dedekind theorem we may now conclude that $\exists$ a point c such that $a \leqslant c$ for all $a \in A$ and $c \leqslant b$ for all $b \in B$.

We now prove that the point c is a limit point of S. Assuming the contrary, there would then exist a neighborhood (α,β) of c such that $(\alpha,\beta) - \{c\}$ does not contain any point of S. Choose a_0 such that $\alpha < a_0 < c$. This implies $a_0 \in (\alpha,\beta) \cap A$. Since $c < \beta$, it follows that $\beta \in B$, so that there must exist infinitely many points of S less than β. None of these points, however, can be in $(\alpha,\beta) - \{c\}$ so that each must be less than a_0, implying that $a_0 \in B$. Thus $a_0 \in A \cap B$, a contradiction. Therefore, c is a limit point of S, which completes the proof. |

It must be observed that the Bolzano-Weierstrass theorem does not hold in the system of rational numbers. For example, the set of distinct points 1, $(1 + 1), \left(1 + 1 + \dfrac{1}{2!}\right), \dots \left(1 + 1 + \dfrac{1}{2!} + \dfrac{1}{3!} + \dots\right), \dots$ is bounded, yet has no limit point in the set of rational numbers. (The set, of course, has the limit point e in the system of real numbers.)

Bolzano-Weierstrass theorem can be proved directly from theorem of nested intervals without using Dedekind Cuts (Exercise 5.1 (9)).

As an immediate consequence of the Bolzano-Weierstrass theorem, we prove the following theorem, which was mentioned in Chapter 4, and which says "Every bounded sequence has a convergent subsequence."

Proof of Theorem 4.2.2. Let S be the associated set of the bounded sequence $\{a_n\}$. If S is finite, then at least one term, say a_m, must occur denumerably many times in $\{a_n\}$. Obviously, $\{a_m, a_m, a_m \ldots\}$ is a subsequence of $\{a_n\}$ and is convergent.

If S is infinite, then, being bounded, it must have a limit point, say x_0. By Theorem 5.1.3 there is a sequence $\{x_n\}$ of distinct points of S whose limit is x_0. The sequence $\{x_n\}$ may not be a subsequence of $\{a_n\}$. We can alter this situation, however. Let $x_1 = a_{n_1}$. Then, if $x_2 = a_p$, and $p < n_1$, ignore x_2. If $p > n_1$, let $p = n_2$; i.e., $x_2 = a_{n_2}$. Proceeding accordingly, and ignoring certain terms, we get a sequence $\{a_{n_1}, a_{n_2}, \ldots a_{n_k}, \ldots\}$ such that $n_i < n_j$ if $i < j$. The new sequence is a subsequence of $\{a_n\}$ and converges to x_0.

(As an illustration of the formation of such a subsequence, suppose $x_1 = a_2, x_2 = a_1, x_3 = a_4, x_4 = a_3, \ldots$. In this case we would obtain

$$a_2, a_4, a_6, \ldots, a_{2n}, \ldots$$

as our required subsequence.) In this way we complete the proof of the theorem. |

Exercise 5.1

1. Prove that the intersection of a nest of open intervals is either a single point or is empty. Give examples for both cases.
2. Prove that the intersection of a finite number of closed intervals is either a closed interval, a single point, or is empty. Give examples.
3. Prove that the intersection of a finite number of open intervals is either an open interval or is empty. Give examples.
4. Prove that the intersection of a finite number of semiopen intervals (open at the same end) is either a semi-open interval or is empty.
5. Give an example of an infinite set with no limit point.
6. Prove Theorem 5.1.3.
7. In the proof of Theorem 5.1.4, where does the point c lie?
8. Construct a denumerable set with three distinct limit points.
9. Prove Bolzano-Weierstrass theorem without using Dedekind cut and using the theorem on nested intervals. (*Hint:* S can be enclosed in $[a,b]$. Why? Divide $[a,b]$ into two closed intervals. Select one which contains infinitely many points of S. Repeat this process and obtain a nest of closed intervals.)

5.2. Closed Sets and Open Sets

Definition 5.2.1. *The set of all limit points of a set S is called the derived set of S and is denoted by* S'.

Definition 5.2.2. *A point* x_0 *is called a point of closure of S if every neighborhood of* x_0 *contains at least one point of S.*

It follows from this definition that a point of closure of S is either a member of S or a limit point of S.

Definition 5.2.3. *The set of all points of closure of a set S is called the "closure of S" and is denoted by $\overline{S}$.*

It is obvious from Def. 5.2.2 and Def. 5.2.3 that $\overline{S} = S \cup S'$.

Definition 5.2.4. *A set S is said to be closed if it contains all of its limit points*; i.e., *if $S \supset S'$.*

Thus S is closed if and only if $S = \overline{S}$.

A closed interval is clearly a closed set, but not every closed set is a closed interval; for example, a set consisting of a single point is closed for the simple reason that such a set has no limit points. As a matter of fact, a finite set is always closed. Indeed, any set with no limit point is closed.

Theorem 5.2.1. Let S be a nonempty, closed, bounded set. Then the greatest lower bound and the least upper bound of S are always contained in S.

Proof. We shall prove the result for the greatest lower bound, the argument being entirely analogous for the least upper bound.

Let a be the greatest lower bound of S. Then for every $\epsilon > 0$, $\exists\, a_0 \in S$ such that $a_0 < a + \epsilon$; moreover, $a_0 > a \Longrightarrow a_0 > a - \epsilon$. Thus, $a - \epsilon < a_0 < a + \epsilon$, and we conclude that an arbitrary neighborhood $(a - \epsilon, a + \epsilon)$ of a contains a point a_0 of S. Accordingly, a is a point of closure of S. Since S is closed, $a \in S$, completing the proof.

We next introduce some notation which will be useful in the next chapter and in Chapter 12. If F is a closed, bounded set containing at least two elements, then it is clear that $F \subset [a,b]$, where a is the glb of F and b is the lub of F. Furthermore, it is clear that if $[c,d]$ is any closed interval containing F then $[a,b] \subset [c,d]$. In this sense the interval $[a,b]$ is the smallest closed interval containing F and will be denoted in this book by *sci* (F).

Definition 5.2.5. *A point x_0 is called an interior point of a set S if there exists a neighborhood of x_0 which is a subset of S.*

Definition 5.2.6. *A point x_0 is called a boundary point, or a frontier point, of S if every neighborhood of x_0 contains at least one point of S and at least one point which does not belong to S.*

Definition 5.2.7. *A point x_0 is called an exterior point of a set S if there exists a neighborhood of x_0 entirely contained in cS.*

It follows from the last three definitions that a point x_0 is either an interior point, a boundary point or an exterior point of a given set S, and that the three possibilities are mutually exclusive.

An exterior point of a set is simply an interior point of its complement, and for this reason the concept of exterior point will be seldom used.

By Def. 5.2.5 it follows that if a boundary point of a set is not in the set then it is a limit point of the set. It may happen, of course, that a boundary point may belong to a set and still be a limit point of that set (for example, the endpoints of $[a, b]$; on the other hand, every point of a finite set is a boundary point but certainly not a limit point of the set.

The next concept which we introduce will be discussed more exhaustively later, yet it is introduced now for emphasis and continuity.

Definition 5.2.8. *A point x_0 is called a point of condensation of a set S if every neighborhood of x_0 contains non-denumerably many points of S.*

A point of condensation is always a limit point, but not vice versa. For example, 0 is a limit point of

$$\left\{1, \frac{1}{2}, \frac{1}{3}, \ldots, \frac{1}{n}, \ldots\right\},$$

but not a point of condensation.

Theorem 5.2.2.* An interior point of a set S of *Real Number System* is always a point of condensation of S and *a fortiori* a limit point of S.

Proof. By definition, if x_0 is an interior point of S then there is a neighborhood (α, β) of x_0 such that $(\alpha, \beta) \subset S$. Now an arbitrary neighborhood U of x_0 contains an open interval I, containing x_0, which must then intersect (α, β) in an open interval (a, b) (cf. Exercise 5.1(3)). Thus $(a, b) = [I \cap (\alpha, \beta)] \subset (I \cap S) \subset I \subset U$, but there are nondenumerably many points in (a, b), and each one of them is in S. Hence, x_0 is a point of condensation of S.

That x_0 is a limit point of S follows trivially from the definitions. The proof is complete. ∎

An isolated point of a set S of the real-number system is, of course, not an exterior point of S, and by the preceding theorem is not an interior point of S. Hence, we can say that "An isolated point of a set S of Real Number System is always a boundary point of S." Again, this result is not true in all the systems. (cf. Exercise 5.3(5)).

We next turn to the notion of "open sets" and establish their relationship to closed sets.

Definition 5.2.8. *The set of all interior points of a set S is called the interior of S, and is denoted by S_i.*

---

*It must be carefully noted that this theorem is *not* true in all metric spaces (Cf. Exercise 5.3(2)). It is true in R^n, however.

For example, the interior of $[a,b]$ is (a,b) and the interior of the finite set is empty.

Definition 5.2.9. *A set S is said to be open if every point of S is an interior point.*

It follows immediately from the two preceding definitions that a set S is open if and only if $S = S_i$.

Every open interval is an open set, as is the set of all real numbers.

It must be remarked here that the terms "open" and "closed" are neither inclusive nor mutually exclusive. For example, the void set and the set of all real numbers are both open as well as closed; on the other hand, a semiopen interval is neither open nor closed. There is a relationship, however, between open sets and closed sets as a result of the following theorem.

Theorem 5.2.3. The complement of an open set is closed and the complement of a closed set is open.

Proof. A closed set F contains all its limit points, so that a point $x \in cF$ cannot be a point of closure of F. Consequently, $\exists$ a neighborhood I_x of x containing no point of F; i.e., $I_x \subset cF$. This implies that x is an interior point of cF, and, as such, cF is open.

Now if G is an open set, then $G = G_i$. If y is a limit point of cG, then every neighborhood of y contains infinitely many points of cG which means that y cannot be an interior point of G; i.e., $y \notin G_i = G \Longrightarrow y \in cG \Longrightarrow cG$ is closed. ∎

Theorem 5.2.4. The union of any number of open sets is open.

Proof. Let $\{G_\alpha : \alpha \in A\}$, where A is any index set, be a class of open sets. Let $S = \bigcup_{\alpha \in A} G_\alpha$. Now if $x \in S$, then $x \in G_\alpha$ for some $\alpha \in A$. Since G_α is open, there exists a neighborhood I_x of $x \ni I_x \subset G_\alpha \Longrightarrow I_x \subset S \Longrightarrow x$ is an interior point of $S \Longrightarrow S$ is open, which completes the proof. ∎

Corollary. The intersection of any number of closed sets is closed.

Proof. If $\{F_\alpha : \alpha \in A\}$ is any class of closed sets, then cF_α is open for every α. Now $\bigcap_{\alpha \in A} F_\alpha = \bigcap_{\alpha \in A} c(cF_\alpha) = c \left(\bigcup_{\alpha \in A} \right) cF_\alpha$ and by the last theorem, $\bigcup_{\alpha \in A} cF_\alpha$ is open; hence, $\bigcup_{\alpha \in A} F_\alpha$ is closed by Theorem 5.2.3. The corollary is established.

Theorem 5.2.5. The intersection of a finite number of open sets is open.

Proof. Let $\{G_k : k = 1, 2, \ldots, n\}$ be a finite number of open sets. Let $T = \bigcap_{k=1}^{n} G_k$. If T is empty, then T is open. We now assume that T is nonempty. Let

$x \in T$; then $x \in G_k$ for every k (= 1,2, ... , n), and, as such, x is an interior point of every G_k. This means that there exist neighborhoods I_k (k = 1,2, ... , n) of x such that $I_k \subset G_k$. Since $I = \bigcap_{k=1}^{n} I_k$ is not empty (x belongs to each I_n), I must be an open interval containing x. (cf. Exercise 5.1(3)). Now $I \subset I_k \subset G_k$ for every k, which means $I \subset T$, implying that x is an interior point of $T \Longrightarrow T$ is open. This completes the proof.

Corollary. The union of a finite number of closed sets is closed.

The proof of this corollary is analogous to that of the last corollary.

It must be remarked here that the intersection of an arbitrary number of open sets may not be open; for example, $\bigcap_{n=1}^{\infty} \left(-\frac{1}{n}, \frac{1}{n} \right) = \{0\}$, which is not open. Moreover, the union of any number of closed sets may not be closed; for example, $\bigcup_{n=1}^{\infty} \left[\frac{n}{n+1}, \frac{n+1}{n+2} \right] = \left[\frac{1}{2}, 1 \right)$, which is not closed.

The following result is an important generalization of the theorem on nested intervals and is due to George Cantor.

Theorem 5.2.6. If $\{C_n\}$ is a denumerable class of nonempty closed and bonded sets such that $C_1 \supset C_2 \supset C_3 \supset \ldots \supset C_n \supset \ldots$, then $C = \bigcap_{n=1}^{\infty} C_n$ is non-empty.

Proof. Let b_n = lub (C_n). Then $b_n \geqslant b_{n+1}$ for every n. By Theorem 5.2.1, each $b_n \in C_n$. Now each lower bound of C, is a lower bound for every C_n, and, as such, a lower bound of $\{b_n\}$. Thus $\{b_n\}$ is a nonincreasing sequence, is bounded below, and is consequently convergent by Theorem 4.1.9. Let $b = \lim_{n \to \infty} b_n$. We shall show that $b \in \bigcap_{n=1}^{\infty} C_n$.

By the corollary of Theorem 5.2.4, $\bigcap_{n=1}^{\infty} C_n$ is closed; hence, assuming by way of contradiction that $b \notin \bigcap_{n=1}^{\infty} C_n$, there would then be a neighborhood $U = (b - \epsilon, b + \epsilon)$ which contains no point of $\bigcap_{n=1}^{\infty} C_n$. This means that U contains no point of C_m for some m. But then U contains no point of C_k for $k \geqslant m$; hence, $b_k \notin U$ for $k \geqslant m$, contradicting the fact that the neighborhood U must contain all but a finite number of terms of the sequence $\{b_n\}$. Hence, $b \in \bigcap\limits_{n=1}^{\infty} C_n$, proving the theorem.

EXAMPLE 5.2.1. The intersection of the closed and bounded sets $\left[-\dfrac{1}{n},0\right]$; $n = 1,2,\ldots$, is $\{0\}$, hence it is nonvoid. On the other hand, $\bigcap\limits_{n=1}^{\infty}\left(\dfrac{1}{n},0\right)$ is empty—the sets are not closed.

The intersection of the following sequence of closed sets $\mathfrak{N} \supset \mathfrak{N} - \{1\} \supset \mathfrak{N} - \{1,2\} \supset \cdots \supset \mathfrak{N} - \{1,2,\ldots,n\} \supset \cdots$ is empty, since these closed sets are not bounded. Reconcile it with Theorem 5.2.6.

Definition 5.2.10. *Let S be any point set. A class of intervals $\{I_\alpha\}$ is called a covering of S if each $x \in S$ is contained in some I_α. If a subclass $\{I_\beta\}$ of $\{I_\alpha\}$ is also a covering of S, then $\{I_\beta\}$ is called a subcovering. If every interval in a covering $\{I_\alpha\}$ is open, then $\{I_\alpha\}$ is called an open covering.*

The following theorem is due to Lindeloff and is very useful in point-set topology. Lindelöff (1870–1946) has made some important contributions in "Topology."

Theorem 5.2.7. (Lindelöff Covering Theorem). Every open covering of a set of real numbers has a countable subcovering.

Proof. Let $\{I_\alpha\}$ be an open covering of the set S, and let $x \in S$. Then, $\exists\ I_\alpha \ni x \in I_\alpha$. Suppose $I_\alpha = (a_\alpha, b_\alpha)$. Now there are rational numbers p_α and $q_\alpha \ni a_\alpha \leqslant p_\alpha \leqslant x \leqslant q_\alpha \leqslant b_\alpha$; i.e., $x \in (p_\alpha, q_\alpha) \subset I_\alpha$. Now the class of open intervals with rational endpoints is denumerable; consequently, the class of intervals of the form (p_α, q_α) is denumerable. Since the set S can be covered by intervals of this form, it follows that S can be covered by a countable subclass of $\{I_\alpha\}$, which completes the proof. ∎

We are now prepared to prove the very important Heine-Borel theorem alluded to at the beginning of this section. E. Heine (1821–81) was also one of Weierstrass' students and became famous because of this theorem. Borel, on the other hand, made many contributions in Mathematics. We mentioned him in Chapter 3.

In the proof of Heine-Borel theorem, we make use of the fact that if F is a closed set and G is open, then $F - G$ $(= F \cap cG)$ is closed.

Theorem 5.2.8. (Heine-Borel Theorem). Every open covering of a closed and bounded set has a finite subcovering.

Proof. Let $\{I_\alpha\}$ be an open covering of a closed and bounded set C. If there are finitely many intervals in $\{I_\alpha\}$, then the conclusion of the theorem becomes obvious. We therefore assume that the class $\{I_\alpha\}$ is infinite. By the Lindeloff covering theorem there exists a countable subclass of $\{I_\alpha\}$ which is a subcovering of C. Again if this countable class is finite, then the result is obvious. We therefore assume that we have a denumerable subcovering $\{I_1, I_2, \ldots, I_n, \ldots\}$ of C. We shall show that it is possible to cover C by a finite number of these intervals.

Suppose by way of contradiction that it is impossible to cover C by a finite number of these intervals; then $C - \left(\bigcup_{k=1}^{n} I_k \right)$ is always nonempty. Moreover, since $\bigcup_{k=1}^{n} I_k$ is open and C is bounded and closed, it follows that $C - \left(\bigcup_{k=1}^{n} I_k \right)$ is closed and bounded. Furthermore, if we let $C_1 = C - I_1$, $C_2 = C - (I_1 \cup I_2)$, ..., $C_n = C - \left(\bigcup_{k=1}^{n} I_k \right), \ldots$ then $C_1 \supset C_2 \supset \cdots \supset C_n \supset \cdots$. Now, using Cantor Theorem (Theorem 5.2.6), $\bigcap_{n=1}^{\infty} C_n$ is not empty. Let $x \in \bigcap_{n=1}^{\infty} C_n$. Since x is in every C_n, x cannot belong to any I_n; yet x belongs to C. This is a contradiction, since $\{I_n\}$ is a covering of C. Hence the result is established. $\blacksquare$

This theorem can be proved directly from Theorem 5.1.1. (Theorem on nested intervals) without using Lindelöff theorem (cf. Exercise 5.2(1)).

If a set S is such that every open covering of S has a finite subcovering, then it is said to have the "Heine-Borel property."

Definition 5.2.11. *A set is called compact if it has the Heine-Borel property.*

The Heine-Borel theorem has more far reaching consequences in analysis than a beginning reader may suspect.

The theorem is not true if the set is not closed and bounded, as is shown by the following examples.

EXAMPLE 5.2.2.

(i) If the set is unbounded (it may be closed) then for an arbitrary open covering of a set it may not be possible to find a finite subcovering. Let us consider the set of all natural numbers. The covering $\left(n - \frac{1}{2}, n + \frac{1}{2} \right)$ of $\mathfrak{N}$ $\left[\text{each } n \text{ is contained in } \left(n - \frac{1}{2}, n + \frac{1}{2} \right) \right]$, cannot possibly have a finite subcovering.

(ii) Now let us consider a bounded set which is not closed. Let $C = \left\{ 1, \frac{1}{2}, \frac{1}{3}, \ldots, \frac{1}{n}, \ldots \right\}$ Let $\frac{1}{n}$ be covered by $\left(\frac{1}{n + \frac{1}{2}}, \frac{1}{n - \frac{1}{2}} \right)$. Now it is impossible to choose a finite number of these intervals which will cover C.

Notice in Example 5.2.2 (ii) that C is not closed, since 0 is a limit point of C but does not belong to C. If we let $C_1 = C - \left(\frac{2}{3}, 2\right), C_2 = C - \left(\frac{2}{3}, 2\right) \cup \left(\frac{2}{5}, \frac{2}{3}\right),$ and so on, then $C_1 = C - \{1\}, C_2 = C - \left\{1, \frac{1}{2}\right\}, \ldots$ and $\bigcap_{n=1}^{\infty} C_n$ is empty (and the conclusion of Cantor's theorem does not hold).

Now if we adjoin the point 0 to C, we obtain $F = \overline{C}$, a closed set; i.e., $F = \left\{0, 1, \frac{1}{2}, \ldots, \frac{1}{n}, \ldots\right\}$. But now we must adjoin an interval, say $\left(-\frac{1}{10}, \frac{1}{10}\right)$, to the covering $\left(\frac{1}{n + \frac{1}{2}}, \frac{1}{n - \frac{1}{2}}\right)$ in order to cover 0. From this new covering it is possible to choose 11 intervals to cover F, since $\left(-\frac{1}{10}, \frac{1}{10}\right)$ covers all but 1, $\frac{1}{2}, \ldots, \frac{1}{10}$.

The Heine-Borel theorem has shown that every closed and bounded set of R has the so-called Heine-Borel property. (This is not true in the system of rational numbers.) The converse of this result, that is, "Every set of R which has the Heine Borel property must be closed and bounded" is also valid. That can be easily proved. However, we shall prove the converse for a more general case, that of R^n in Sec. 5.6 (cf. Theorem 5.6.4).

We end this section by introducing "relatively open" and "relatively closed" subsets of a set.

Definition 5.2.12. *A subset O of a set S is called relatively open in S (or simply open in S) if there is an open set G such that $O = G \cap S$.*

For example, $\left[0, \frac{1}{2}\right)$ is relatively open in $S = [0, 1]$ since we may write $\left[0, \frac{1}{2}\right) = \left(-\frac{1}{2}, \frac{1}{2}\right) \cap [0, 1]$. Also, $\left(0, \frac{1}{2}\right)$ is relatively open in S. But $\left(0, \frac{1}{2}\right]$ is not relatively open in S.

Definition 5.2.13. *A subset C of a set S is called relatively closed in S if there is a closed set F such that $C = F \cap S$.*

For example, $\left(0, \frac{1}{2}\right]$ is relatively closed in $(0, 1]$. But $\left(0, \frac{1}{2}\right)$ is not relatively closed in $(0, 1]$.

By these definitions it follows that a set S is "relatively open" as well as "relatively closed" in S itself, since we can write $S = S \cap R$; and R is open as well as closed.

For practical purposes, we may say that a set A is relatively open in S ($A \subset S$) if every point of A is an interior point of A or a boundary point of S (prove it); and A is relatively closed in S if every limit point of A which is in S also belongs to A.

Exercise 5.2

1. Prove the Heine-Borel theorem without using the Lindeloff theorem, instead using the theorem of nested intervals. (*Hint:* See Exercise 5.1(9).)
2. Prove that if $[a,b]$ = sci (F), where F is a nonempty closed set, then $[a,b]$ − F is open.
3. Prove that $\overline{S}$ (closure of S) is the intersection of all closed sets which contain S.
4. Prove that $\overline{(A \cup B)} = \overline{A} \cup \overline{B}$.
5. Show that $\overline{(A \cap B)} \subset \overline{A} \cap \overline{B}$. Give an example when $\overline{(A \cap B)} \neq \overline{A} \cap \overline{B}$.
6. Prove that S_i (interior of S) is the union of all open sets contained in S.
7. Show that $(A \cap B)_i = A_i \cap B_i$.
8. Show that $(A \cup B)_i \supset A_i \cup B_i$. Give an example when $(A \cup B)_i \neq A_i \cup B_i$.
9. Find a subset of R whose interior is empty and whose closure is R. (There are more than one.)
10. Let $\mathfrak{B}(S)$ denote the set of all boundary points of S. Show that $\mathfrak{B}(S) = \overline{S} \cap \overline{cS}$.
11. Let $S = [0,1]$. Find a set which is
    (a) relatively open in S but not open,
    (b) not closed but relatively closed in S,
    (c) neither relatively open nor relatively closed in S.
12. Prove that a set A is relatively open in S if and only if every point of A is either an interior point of A or a boundary point of S.
13. Give an example of an open covering of $[1,2)$ which has no finite subcovering.
14. Let $\left\{ \left(-\dfrac{1}{2}, \dfrac{3}{11} \right), \left(\dfrac{1}{2}, \dfrac{3}{2} \right), \left(\dfrac{1}{3}, \dfrac{4}{3} \right), \ldots, \left(\dfrac{1}{n}, \dfrac{n+1}{n} \right), \ldots \right\}$ be a covering of $[0,1]$. Find its smallest finite subcovering.
15. Show that x_o is a point of closure of a set S if and only if there is a sequence of points of S (not necessarily distinct) converging to x_o.

5.3. Metric Spaces

The final sections of this chapter will be devoted to the generalization of the various properties of the real number system to spaces of a more abstract nature, namely, metric spaces. We attempt to examine those properties of an abstract mathematical system in which a certain intuitive concept is maintained, in this case, the idea of distance.* In doing so, we would be sacrificing some of the algebraic properties of real numbers like field properties, linear order relations, etc. Yet we would retain those properties which depend upon the idea of distance alone, like neighborhoods, open sets, closed sets, convergence and limit of a sequence, and also continuity of functions (which would be discussed in Chapter 7).

We now consider important basic properties of the usual distance between two real numbers and use them as defining postulates. Thus, we introduce

Definition 5.3.1. *Let M be a nonempty set and suppose there exists a function $d: M \times M \longrightarrow R$ satisfying the following conditions*:

M.1. $d(x,y) \geqslant 0$
M.2. $d(x,y) = 0 \Longleftrightarrow x = y$
M.3. $d(x,y) = d(y,x)$
M.4. $d(x,y) + d(y,z) \geqslant d(x,z).$

Then M, together with the function d, is called a metric space, and is denoted by $\{M,d\}$.

Condition M.4 is called the triangle inequality, and d is called the "distance function," or "metric." It is important to note that d actually determines the structure of the metric space, as it is possible to obtain different metric spaces from the same set M by merely changing the metric (cf. Exercise 5.3(1,2)). The term "distance" is undefined here.

By slightly modifying M.4 (cf. Exercise 5.3(3)), we can show that M.1 and M.3 follow from M.2 and M.4. In other words, we can define a metric space with only two postulates for the metric.

The set of all real numbers with $d(x,y) = |x - y|$ (called the usual metric) is a metric space, as is the set of all rational numbers with the same metric. These assertions are easily verified.

Definition 5.3.2. *Let $\{M,d\}$ be a metric space. By an ϵ-neighborhood of a point $x_0 \in M$ is meant the set $\{x \in M : d(x,x_0) < \epsilon\}$; moreover, by an arbitrary neighborhood of x_0 is meant the set $\{x \in M : d(x,x_1) < \epsilon\}$ containing x_0, and $x_1 \in M$. This would imply $d(x_0,x_1) < \epsilon$.*

It is easily seen that the idea of neighborhood in a metric space is a natural generalization of that on the real line—an interval $(x_0 - \epsilon, x_0 + \epsilon)$ being replaced by the set $\{x : d(x,x_0) < \epsilon\}$. In this context we can introduce all those concepts which can be defined in terms of neighborhoods on the real line. For example, convergence and limit of a sequence, limit point of a set, interior, exterior and boundary points of a set, closed sets, open sets, compact sets, covering, and so on. All this can be accomplished by a simple change of terminology.

---

*"Distance" does not necessarily mean Euclidean distance.

To give an example, a sequence $\{x_n\}$ of points in a metric space $\{M,d\}$ would be called a Cauchy sequence if for any $\epsilon > 0$ ∃ a natural number n_0 such that $d(x_n,x_m) < \epsilon$ for $n, m \geqslant n_0$.

For our most important results to hold, we would require one more condition on the metric space; namely, "completeness."

Definition 5.3.3. *A metric space $\{M,d\}$ is said to be complete if every Cauchy sequence in $\{M,d\}$ has a limit which belongs to M.*

In a complete metric space, then, the Bolzano-Weierstrass theorem, the Heine-Borel theorem and Cantor's theorem on compact sets (Theorem 5.2.6) are valid.

The reader will note, indeed, that $\{R,d\}$, where d is the usual metric, is a complete metric space.

Let $\{M,d\}$ be a metric space, and let S be a subset of M. If we consider S with the metric d of $\{M,d\}$, then $\{S,d\}$ itself would be a metric space, and is called a subspace of the original metric space $\{M,d\}$.

For example, $[0,1]$ is a metric space and a subspace of $\{R,d\}$, and so is $(0,1) \cup (2,3)$.

If $\{S,d\}$ is a subspace of $\{M,d\}$ then open sets and closed sets of $\{S,d\}$ will be those subsets of S which are "relatively open in S" and "relatively closed in S," respectively. (Cf. Defs. 5.2.12 and 5.2.13.)

We now give an example of an interesting metric space with some unusual properties.

EXAMPLE 5.3.1. Consider the metric space $\{M,d\}$ (M being nonempty set) defined as follows: $d(x,y) = 1$ if $x \neq y$,

$d(x,y) = 0$ if $x = y$.

It can be easily shown that this is a metric space [Cf. Exercise 5.3(2)]. This metric space is called "discrete."

Every point of this metric space is an isolated point, no matter what set it belongs to. For $\epsilon \leqslant 1$ the ϵ-neighborhood of x_0 $\{x : d(x,x_0) < \epsilon\}$ contains x_0 alone. It is also interesting to note that every point of this metric space will also be an interior point of every set it belongs to. (Prove it.) This is in contradiction to the conclusions of Theorem 5.2.2 and its corollary (which were valid for real number system). As we said earlier, this is an unusual metric space.

Exercise 5.3

1. Prove that R^n is a metric space with $d(X, Y) = \| X - Y \| = \sqrt{\sum_{i=1}^{n} (x_i - y_i)^2}$,

where $X = (x_1, x_2, \ldots x_n)$ and $Y = (y_1, y_2, \ldots y_n); x_i, y_i \in R, i = 1, 2, \ldots n$.

2. Let M be a nonvoid set and define $d(x,y) = 1$ if $x \neq y$, $d(x,y) = 0$ if $x = y$. Show that d defines a metric on M.
3. Show that it is possible to replace $M.1-4$ by the following:

$$M_1' : d(x,y) = 0 \Longleftrightarrow x = y$$
$$M_2' : d(x,y) + d(z,y) \geqslant d(x,z).$$

5.4. Connectedness

Intuitively, we may say that a space is connected if we can move from one point to another without going out of the space. If this is not always possible, then we say the space is disconnected. In mathematics it is easier to define disconnectedness and then the negation of disconnectedness is termed as connectedness.

Definition 5.4.1. *A metric space $\{M,d\}$ is said to be disconnected if there exist two nonempty disjoint open sets G_1 and G_2 such that $G_1 \cup G_2 = M$; G_1 and G_2 are said to form a disconnection or a partition for $\{M,d\}$.*

A metric space $\{M,d\}$ is called *connected* if it is not disconnected, of course.

If G_1 and G_2 form a partition of a disconnected space $\{M,d\}$ then these two open sets are obviously complementary. As such both of them are closed also. Furthermore, none of them is empty or equal to M. Conversely, if there is a nonempty set S not equal to M, and is open as well as closed, then so will be its complement. So S and cS would form a disconnection of M. Thus we have proved the following theorem.

Theorem 5.4.1. $\{M,d\}$ is connected $\Longleftrightarrow$ the only nonempty set in M which is open as well as closed is M itself.

Having defined connectedness of a metric space, we now introduce connectedness of a set. If S is a set in a metric space $\{M,d\}$ then regarding $\{S,d\}$ as a metric space, the following would be the natural definition of disconnectedness.

Definition 5.4.2. *A set S of a metric space $\{M,d\}$ is said to be disconnected if there are two nonempty disjoint subsets S_1 and S_2 of S such that they are open (relatively) in S and $S_1 \cup S_2 = S$. S_1 and S_2 form a disconnection of S.*

If S_1 and S_2 form a partition of a disconnected *set* S, then there are open sets G_1 and G_2 such that $S_1 = S \cap G_1$ and $S_2 = S \cap G_2$. Furthermore, $S_1 \cap S_2 = \phi \Longrightarrow (G_1 \cap G_2) \cap S = \phi$. Also since $S_1 \cup S_2 = S \Longrightarrow S \subset G_1 \cup G_2$. Therefore, we have the following result.

Theorem 5.4.2. A set S is disconnected $\Longleftrightarrow$ there exist two nonempty open sets G_1 and G_2 (not necessarily disjoint) such that $(G_1 \cap G_2) \cap S = \phi, S \cap G_1 \neq \phi, S \cap G_2 \neq \phi$, and $S \subset G_1 \cup G_2$.

In this theorem, we could replace the term "open" by "closed" (why?). That is, a set S is disconnected $\Longleftrightarrow$ there are two nonempty closed sets F_1 and F_2 (not necessarily disjoint) such that $F_1 \cap S \neq \phi, F_2 \cap S \neq \phi, (F_1 \cap F_2) \cap S = \phi$, and $S \subset F_1 \cup F_2$.

In Miscellaneous Exercises for Chapter 2 (Exercise 1), we introduced an important notion of path-connectedness, which we repeat here. A subset S of R is path-connected if for every two points a and b in S all the points between a and b are in S. Furthermore, S is path-connected if and only if S has any one of the following forms: $(a,b), (a,b]$ $[a,b]$, $[a,b), (-\infty,a), (-\infty,a], (a,\infty), [a,\infty),$ $(-\infty,\infty)$.

The following theorem establishes the equivalence of connectedness and path-connectedness in the real number system. This is not true in every metric space.

Theorem 5.4.3. A subset S of R consisting of more than one point is connected $\Longleftrightarrow S$ is path-connected.

Proof. To prove $\Longrightarrow$, we assume by way of contradiction that S is not path-connected, which means there are two distinct points a and b in $S, \ni a < b$, and there exists at least one point c between a and b such that $c \notin S$.

Now letting $G_1 = (-\infty, c)$ and $G_2 = (c, \infty)$, we notice $S \cap G_1 \neq \phi$, since it contains a. Similarly, $G_2 \cap S \neq \phi$. Furthermore, $S \subset G_1 \cup G_2$ and $(G_1 \cap G_2) \cap S = \phi$. Using Theorem 5.4.2 it follows that S is disconnected.

To prove $\Longleftarrow$ we again use indirect method and assume that S is disconnected. In that case there are two closed sets F_1 and F_2, such that $F_1 \cap S \neq \phi$, $F_2 \cap S \neq \phi, (F_1 \cap F_2) \cap S = \phi$, and $F_1 \cup F_2 \supset S$. Let a and $b \in S$, and let $a < b$. We shall show there is at least one point between a and b which does not belong to S. Without any loss of generality (with the change of notation if necessary) we may assume $a \in F_1$ and $b \in F_2$. Let $c = \text{glb} \ \{F_2 \cap [a,b]\}$. Obviously, $F_2 \cap [a,b]$ is closed and c must lie in it, implying c must belong to F_2. Furthermore, a is a lower bound of $F_2 \cap [a,b]$. Thus $a \leqslant c$. But $a \in F_1 \cap S \Longrightarrow a \notin F_2$, and thus $a < c$. Now let $[a,c] \cap F_1 = F_1'$, and let $l = \text{lub} \ (F_1')$, which means $l \in F_1' \Longrightarrow l \in F_1$. Also c is an upper bound of $F_1' \Longrightarrow l \leqslant c$.

If $l = c$, then $c \in F_1$. But $c \in F_2$ which means $c \in F_1 \cap F_2$. Since $(F_1 \cap F_2) \cap S = \phi, c \notin S$, and $a < c < b$, which proves our assertion.

Let $l < c$ and let $l < x_0 < c$. Now $x_0 > \text{lub} \ (F_1')$ and thus $x_0 \notin F_1'$. But $a < l < x_0 < c \Longrightarrow x_0 \in [a,c]$. Thus $x_0 \notin F_1$. By a similar reasoning, we can show that $x_0 \notin F_2$. Therefore, we have found x_0 lying between a and b such that $x_0 \notin S$. The proof is now complete. $\blacksquare$

Corollary. A set S of real numbers consisting of at least two points is connected $\Longleftrightarrow S$ is of any one of the following forms: $(a,b), [a,b], [a,b), (a,b],$ $(-\infty,a), (-\infty,a], (a,\infty), [a,\infty), (-\infty,\infty)$.

Exercise 5.4

1. Show that the set of natural numbers on the real line is not a connected set. Prove the same thing for the set of all rational numbers.
2. Let $\{M,d\}$ be a metric space. Let S be a connected set of M, and let X be a set which is open as well as closed. Show that if $S \cap X$ is not empty then $S \subset X$. (*Hint*: Use indirect method and Theorem 5.4.2.)
3. Let S and T be two connected sets of a metric space. Show that if $S \cap T$ is not empty then $S \cup T$ is connected.
4. Show that the set $A = \left\{ (x,y): y = \sin\dfrac{1}{x}, x \neq 0 \right\} \cup \{(0,y): -1 \leqslant y \leqslant 1\}$ is connected in Euclidean plane R^2 with the usual metric.

5.5. Distance and Separation

In this section we introduce the idea of "distance between a point and a set" and that of "distance between two sets."

Definition 5.5.1. *Let S be a nonempty set in a metric space $\{M,d\}$. Let $x \in M$. The distance between x_0 and S, written as $d(x_0,S)$, is defined as follows:*

$$d(x_0,S) = \text{glb } \{d(x,x_0): x \in S\}.$$

Definition 5.5.2. *If A and B are two nonempty sets in the metric space $\{M,d\}$ then the distance between A and B, written as $d(A,B)$, is defined as follows:*

$$d(A,B) = \text{glb } \{d(x,y): x \in A, y \in B\}.$$

It follows from these definitions that if $x_0 \in S$, then $d(x_0,S) = 0$; moreover, if A and B are not disjoint then $d(A,B) = 0$. The converse of these statements are not necessarily true as indicated by the following examples.

EXAMPLE 5.5.1. (a) The distance between the point 0 and the set $(0,1)$ is zero, but $0 \notin (0,1)$.

(b) The distance between $(0,1)$ and $[1,2]$ is zero, but the two sets are disjoint.

We do, however, have the following relationship between $d(x_0,S)$ and S.

Theorem 5.5.1. *If S is a nonempty set in a metric space $\{M,d\}$ then $d(x_0,S) = 0 \Longleftrightarrow x_0 \in \overline{S}$.*

Proof. Let $d(x_0,S) = 0$, and let $\epsilon > 0$. Using a property of the greatest lower bound and Def. 5.5.1 we can find $x_1 \in S$ such that $d(x_0,x_1) <$

$d(x_0, S) + \epsilon = \epsilon$. In other words, every ϵ-neighborhood of x_0 contains at least one point of S, implying $x_0 \in \overline{S}$.

Conversely, if $x_0 \in \overline{S}$, then every ϵ-neighborhood of x_0 contains at least one point, say x', of S, thus

$$d(x_0, S) \leqslant d(x_0, x') \Longrightarrow d(x_0, S) < \epsilon.$$

Since ϵ is arbitrary and $d(x_0, S) \geqslant 0$, it follows that $d(x_0, S) = 0$. The proof is now complete. ∎

This theorem gives an intuitive significance to the terms "closure" and "point of closure."

An immediate corollary of this theorem is that if F is a closed set, then $d(x_0, F) = 0 \Longleftrightarrow x_0 \in F$.

Theorem 5.5.2. (a) $d(x, A) \leqslant d(x, y) + d(y, A)$
(b) $d(A, B) \leqslant d(A, x) + d(x, B)$.

The proof of this theorem follows directly from the triangle inequality and Definitions 5.5.1 and 5.5.2.

We now prove a theorem which is very important in topology and is popularly known as the Separation Axiom T_4 (or the axiom of normality).

Theorem 5.5.3. Let F_1 and F_2 be disjoint closed sets in a metric space $\{M, d\}$. Then there exist two disjoint open sets G_1 and G_2 containing F_1 and F_2, respectively.

Proof. If one of F_1 and F_2 is empty, then we let G_1 be empty and $G_2 = M$, and the conclusion of the theorem is obvious. Thus we assume that neither F_1 nor F_2 is empty. Let $G_1 = \{x : d(x, F_1) < d(x, F_2)\}$, and let $G_2 = \{x : d(x, F_2) < d(x, F_1)\}$. That G_1 and G_2 are disjoint is obvious from their construction. Furthermore, if $x \in F_1$, then $d(x, F_1) = 0$; but since F_2 is closed and disjoint from F_1, $d(x, F_2) \neq 0 \Longrightarrow d(x, F_2) > 0 \Longrightarrow d(x, F_1) < d(x, F_2) \Longrightarrow x \in G_1$. That proves $F_1 \subset G_1$. Similarly, $F_2 \subset G_2$.

What remains to be shown is that G_1 and G_2 are open. Let $x_0 \in G_1$. Then $d(x_0, F_1) < d(x_0, F_2)$. Hence,

$$\exists \text{ a } \delta > 0 \text{ such that } d(x_0, F_1) + \delta = d(x_0, F_2). \tag{5.1}$$

Now let $I = \left\{ x : d(x, x_0) < \dfrac{\delta}{2} \right\}$; that is, I is a $\dfrac{\delta}{2}$-neighborhood of x_0. If $y \in I$, then $d(y, x_0) < \dfrac{\delta}{2}$. Using Theorem 5.5.2, and (5.1) we have

$$d(y, F_1) \leqslant d(y, x_0) + d(x_0, F_1) < \frac{\delta}{2} + d(x_0, F_2) - \delta = d(x_0, F_2) - \frac{\delta}{2}.$$

Also, $d(x_0 F_2) \leqslant d(x_0,y) + d(y,F_2) < \dfrac{\delta}{2} + d(y,F_2) \Longrightarrow d(y,F_1) < d(y,F_2) +$

$\dfrac{\delta}{2} - \dfrac{\delta}{2} \Longrightarrow y \in G_1 \Longrightarrow I \subset G_1$. We have shown that x_0 is an interior point of G_1. Since $x_0 \in G_1$ was arbitrary, we conclude that G_1 is open. A similar argument establishes that G_2 is open, and the proof is complete. $\blacksquare$

Exercise 5.5

1. If F_1 and F_2 are two nonempty closed sets and if one of them is bounded, prove that there exist $x_1 \in F_1$ and $x_2 \in F_2$ such that $d(x_1,x_2) = d(F_1,F_2)$.
2. Construct two unbounded closed sets F_1 and F_2 which are disjoint and $d(F_1,F_2) = 0$.
3. If S is any nonempty set and $\epsilon > 0$ show that the set $\{x : d(x,S) < \epsilon\}$ is open.

5.6. Metric space of R^n

A special case of a metric space is R^n (cf. Exercise 5.3(1)), which was introduced in Chapter 2. Open sets and closed sets in this space can be defined in the same manner as in a metric space. To generalize these concepts from the real-number system, one simply has to replace open neighborhoods by open balls (or open boxes if convenient). To give some examples the sets $\{\langle x,y \rangle : x^2 + y^2 < 1\}$, $\{\langle x,y \rangle : y^2 < x\}$ are open in R^2 whereas the sets $\{\langle x,y \rangle : x^2 + y^2 = a^2\}$, $\{\langle x,y \rangle : y \geqslant x^3\}$ are closed and sets like $\left\{\langle x,y \rangle : y = \sin \dfrac{1}{x}\right\}$ are neither open nor closed.

In this section we wish to reemphasize the property of completeness of R^n which was established in Sec. 4.4. It may be recalled that there we made use of the "completeness" of R. The completeness of R is equivalent to various other properties listed in the Appendix. However, in Chapter 4 to prove this property, we used Theorem 4.2.2 and that theorem depends upon the Bolzano-Weierstrass theorem. Since we have already established the counterpart of Theorem 4.2.2 in R^n which is Theorem 4.4.5, we can use it (Theorem 4.4.5) to prove the Bolzano-Weierstrass property in R^n.

Theorem 5.6.1. Every infinite bounded set of R^n has at least one limit point.

Proof. Let S be an infinite bounded set of R^n. Then S must contain a denumerable subset, say D. Let us write it in the form of a sequence $D = \{P_1, P_2, \ldots, P_m, \ldots\}$, $P_i \neq P_j$. This sequence of points must have a convergent subsequence (Theorem 4.4.5), say $\{P_{m_k}\}$, and let L be its limit. Now every open ball centered at L would contain all but a finite number of terms of this subsequence. Since this subsequence consists of distinct points of S, every such open ball would contain infinitely many points of S which implies that L is a limit point of S. The proof is now complete. $\blacksquare$

Because of the dependence of the last theorem on Theorem 4.4.5 and that of Theorem 4.2.2 on Theorem 5.1.4, one may observe that the Bolzano-Weierstrass property is logically equivalent to the property: "Every bounded sequence has a convergent subsequence."

The Heine-Borel theorem in R^n is similar to its counterpart in R. Thus it may be stated as follows.

Theorem 5.6.2. Every open covering of a closed and bounded set in R^n has a finite subcovering.

The proof is left as an exercise.

In the next theorem we seek a generalization of the theorem on nested intervals in R (Theorem 5.1.1). The proof is based on the completeness property of R^n.

Theorem 5.6.3. Let $\{F_m\}$ be a sequence of nonempty closed sets such that $F_m \supset F_{m+1}$ and diam(F_m) approaches zero as $m \longrightarrow \infty$. Then $\bigcap_{m=1}^{\infty} F_m$ is a single point.

Proof. Let $P_1 \in F_1$, $P_2 \in F_2, \ldots, P_m \in F_m, \ldots$. For $\epsilon > 0$, there is m_0 such that diam $(F_{m_0}) < \epsilon$. Now if $m \geqslant m_0$ and $k \geqslant m_0$, then $F_m \subset F_{m_0}$, $F_k \subset F_{m_0}$, and this would imply that P_k and P_m are in F_{m_0}.

Therefore, $\|P_k - P_m\| < \epsilon$ for m, $k \geqslant m_0$ which shows that $\{P_m\}$ is a Cauchy sequence. From the completeness of R^m, it follows that the sequence $\{P_m\}$ has a limit. Let $\lim_{m \to \infty} P_m = L$.

Now since $F_k \supset F_{k+1} \supset F_{k+2} \ldots$, the sequence $\{P_k, P_{k+1}, \ldots, P_{k+m}, \ldots\}$ is a sequence of points of F_k. This implies that L is a point of closure of F_k, and because F_k is closed, $L \in F_k$. Since k is arbitrary, $L \in \bigcap_{k=1}^{\infty} F_k$.

Now if $L' \neq L$ is another point of $\bigcap_{k=1}^{\infty} F_k$, then by letting $\epsilon = \frac{1}{2} \|L' - L\|$, we can find F_m such that diam $(F_m) < \epsilon$, but that implies that F_m cannot contain both L and L' and hence $\bigcap_{k=1}^{\infty} F_k$ is a single point. The proof is now complete. $\blacksquare$

If we have sequences of nonempty, bounded, closed sets $\{F_m\}$ with $F_m \supset F_{m+1}$, then the $\bigcap_{k=1}^{\infty} F_k$ has at least one point (whether or not diam $(F_m) \longrightarrow 0$).

In Sec. 5.2, the Heine-Borel theorem established that if a set of real numbers is closed and bounded then it is compact and we stated that the converse was also true. We now prove the converse for R^n. (Indeed, the proof is valid for R also).

Theorem 5.6.4. Every compact set (having the Heine-Borel property) of R^n is closed and bounded.

Proof. Let C be a compact set of R^n, and $B_k = \{X : \|X\| < k\}$ k being a natural number. The class $B_k : \{k = 1, 2, \ldots\}$ of open balls is an open covering for the entire R^n, and *a fortiori*, an open covering for C. Because of the Heine-Borel property, there is a finite subcovering say $B_1, \ldots B_m$ for C. In that case B_m (the largest of these open balls) covers C which means that C is bounded.

To show that C is closed, assume by way of contradiction that there is a limit point P of C such that $P \notin C$. Now for $X \in C$ and for $r_X = \dfrac{1}{2} \|P - X\|$, construct the open ball B_X with center at X and radius r_X. The class of open balls $\{B_X : X \in C\}$ is an open covering for C and there would exist a finite subcovering, say $\{B_{X_1}, \ldots, B_{X_n}\}$. Let $r = \min \{r_{X_1}, \ldots, r_{X_n}\}$. Then for $0 < \epsilon \leqslant r$, the open ball $\{X : \|X - P\| < \epsilon\}$ does not contain any point of C for if it does, say Y, then Y must be in some B_{X_k} (of the finite subcovering), and $\|Y - P\| < \epsilon \leqslant r \leqslant r_k$. In that case,

$$r_k = \frac{1}{2} \|X_k - P\| = \frac{1}{2} \|X_k - Y + Y - P\| \leqslant \frac{1}{2} (\|X_k - Y\| + \|Y - P\|)$$

$$< \frac{1}{2} (r_k + \epsilon) \leqslant \frac{1}{2} (r_k + r_k) = r_k$$

which means that $r_k < r_k$ and we have a contradiction. Thus P could not be a limit point of C. Hence C is closed; and that completes the proof. $\blacksquare$

Exercise 5.6

1. Prove Theorem 5.6.2.
2. Prove that the intersection of sequence $\{F_m\}$ of nonempty, closed and bounded sets of R^n with $F_m \supset F_{m+1}$ (the diameter of F_m may or may not approach zero as $m \longrightarrow \infty$) is nonempty.
3. Give an example of a sequence $\{F_m\}$ of nonempty closed sets in R^2 with $F_m \supset F_{m+1}$ such that $\bigcap_{m=1}^{\infty} F_m$ is empty.

4. Determine which of the following sets are open, closed, neither.
   (a) $\{\langle x,y,z\rangle : x^2 + y^2 < z^2\} \cup \{\langle x,y,z\rangle : x^2 + y^2 = 1\}$ in R^3,
   (b) $\{\langle x,y\rangle : y < x^2\} \cap \{\langle x,y\rangle : y \geq x\}$ in R^2,
   (c) $\{\langle x,y\rangle : y^2 < x\} \cap \{\langle x,y\rangle : y \geq x^3\}$ in R^2.

5. Prove that if C is a compact subset of R^n then C is compact when regarded as a subset of R^{n+1}. In particular, show that the closed interval $[a,b]$ is closed in R^2.

6. Show that the open interval (a,b) is not open in R^2.

7. Show that the set $\left\{\langle x,y\rangle : y = \sin \dfrac{1}{x}\right\}$ is neither open nor closed in R^2.

8. Let $\mathcal{Q}^n$ be the set of all points which have rational coordinates. Show that every point of R^n is a limit point of $\mathcal{Q}^n$.

9. What is wrong with the following proof of the Bolzano-Weierstrass theorem in R^n?

   "Let S be an infinite bounded set of R^n with coordinate sets $\{S_i : i = 1, \dots, n\}$. At least one of S_i's must be infinite and since it is bounded it must have a limit point (Bolzano-Weierstrass property of R), say l. There must exist a sequence of distinct real number $\{p_i^m\}$ of S_i such that $\lim\limits_{m\to\infty} p_i^m = l$. Let $a_1 \in S_1, \dots, a_{i-1} \in S_{i-1}, a_{i+1} \in S_{i+1}, \dots, a_n \in S_n$. Now the sequence of distinct points $\{\langle a_1, \dots, a_{i-1}, p_i^m, a_{i+1}, \dots, a_m\rangle : m = 1, 2, \dots\}$ of S has the limit $\langle a_1, \dots, a_{i-1}, l, a_{i+1}, \dots, a_n\rangle$, and this limit is a limit point of S."

Miscellaneous Exercises for Chapter 5

1. Let $M = R \times R$, and $d(a_1,a_2) = \sqrt{(x_2 - x_1)^2 + (y_2 - y_1)^2}$, $a_1 = (x_1,y_1)$, $a_2 = (x_2,y_2)$. $\{M,d\}$ is a metric space of Euclidean plane. Determine which ones of the following sets are open, closed, neither.
   (a) $\{(x,y) = y^2 \leq x\}$
   (b) $\{(x,y) : x^2 + y^2 = 4\}$
   (c) $\{(x,y) : y < 0\} \cup \{(x,y) : x^2 + y^2 \leq 1\}$
   (d) $\{(x,y) : x < 0\} \cap \{(x,y) : y < x^3\}$
   (e) $(R \times R) - $ (the closed interval $[0,1]$).

2. Determine which ones of 1 (a)–(e) are connected.

3. Describe the interiors and boundaries of the sets in 1 (a)–(e).

4. Prove that the set of all limit points of a set is always closed.

5. Show that (a,b) considered as a subspace of R is not a complete metric space.

6. Show that (a,b) with discrete metric—that is, $d(x,y) = 1$ if $x \neq y$, $d(x,x) = 0$ is a complete metric space.

7. Prove that a set S of real numbers is compact if and only if every sequence of points of S has convergent subsequence whose limit is in S.

8. Give an example of a connected set of R which is not compact, and an example of a compact set which is not connected.
9. Show that a closed set F in R has an empty interior if and only if every real number is a limit point of its complement.
10. Prove that every compact subset of a complete metric space is complete.

6

Open Sets, Closed Sets, Perfect Sets, Borel Sets

This chapter is devoted to the study of various types of sets mentioned in the title. We are going to deal with the metric space of real numbers (and R^n in Sec. 6.6) unless otherwise specified. We start with the structure of open sets and closed sets in R.

6.1. The Structure of Open Sets and Closed Sets

In an attempt to gain a deeper understanding of the form and structure of certain sets of *real numbers*, we introduce the concept of component intervals, which serves in this capacity as a fundamental structure or building block.

Definition 6.1.1. (a) *An open interval (a,b) is called a component (or a component interval) of an open set G if $(a,b) \subset G$, $a \notin G$ and $b \notin G$.*

(b) *Also $(-\infty,a)$ is called a component of the open set G if $(-\infty,a) \subset G$, but $a \notin G$.*

(c) *(a,∞) is said to be a component of G, if $(a,\infty) \subset G$, and $a \notin G$.*

(d) *$(-\infty,\infty)$ is the component of R.*

In other words, a component of an open set G is a path connected open set (α,β) (α being a real number or $-\infty$, and β being a real number or ∞), such that $(\alpha,\beta) \subset G$, yet $\alpha \notin G$ and $\beta \notin G$.

EXAMPLE 6.1.1.

(i) If we delete all natural numbers from the set R, then we have an unbounded open set (Why?) whose components are $(-\infty,1)$ and intervals of the type $(n,n+1)$, n being a natural number.

(ii) If we delete $[0,1]$ from R, we have an open set whose components are $(-\infty,0)$ and $(1,\infty)$.

One may observe that the term "component" is quite descriptive.

Looking now at two distinct components of an open set G, it is clear that these components cannot intersect; for if they overlap, then an endpoint of one of them must lie in the interior of the other, and as such this endpoint will belong to G, a situation which is not possible according to Def. 6.1.1. This proves the following result:

Theorem 6.1.1.   Two distinct components of an open set are disjoint.

The next theorem describes the structure of an open set.

Theorem 6.1.2.   Every nonempty open set is the union of a countable number of components.

Proof.   Let G be a nonempty open set. We first prove that there is at least one component of G and that every point of G is in a component. Let $x_0 \in G$. We shall show that there exists a component (α,β) of G such that $x_0 \in (\alpha,\beta)$.

Consider the set $F = cG \cap (-\infty, x_0]$.

If F is empty, then $(-\infty, x_0] \subset G$, and we let $\alpha = -\infty$.

If F is nonempty, then since it is bounded above by x_0, it must have the lub. Let it be α. Since F is closed, $\alpha \in F \Longrightarrow \alpha \in cG$ which imples $\alpha \neq x_0$. Now x_0 being an upper bound of F, must be greater than α. If $x \in (\alpha, x_0]$ then $x \notin F$, yet $x \in (-\infty, x_0]$, and therefore, $x \in G$. Thus $(\alpha, x_0] \subset G$.

In either case we have found α (a real number or $-\infty$) such that $(\alpha, x_0] \subset G$, and $\alpha \notin G$.

In a similar fashion we can find a β (a real number or ∞) such that $x_0 < \beta$, and $[x_0, \beta) \subset G$, yet $\beta \notin G$.

Now $(\alpha, \beta) = (\alpha, x_0] \cup [x_0, \beta) \subset G$, and $\alpha \notin G$, $\beta \notin G$. Thus we have found a component of G containing x_0.

Finally, since any two distinct components of G are disjoint, we can associate each component with a rational number belonging to that component. This gives us an injective mapping of the class of components of G into the set of rational numbers; therefore, there are only countably many component intervals of G. Since we have already shown that every point of G is in some component of G, our theorem is proved. $\blacksquare$

We next discuss the structure of closed sets. First consider a bounded closed set F. F may be empty or may consist of a single point; but if F consists of more than one point let $[a,b] = \text{sci}\,(F)$, the smallest closed interval containing F (a and b are the glb and lub, respectively, of F). Furthermore, $[a,b] - F$ is open. (Cf. Exercise 5.2(2)) It is empty if $F = [a,b]$. Otherwise, $[a,b] - F$ is a nonempty bounded open set and as such is the union of a countable number of open intervals. This proves the following result.

Theorem 6.1.3.   A nonempty bounded closed set is either a single point, a closed interval or can be obtained from a closed interval by removing a countable number of mutually disjoint open intervals.

The open intervals removed in the latter case will be called complementary intervals of F.

If a closed set F is unbounded, then cF is still open, and Theorem 6.1.2 yields the following result.

Theorem 6.1.4. An unbounded closed set is either the entire set of reals, R, or is obtained from R by removing a countable number of open intervals (of finite or of infinite length) which are mutually disjoint.

6.2. Perfect Sets

Definition 6.2.1. *A set S is called "dense in itself" if every point of S is a limit point of S.*

In other words, S is dense in itself if and only if $S \subset S'$. Any interval is dense in itself, as are the set of all rationals and the set of all irrationals. Moreover, by virtue of Theorem 5.2.2, every open set (of reals) is dense in itself.

Definition 6.2.2. *A set S is called perfect if it is both closed and dense in itself.*

Now S is closed if and only if $S' \subset S$; S is dense in itself if and only if $S \subset S'$; therefore, S is perfect if and only if $S = S'$.

Clearly, every closed interval, the void set and the entire set of reals are all perfect sets; but these are not the only perfect sets, as illustrated by the following most fascinating example.

EXAMPLE 6.2.1. The *Cantor Set*.

From the interval $[0,1]$ remove the following sequence of open intervals: the open interval $\left(\dfrac{1}{3}, \dfrac{2}{3}\right)$, which is the "open middle third" of $[0,1]$, leaving $\left[0, \dfrac{1}{3}\right]$ and $\left[\dfrac{2}{3}, 1\right]$; next, the intervals $\left(\dfrac{1}{9}, \dfrac{2}{9}\right)$ and $\left(\dfrac{7}{9}, \dfrac{8}{9}\right)$, the open middle thirds

of $\left[0, \dfrac{1}{3}\right]$ and $\left[\dfrac{2}{3}, 1\right]$ respectively, leaving $\left[0, \dfrac{1}{9}\right]$, $\left[\dfrac{2}{9}, \dfrac{1}{3}\right]$, $\left[\dfrac{2}{3}, \dfrac{7}{9}\right]$ and $\left[\dfrac{8}{9}, 1\right]$; at each succeeding stage remove the open middle third of each remaining closed interval. If this process is carried out denumerably many times, the result is the Cantor set, sometimes called the *Cantor ternary set* or the *Cantor middle-third set*.

Interestingly enough, the sum of the lengths of the intervals removed is

$$\frac{1}{3} + \left(\frac{1}{9} + \frac{1}{9}\right) + \left(\frac{1}{27} + \frac{1}{27} + \frac{1}{27} + \frac{1}{27}\right) + \cdots = \frac{1}{3} + \frac{2}{9} + \frac{4}{27} \cdots = \sum_{n=0}^{\infty} \frac{1}{3}\left(\frac{2}{3}\right)^n =$$
$$\frac{1}{3}\frac{1}{1 - \frac{2}{3}} = 1.$$

Intuitively, it may appear that the only points left in the Cantor set are the end-points $\frac{1}{3}, \frac{2}{3}, \frac{1}{9}, \frac{2}{9}, \frac{7}{9}, \frac{8}{9}, \ldots$, and these are denumerable in number. Closer scrutiny reveals that this impression is false, and that the Cantor set is actually nondenumerable. We formalize this important result as follows:

Theorem 6.2.1.   *The Cantor* set is nondenumerable.

Proof.   Let us express all real numbers in $[0,1]$ in their ternary expansions; e.g., $\frac{2}{9}$ becomes $(.02)_3$; $\frac{5}{27}$ has the ternary expansion $(.012)_3$, etc.

When we removed the interval $\left(\frac{1}{3}, \frac{2}{3}\right)$ in the construction of the Cantor set, we actually removed all those real numbers between .1 and .2 (in ternary form); i.e., all numbers having the digit 1 in the first place of their ternary expansion. Removing the intervals $\left(\frac{1}{9}, \frac{2}{9}\right)$ and $\left(\frac{7}{9}, \frac{8}{9}\right)$ means discarding all those numbers having ternary expansion between .01 and .02 or between .11 and .12. Each of these numbers has the digit 1 in the second place (in ternary form).

The next (third) removal discards all those numbers with 1 in the third place; the fourth removal discards those numbers with 1 in the fourth place, etc. We see, then, that the points in $[0,1]$ which do not belong to the Cantor set have the digit 1 in at least one place in their ternary expansion. The Cantor set must then consist of those numbers in the interval $[0,1]$ which contain only the digits 0 and 2 in their ternary expansion. There is an obvious 1-1 correspondence between these numbers and the set of all numbers in $[0,1]$ expressed in binary form. This 1-1 correspondence merely interchanges the digits 1 and 2. For example,

$$(.02)_3 \longleftrightarrow (.01)_2 \quad \text{and} \quad (.0202)_3 \longleftrightarrow (.0101)_2$$

In other words, the set of all reals in $[0,1]$ can be put in 1-1 correspondence with a subset of the Cantor set, and by the Schroeder-Bernstein theorem it follows that the Cantor set has power c; i.e., it is nondenumerable. This completes the proof. |

It must be remarked here that the end-points of the removed intervals in the construction of the Cantor set have two different ternary representations, one containing the digit 1 at a certain place, and the other having the digits 0 or 2 recurring; for example,

$$\frac{1}{3} = (.1000\ldots)_3 = (.0222\ldots)_3$$

and

$$\frac{2}{3} = (.1222\ldots)_3 = (.200\ldots)_3$$

If we agree to use that representation which does not contain the digit 1, then all the points in the Cantor set can be expressed in ternary expansion by the digits 0 and 2, as claimed in the above proof.

Theorem 6.2.2. The Cantor set is closed.

Proof. This follows immediately from the construction, since the Cantor set is the complement of the union of the removed open intervals and the intervals $(-\infty, 0)$ and $(1, +\infty)$.

Theorem 6.2.3. The Cantor set is dense in itself.

Proof. Let x_0 be any point of the Cantor set. If we write x_0 in ternary expansion, say

$$x_0 = .t_1 t_2 t_3 \ldots t_n \ldots,$$

then $t_1 = 0$ or 2. We now construct a sequence of points in the following manner:

$$x_1 = .t_1' t_2 t_3 t_4 \ldots t_n \ldots$$
$$x_2 = .t_1 t_2' t_3 t_4 \ldots t_n \ldots$$
$$x_3 = .t_1 t_2 t_3' t_4 \ldots t_n \ldots$$
$$\vdots$$
$$x_n = .t_1 t_2 t_3 t_4 \ldots t_{n-1} t_n' t_{n+1} \ldots$$
$$\vdots$$

where $t_n' = 0$ if $t_n = 2$ and $t_n' = 2$ if $t_n = 0$. In this way, we obtain a sequence of distinct points, $\{x_n\}$, all belonging to the Cantor set, such that x_n differs from x_0 in the nth place in the ternary expansions. The reason for constructing this sequence is, of course, that $\lim_{n \to \infty} x_n = x_0$, and we may conclude from Theorem 5.1.3 that x_0 is a limit point of the Cantor set. Thus every point of the Cantor set is a limit point, and the set is therefore dense in itself.

As a consequence of Theorems 6.2.2 and 6.2.3 we have:

Theorem 6.2.4. The Cantor set is perfect.

It is not difficult to perceive the structure of a perfect set, realizing that a perfect set is a closed set without any isolated points. The following definition will be useful in the discussion of perfect sets.

Definition 6.2.2. *Two intervals are said to be "adjacent" (or "abutting") if the left end-point of one is the same as the right-end point of the other; otherwise, they are called nonadjacent (or nonabutting).*

We now prove the following theorem for perfect sets.

Theorem 6.2.5. A nonempty perfect set is either the whole set R, a closed interval $[a,b]$, or can be obtained from a closed interval (or from R) by removing a countable number of nonadjacent disjoint open intervals (or path connected open sets) which have neither a nor b as an end-point. Conversely, any set obtained in such a manner is perfect.

Proof. Let P be a nonempty perfect set. Then P is closed, and we may apply Theorem 6.1.3 or Theorem 6.1.4 depending upon whether P is bounded or unbounded.

Suppose, first, that P is bounded. Let $[a,b]$ = sci (P). If $P = [a,b]$ the theorem is proved. If $P \neq [a,b]$, then, by Theorem 6.1.3, P is obtained from $[a,b]$ by the removal of a countable number of disjoint open intervals. No two of these intervals can be adjacent. To see this, suppose that (α,β) and (β,γ) are two adjacent intervals which are removed. Then $\beta \in [a,b]$, $\beta \notin [a,b] - P$ and therefore $\beta \in P$. But then (α,γ) is a neighborhood of β containing no other point of P, implying that β is an isolated point of P, a contradiction. Furthermore, neither a nor b can be the end point of a complementary interval. For suppose (a,c) is a complementary interval and let a_1 be any point less than a. Then the neighborhood (a_1,c) contains no other point of P, making a an isolated point of P, again a contradiction [$a \in P$ by virtue of $[a,b]$ being the sci (P)].

Next consider the converse for the case when P is bounded; i.e., let P be obtained from $[a,b]$ by removing a countable number of nonadjacent disjoint open intervals having neither a nor b as an end point. Such a set is clearly closed. Thus in this case if we assume that P is not perfect, then P must contain at least one isolated point, say x_0. If $a < x_0 < b$, then $\exists$ a neighborhood (α_0,β_0) of x_0 containing no other point of P, and since a, $b \in P$, then $(\alpha_0,\beta_0) \subset [a,b]$. Now (α_0,x_0) and (x_0,β_0) are contained in two distinct complementary intervals, say I_1 and I_2. These intervals are disjoint and do not contain x_0; hence, x_0 is a common end point of both; i.e., I_1 and I_2 are adjacent, in contradiction to our assumption. If $a = x_0$, an isolated point of P, then it can be shown quite easily that a would be a left-end point of a complementary interval of P—a contradiction. In the same manner we can show that b cannot be an isolated point of P.

The theorem can be proved for the case when P is unbounded in an analogous fashion, using R instead of $[a,b]$ and extending the open intervals under discussion to include $(-\infty,a)$, $(a,+\infty)$ and $(-\infty,+\infty)$. ❚

Using this theorem, it follows immediately that the Cantor set is perfect, and Theorem 6.2.3 (also Theorem 6.2.1 and Theorem 6.2.4) become redundant. The proof given for Theorem 6.2.3 does, however, hold special interest in itself.

6.3. Nowhere Dense Sets and Sets of First and Second Category

In this section we discuss nowhere dense sets and the sets which are obtained from nowhere dense sets. We also prove the famous Baire category theorem, the most important result of this section.

Definition 6.3.1. *A set S is said to be nowhere dense if every point of closure of S is a boundary point of $\overline{S}$ (and, a fortiori, a boundary point of S).*

In other words, S is nowhere dense if and only if its closure, $\overline{S}$, has no interior point.

Theorem 6.3.1. S is nowhere dense $\Longleftrightarrow$ there does not exist an open interval contained in $\overline{S}$.

Proof. If $x \in \overline{S}$, and S is nowhere dense, then x is a boundary point of $\overline{S}$; therefore, every open interval containing x contains at least one point which does not belong to $\overline{S}$; i.e., ∄ an open interval contained in $\overline{S}$.

Conversely, suppose ∄ an open interval contained in $\overline{S}$. If $x \in \overline{S}$ and I_x is an arbitrary neighborhood of x, then I_x contains at least one point not in S. Since I_x certainly contains at least one point which does lie in S (x is a point of closure of S), we may conclude that x is a boundary point of $\overline{S}$; thus S is nowhere dense and the proof is complete. ❚

It follows that a set S is nowhere dense if and only if $\overline{S}$ is nowhere dense. Thus, a closed set F is nowhere dense if and only if F has no interior point; for example, a finite set is nowhere dense. A set without a limit point is nowhere dense, being both closed and lacking any interior point. The set of natural numbers is such a set.

The set of rational numbers, on the other hand, is not nowhere dense, since its closure is the set of all real numbers. Thus we have a denumerable set which is *not* nowhere dense.

One can easily prove that the Cantor set is nowhere dense (and, of course, nondenumerable), adding to the list of properties of this singularly interesting set.

There is a slight connection between the notions of "nowhere dense" and "dense in itself." As might be expected, if S is nowhere dense, then cS is dense in itself. The set of rationals, however, serve to illustrate that the converse is not necessarily true.

Definition 6.3.2. *A set which is the union of a denumerable number of nowhere dense sets is called a set of first category or a meagre set.*

A nowhere dense set is always "meagre" since it can always be expressed as the union of itself and denumerably many empty sets.

A set of first category is not necessarily nowhere dense, the set of rational numbers being an example.

We now state a theorem whose proof follows trivially from Def. 6.3.2.

Theorem 6.3.2. The union of two sets of first category is a set of first category.

We now prove a theorem which is due to the mathematician, R. Baire (1874–1932) who made some contributions in analysis.

Theorem 6.3.3. (Baire category theorem): The set of all real numbers is not of first category.

Proof. By way of contradiction we assume that the set of all real numbers R is of first category. We can then write $R = \bigcup_{n=1}^{\infty} A_n$, where each A_n is nowhere dense, and so is each $\overline{A}_n$. But $R = \bigcup_{n=1}^{\infty} A_n \subset \bigcup_{n=1}^{\infty} \overline{A}_n \subset R$. Thus $R = \bigcup_{n=1}^{\infty} \overline{A}_n$. Observe each $\overline{A}_n$ is closed.

Since $\overline{A}_1$ is a nowhere dense $\overline{A}_1 \neq R$, and thus there exists a real number x_1 such that $x_1 \notin \overline{A}_1$. Furthermore, x_1 cannot be a limit point of $\overline{A}_1$ which means there exists an open interval I_1 containing x_1 such that $I_1 \cap A_1 = \phi$. It is possible to construct a closed interval $F_1 \subset I_1$ which implies $F_1 \cap \overline{A}_1 = \phi$.

Now $\overline{A}_2$ is nowhere dense, and therefore it cannot contain interior of F_1 (which is an open interval). Thus there exists x_2 in the interior of F_1 such that $x_2 \notin \overline{A}_2$. Again there exists an open interval I_2 containing x_2 such that $I_2 \cap \overline{A}_2 = \phi$. Let $I_2' = I_2 \cap (\text{Interior of } F_1)$. I_2' is an open interval containing x_2 and contained in $I_2 \implies I_2' \cap \overline{A}_2 = \phi$. Construct a closed interval $F_2 \subset I_2'$. This means $F_2 \cap \overline{A}_2 = \phi$. Furthermore, $F_2 \subset F_1$ which means $F_2 \cap \overline{A}_1 = \phi$.

Proceeding in this manner with $\overline{A}_3, \overline{A}_4, \ldots$, we can construct a sequence of closed intervals:

$$F_1 \supset F_2 \supset F_3 \supset \cdots \supset F_n \supset \ldots$$

such that $F_n \cap \overline{A}_n = \phi$.

Now by Cantor theorem there exists at least one point $x_0 \in \bigcap_{n=1}^{\infty} F_n$. Now

$x_0 \notin \overline{A}_n$ for every n (for if it does belong to $\overline{A}_m$ then $\overline{A}_m \cap \overline{F}_m \neq \emptyset$) which means $x_0 \notin \bigcup\limits_{n=1}^{\infty} \overline{A}_n = R$, which is a contradiction. This proves the theorem. ▌

It must be observed from the last part of the proof that this theorem heavily depends upon Cantor theorem which is equivalent to the "Property of Completeness" (the set of rational numbers does not obviously satisfy this theorem). For this reason, the Baire category theorem can be generalized for any complete metric space as follows:

Theorem 6.3.4. No complete metric space is of first category.
A very important corollary of Theorem 6.3.3 is as follows:

Corollary. The set of irrationals is not of first category.
This follows obviously from Theorems 6.3.2 and 6.3.3.

6.4. Borel Sets

An interesting class of sets which plays an important role in analysis in general and measure theory in particular was studied by Borel; the sets in this class are, consequently, popularly known as Borel sets. Recall that the union of any number of open sets is open, whereas the intersection of closed sets is always closed. Borel sets are constructed basically by considering unions of closed sets and intersections of open sets. This can be done in any *metric space.*

Definition 6.4.1. *A set is said to be of type F_σ if it can be obtained as the union of a denumerable number of closed sets.*

Definition 6.4.2. *A set is said to be of type G^δ if it can be obtained as the intersection of a denumerable number of open sets.* *
It follows immediately that every closed set F is of type F_σ and that every open set G is of type G^δ for we can write

$$F = F \cup \emptyset \cup \emptyset \cup \cdots \quad \text{and} \quad G = G \cap R \cap R \cap \cdots$$

However, not every set of type F_σ need be closed, nor must every set of type G^δ

---

*It may be remarked here that an apparent reason for the usage of the symbols σ and δ is that they are Greek abbreviations for the German words *Summe* (sum) and *Durchschnitt* (intersection), respectively. There is a slight indication that F and G are abbreviations for the French word *ferme* (closed) and the German word *Gebiet* (region), respectively. The fact that F and G are also the first letters of French and German, respectively, is purely a coincidence.

be open. For example, the set of all rational numbers is of type F_σ, being the denumerable union of its elements (each singleton is closed); moreover, every semiopen interval is of type G^δ since we can write

$$(a,b] = \bigcap_{n=1}^{\infty} \left(a, b + \frac{1}{n} \right)$$

As a matter of fact, every interval is of type F_σ and G^δ, for

$$(a,b) = \bigcup_{n=1}^{\infty} \left[a + \frac{1}{n}, b - \frac{1}{n} \right]$$

and

$$[a,b] = \bigcap_{n=1}^{\infty} \left(a - \frac{1}{n}, b + \frac{1}{n} \right)$$

As a consequence of De Morgan's law and Theorem 5.2.3, we have the following immediate result:

Theorem 6.4.1. The complement of a set of type F_σ is a set of type G^δ; the complement of a set of type G^δ is a set of type F_σ.

An immediate consequence of this theorem is that the set of irrationals is of type G^δ. Employing the corollary of Theorem 6.3.4, however, we will establish the following:

Theorem 6.4.2. The set of irrationals is not of type F_σ.

Proof. Let Y denote the set of irrationals, and assume by way of contradiction that Y is of type F_σ. Then we can write $Y = \bigcup_{n=1}^{\infty} F_n$, with each F_n closed. Now F_n cannot contain any interval since every interval must contain both rational and irrational points; furthermore, F_n is closed and thus F_n is nowhere dense. Y then is a set of first category, contradicting the aforementioned corollary, and the theorem is established. ∎

Combining Theorems 6.4.1 and 6.4.2 we have the following corollary.

Corollary. The set of rational numbers is not of type G^δ.

Theorem 6.4.3. Every open set is of type F_σ.

Proof. We can write

$$(-\infty,a) = \bigcup_{n=1}^{\infty} \left[-n, a - \frac{1}{n} \right]$$

$$(a,+\infty) = \bigcup_{n=1}^{\infty} \left[a + \frac{1}{n}, n\right]$$

and

$$(-\infty,+\infty) = \bigcup_{n=1}^{\infty} [-n,n]$$

The empty set, being closed, is of type F_σ. Finally, a nonempty open set, G, is the union of a countable number of open intervals and/or sets of the form $(-\infty,\infty)$ $(-\infty,a)$ and $(a,+\infty)$; each of these components is of type F_σ, and it is obvious that the countable union of sets of type F_σ is again of type F_σ, and the proof is complete. ∎

Notice that the set of all real numbers $(-\infty,\infty)$ is of type F_σ.

Theorem 6.4.4. Every closed set is of type G^δ.

Proof. This is an immediate consequence of Theorems 6.4.1 and 6.4.2.

By allowing the denumerable intersections and unions of sets of type F_σ on G^δ, we construct other Borel sets.

Definition 6.4.3. *A set S is of type F_σ^δ if it is the intersection of a denumerable class of sets of type F_σ; S is of type $F_{\sigma\sigma}^\delta$ (or $F_\sigma^{\delta^2}$) if it is the union of a denumerable class of sets of type F_σ; and so on.*

Definition 6.4.4. *A set S is of type G_σ^δ if it is the union of a denumerable class of sets of type G^δ; S is of type $G_\sigma^{\delta\delta}$ (or $G_\sigma^{\delta^2}$) if it is the intersection of a denumerable class of sets of type G_σ^δ; and so on.*

It is an easy matter to show that the complement of a set of type $F_{\sigma i}^{\delta j}$ is a set of type $G_{\sigma i}^{\delta j}$; and vice versa.

As a matter of convenience, we say that a closed set is of type $F_{\sigma 0}^{\delta 0}$ and an open set is of type $G_{\sigma 0}^{\delta 0}$.

We can construct a hierarchy of Borel sets if we observe that every set of type F_σ is of type F_σ^δ (since $F_\sigma = (F_\sigma \cap R \cap R \cap \cdots)$), and that every set of type F_σ^δ is of type $F_{\sigma\sigma}^\delta$, and so on. Similarly, every set of type G^δ is of type G_σ^δ, and so on. Then, using induction we can establish the following theorem:

Theorem 6.4.5. Every set of type $F_{\sigma i}^{\delta j}$ is of type $F_{\sigma l}^{\delta k}$ and of type $G_{\sigma l}^{\delta k}$, where $k \geqslant j$ and $l \geqslant i$. A similar result is true for sets of type $G_{\sigma i}^{\delta j}$.

Indeed, we can construct the following chart:

$$
\begin{array}{c}
F \;\longrightarrow\; F_\sigma \;\longrightarrow\; F_\sigma^\delta \;\longrightarrow\; F_{\sigma\sigma}^\delta \;\longrightarrow\; F_{\sigma\sigma}^{\delta\delta} \;\longrightarrow\; \cdots \\[2pt]
G \;\longrightarrow\; G^\delta \;\longrightarrow\; G_\sigma^\delta \;\longrightarrow\; G_\sigma^{\delta\delta} \;\longrightarrow\; G_{\sigma\sigma}^{\delta\delta} \;\longrightarrow\; \cdots
\end{array}
$$

where $A \longrightarrow B$ means that every set of type A is also of type B.

Definition 6.4.5. *A class of sets* $\{A_\alpha\}$ *is called a σ-field* (or σ-algebra or Borel field) if* (a) *the entire set R and the void set ϕ are members of the class,* (b) *the union of a countable number of sets of the class is in the class, and* (c) *the complement of every set of the class is in the class.*

From Def. 6.4.5 it follows easily that the intersection of a countable number of sets in a Borel field is a member of that Borel field.

An important result which can be shown without much difficulty is that the class of Borel sets is the smallest Borel field containing open sets (actually, open intervals). This means that any other Borel field containing open intervals will contain the class of Borel sets.

6.5. Condensation Points

In Chapter 5 we briefly discussed the concept of point of condensation. Here we explore this topic more thoroughly, to enable us to present yet another structural theorem on closed sets. This section may be skipped without any loss of continuity.

Recall that a point x_0 is a condensation point of a set S if every neighborhood of x_0 contains nondenumerably many points of S. The following result is significant.

Theorem 6.5.1. Every nondenumerable set S has at least one point of condensation in S.

Proof. Suppose by way of contradiction that no point of S is a condensation point of S. Then for each $x \in S$, $\exists$ an open interval I_x, containing x, such that $I_x \cap S$ is countable. Let J_x be an open interval contained in I_x, and containing x but having rational end points; indeed, $J_x \cap S$ is again countable. Moreover, the class of all such intervals J_x is necessarily countable and we may enumerate as follows:

$$J_1, J_2, J_3, \ldots J_n, \ldots$$

Since each point of S is in some J_k, we may conclude that $S = \bigcup_k (J_k \cap S)$. The countable union of countable sets being countable, we have obtained a contradiction. ∎

This theorem is analogous to the Bolzano-Weierstrass theorem, but is stronger in the sense that the condition of boundedness is not required as is the case with

---

*The reader should be cautioned that a σ-field is not a field in the usual sense. This misnomer is analogous to the fact that quicksilver is not really silver in the classification of chemical elements and that mud pies are not generally considered to be palatable cuisine in the classification of digestible matter.

the Bolzano-Weierstrass theorem; furthermore, whereas the Bolzano-Weierstrass theorem assures only the existence of a limit point of a bounded, infinite set, the present theorem guarantees not only that a nondenumerable set has a point of condensation, but that some point of condensation must actually be contained in the set.

We next introduce two concepts which are analogous to "derived set" and "dense in itself," respectively.

Definition 6.5.1. *The set of all points of condensation of a set S will be called the condensed set of S and will be denoted by* S_c.

Definition 6.5.2. *If every point of a set S is a point of condensation of S,* i.e., *if* $S \subset S_c$, *then S will be called "condensed in itself."*

The condensed set of (a,b) is $[a,b]$, and that of $(0,1] \cup \{2,3,4,\ldots,n,\ldots\}$ is $[0,1]$. The condensed set of the set of irrationals is R, but the condensed set of the rationals is void.

Every open set of reals is condensed in itself, as is every interval. The set of irrationals is also condensed in itself.

The following result is useful in a subsequent theorem.

Theorem 6.5.2. $S - S_c$ *is countable.*

Proof. Let $x_0 \in S - S_c$. Then x_0 is not a point of condensation of S; as such, there exists a neighborhood of x_0 which contains only countably many points of S and *a fortiori* countably many points of $S - S_c$. Thus, no point of $S - S_c$ is a point of condensation of $S - S_c$. Using Theorem 6.5.1, we see that $S - S_c$ must be countable, and that completes the proof. |

Since we can write $S = (S - S_c) \cup (S \cap S_c)$, we have the following.

Corollary. If S is nondenumerable, then $S \cap S_c$ is nondenumerable.

Theorem 6.5.3. *For any set S, the set* S_c *is perfect.*

Proof. If S_c is empty, then it is perfect. If S_c is not empty, then S is nondenumerable. Let $x_0 \in S_c$, and let I_{x_0} be an arbitrary neighborhood of x_0. Then I_{x_0} contains nondenumerably many points of S. Therefore, I_{x_0} contains nondenumerably many points of S_c, for otherwise I_{x_0} would contain nondenumerably many points of $S - S_c$. This proves that x_0 is a point of condensation of S_c and *a fortiori* a limit point of S_c. Therefore, S_c is dense in itself.

Now let y_0 be a point of closure of S_c. Then an arbitrary neighborhood, I_{y_0}, of y_0 contains at least one point x of S_c. Now I_{y_0} is also a neighborhood of x and must contain nondenumerably many points of S. Thus an arbitrary neighborhood of y_0 contains nondenumerably many points of S. This implies that $y_0 \in S_c$. Hence S_c is closed.

Since S_c is dense in itself and closed it is perfect, and the proof is complete. |

The above theorem may be compared with the result that the derived set of S is always closed.

In proving the first part of the theorem, we established the following interesting result.

Theorem 6.5.4. The set of points of condensation of a set is condensed in itself.

This result is quite in contrast with the fact that the derived set of a set may not be dense in itself.

For open sets, however, we have a very strong result.

Theorem 6.5.5. For any open set G of reals, $\overline{G} = G' = G_c$.

Proof. This is an immediate consequence of Theorem 5.2.2 and Theorem 6.1.2.

We conclude this section with the structural theorem on closed sets as promised.

Theorem 6.5.6. Every closed set is the union of a perfect set and a countable set.

Proof. Let F be a closed set and let F_c be the condensed set of F. By Theorem 6.5.3, F_c is perfect; moreover, by Theorem 6.5.2, $F - F_c$ is countable. Since F is closed, $F_c \subset F' \subset F$. Therefore

$$F = F_c \cup (F - F_c)$$

and the proof is complete. |

6.6. Borel Sets, Nowhere Dense Sets in R^n

In this section we consider the validity of some of the results of the preceding sections in the case of R^n. First it must be pointed out that the open sets of $R^n (n > 1)$ cannot be characterized in the same way as those in R, that is, by means of countable component intervals. In other words, there is no counterpart of Theorem 6.1.2 in R^n. We do have the following result, however.

Theorem 6.6.1. Every nonempty open set of R^n is the union of a countable class of open balls (not necessarily pairwise disjoint) of R^n.

Proof. Let G be a nonempty subset of R^n. Construct a subset S of G that consists of only those points of G which have rational coordinates.

If $P \in S \subset G$, then P is an interior point of G and there exists an open ball with center at P and contained in G. Let m_p be a natural number such that the

$$\circ\; B_p = \left\{ X : \|X - P\| < \frac{1}{m_p} \subset G \right\}$$

(m_p could very well be 1 or very large).

Now every point of G is either in S or a limit point of S (cf. Exercise 5.6(8)). Therefore, it can be easily shown that

$$G = \bigcup_{P \in S} \left\{ X : \|X - P\| < \frac{1}{m_P} \right\}$$

and since S is countable the result is proved. ∎

The next result shows that every open set of R^n is of type F_σ.

Theorem 6.6.2. Every nonempty open set G is the union of a countable class of closed balls, and hence it is of type F_σ.

The proof of this theorem is similar to that of the last theorem and is therefore left as an exercise.

Because of the De Morgan's property the following corollary is obvious.

Corollary. Every closed set of R^n is of type G^δ.

It may be recalled that every open set is of type G^δ and every closed set is of type F_σ.

Next, we generalize the ideas of nowhere dense sets, and sets of first category.

To define a nowhere dense set it would be appropriate to use the criterion of Theorem 6.3.1.

Definition 6.6.1. *A subset S of R^n is nowhere dense if $\overline{S}$ contains no open ball.*

One may describe a nowhere dense set as a set whose closure is not "plump" in the sense that it does not contain any open ball.

In $R^n (n > 1)$ all subsets of R^{n-1} become nowhere dense (prove it). In particular all the subsets of R are nowhere dense in R^2, since the closure of any set of R will be in R and as such would not contain any open ball (or open disk) of R^2. It may also be observed that no nonempty subset of R^{n-1} is open in R^n. Without giving a formal definition, we may recall that: "A set is of first category (meager) if it is a countable union of nowhere dense sets; otherwise it is of second category (nonmeager)." Since R^n is complete, the Baire category theorem holds in R^n.

Theorem 6.6.3. R^n is of second category (nonmeager).

The proof is left as an exercise.

It is interesting to note that the set of all irrational numbers $(R - \mathbb{Q})$ when considered as a subset of R^2 becomes a nowhere dense set and, a fortiori, a set of first category in R^2 in contrast to the conclusion of the corollary of Theorem 6.3.3.

The set Q^n of R^n which consists of all those points having rational coordinates is not nowhere dense, but being denumerable, it is of first category.

An example of an n-dimensional, nondenumerable nowhere dense set in R^n is the n-dimensional Cartesian product of the Cantor set by itself. It may be denoted by K^n where K is the Cantor set (Example 6.2.1).

We conclude this section with a definition of "points of condensation" of a set in R^n.

Definition 6.6.2. *A point of condensation of a set S in R^n is a point x_0 such that every open ball containing x_0 contains nondenumerably many points of S.*

All the results of Sec. 6.5 are valid for R^n.

Exercise 6.6

1. Prove Theorem 6.6.2.
2. Prove Theorem 6.6.3.
3. Show that every subset of R^n is nowhere dense in R^{n+1}.
4. Prove that every nondenumerable set of R^n has a point of condensation.

Miscellaneous Exercises for Chapter 6

1. Construct an open set in R having denumerably many components with lengths $1, \dfrac{1}{2}, \ldots, \dfrac{1}{n}, \ldots$.

2. Construct a generalized Cantor set as follows: let $0 < a < 1$.

   (i) Remove from $[0,1]$ an open interval of length $\dfrac{1}{2} a$ with midpoint $\dfrac{1}{2}$.

   (ii) From the remaining closed interval remove two open intervals each of length $\dfrac{1}{8} a$ with midpoints $\dfrac{1}{4} - \dfrac{1}{4} a$ and $\dfrac{3}{4} + \dfrac{1}{4} a$, respectively. Repeat this process indefinitely as in the case of the Cantor set. Show that the sum of the lengths of intervals removed from $[0,1]$ is "a" and the remaining set is a perfect nowhere dense set. (Such a set is called a Cantor set of positive measure since the "length" of what is left is $1 - a > 0$).

3. Obtain a set in R similar to the Cantor set in construction but consisting of irrational numbers only.

4. Let A be a nonempty set. Call the set $\{(x - y): x,y \in A\}$ the difference set of A. Show that the "zero" on the real line is an interior point of the difference set of the Cantor set.

5. Prove that every complete metric space is of second category.

7

Continuity

One of the most important concepts in the development of modern mathematics is that of "continuity." In a metric space this notion is based on the idea of limit of a function. We shall begin this chapter with the definition of limit of a function before moving on to "continuous functions" and "uniformly continuous functions."

7.1. Limit of a Function

Definition 7.1.1. *Let $\{X,d\}$ and $\{Y,d'\}$ be two metric spaces and let D be a subset of X. Let $f: D \longrightarrow Y$ be a function defined on at least a deleted neighborhood of a point x_0 of D. We say f is said to have a limit $L \in Y$ as x approaches x_0 (written $x \longrightarrow x_0$) if for every neighborhood I_L of L, $\exists$ a deleted neighborhood N_{x_0} of x_0 such that $x \in N_{x_0} \cap D \implies f(x) \in I_L$. We write $\lim\limits_{x \to x_0} f(x) = L$.*

The following examples illustrate the idea of a limit in the real number system.

EXAMPLE 7.1.1. The function $f(x) = \dfrac{2x^2 - 2}{x - 1}$ is defined for all real values of x except for $x = 1$; however, the limit of this function exists as $x \longrightarrow 1$ and is equal to 4. If we consider any neighborhood of 4, say $(4 - \epsilon, 4 + \epsilon)$ then for any deleted neighborhood N_δ of 1 which is contained in $(1 - \delta, 1 + \delta)$, it is true that $x \in N_\delta \implies f(x) \in (4 - \epsilon, 4 + \epsilon)$. For example, if $\epsilon = \dfrac{1}{100}, x = 1.0001$, then $f(x) = 4.0002 \in (4 - \epsilon, 4 + \epsilon)$.

EXAMPLE 7.1.2. The function $f(x) = \dfrac{1}{x - 1}$ does not have a limit as

$x \longrightarrow 1$, for if L is such a limit, then for $\epsilon > 0$ it is always possible to choose an x in any deleted neighborhood of 1 such that $\dfrac{1}{x-1} > L + \epsilon$.

Exercise 7.1

In the following problems, consider the functions $f\colon \{R^2 - \langle 0,0 \rangle\} \longrightarrow R$, that is the domain of f is the set of all points in $R \times R$, except $\langle 0,0 \rangle$, and the range is in R.

1. Let

$$f\langle x,y \rangle = \frac{xy}{|x| + |y|}$$

Show that $\displaystyle\lim_{\langle x,y \rangle \to \langle 0,0 \rangle} f(x,y) = 0$.

$\left(Hint: \left| \dfrac{xy}{|x| + |y|} \right| \leqslant |x| \right)$

2. Let

$$f\langle x,y \rangle = \frac{x}{x^2 + y^2}$$

Show that $\displaystyle\lim_{\langle x,y \rangle \to \langle 0,0 \rangle} f\langle x,y \rangle$ does not exist.

3. Let

$$f\langle x,y \rangle = \frac{x^2 y^2}{x^2 + y^2}$$

Show that $\displaystyle\lim_{\langle x,y \rangle \to \langle 0,0 \rangle} f\langle x,y \rangle = 0$.

(Hint: $x^2 \leqslant x^2 + y^2, y^2 \leqslant x^2 + y^2$.)

7.2. Continuity

Let f be a function defined on a subset D of a metric space $\{X,d\}$ with values in a metric space $\{Y,d'\}$. Let $x_0 \in D$. The continuity of f at x_0 may be defined in any one of the following three ways:

Definition 7.2.1(a). *f is said to be continuous at x_0 if given $\epsilon > 0 \; \exists \; a$ $\delta(x_0, \epsilon) > 0$ such that $d'(f(x), f(x_0)) < \epsilon$ whenever $x \in N_\delta \cap D$, where N_δ is the δ-neighborhood of x_0.*

Definition 7.2.1(b). *f is said to be continuous at x_0 if for every convergent* *sequence $\{x_n\}$ in D having the limit x_0, the sequence $\{f(x_n)\}$ is convergent and* $\lim\limits_{n \to \infty} f(x_n) = f(x_0)$.

Definition 7.2.1(c). *f is said to be continuous at x_0 if for every open set O containing $f(x_0)$ there exists an open set G containing x_0 such that $f(G \cap D) \subset O$.*

We may rephrase Def. 7.2.1(c) as follows:

f is continuous at $x_0 \Longleftrightarrow$ for every open set O containing $f(x_0)$ there is a set S containing x_0 and relatively open in D, such that $f(S) \subset O$.

It is interesting to observe that if x_0 is an isolated point of D then f is always continuous at x_0 no matter what the function is.

To prove the equivalence of Def. 7.2.1(a), Def. 7.2.1(b) and Def. 7.2.1(c) we start by showing that Def. 7.2.1(a) $\Longrightarrow$ Def. 7.2.1(b). If f is continuous at x_0 according to Def. 7.2.1(a) then for $\epsilon > 0$ $\exists$ a $\delta > 0$ such that $d'(f(x), f(x_0)) < \epsilon$ whenever $x \in N_\delta \cap D$. Let an arbitrary sequence $\{x_n\}$ in D converge to x_0, then for $\delta > 0$ $\exists$ a natural number n_0 such that $x_n \in N_\delta$ for $n \geqslant n_0$.

But $x_n \in N_\delta \Longrightarrow d'(f(x_n), f(x_0)) < \epsilon$ for $n \geqslant n_0$ which proves that $\{f(x_n)\}$ converges to $f(x_0)$.

We shall use the indirect method to show that Def. 7.2.1(b) $\Longrightarrow$ Def. 7.2.1(a). Suppose there exists a point $x_0 \in D$ such that f is continuous according to Def. 7.2.1(b) but not according to Def. 7.2.1(a). This means there exists an $\epsilon_0 > 0$ such that for every $\delta > 0$, $d'(f(x), f(x_0)) \geqslant \epsilon_0$ for at least one point x of D in the δ-neighborhood of x_0.

Let $\delta_0 = \dfrac{1}{2}$. We can find $x_1 \in N_{\delta_0} \cap D$ such that $d'(f(x_1), f(x_0)) \geqslant \epsilon_0$.

Notice $x_1 \neq x_0$. Now let $\delta_1 = \min\left\{\dfrac{1}{2}, d(x_1, x_0)\right\}$. We can find x_2 in the δ_1-neighborhood of x_0 such that $d'(f(x_2), f(x_0)) \geqslant \epsilon_0$. Again $x_2 \neq x_0$ and also $x_2 \neq x_1$ for x_1 does not lie in the δ_1-neighborhood of x_0. It must be observed that $\delta_1 \leqslant \dfrac{1}{2}\delta_0$. Proceeding in this manner, we can construct a sequence

$\{x_n\}$ such that $x_n \in \delta_n$-neighborhood of x_0 and $\delta_n \leqslant \dfrac{1}{2}\delta_{n-1}$, $d'(f(x_n), f(x_0)) \geqslant$ ϵ_0 for every n. It is easy to show that $\{x_n\}$ converges to x_0 but $\{f(x_n)\}$ does not converge to $f(x_0)$, which contradicts our hypothesis, hence the result.

To show that Def. 7.2.1(a) $\Longrightarrow$ Def. 7.2.1(c), we proceed as follows. Suppose f is continuous at x_0 according to Def. 7.2.1(a) and that O is an open set containing $f(x_0)$. Then, $\exists$ a neighborhood N_ϵ of $f(x_0)$ such that $N_\epsilon \subset O$, where $N_\epsilon = \{y : d'(y, f(x_0)) < \epsilon\}$. According to Def. 7.2.1(a) $\exists$ a $\delta > 0$ such that $x \in N_\delta \cap D \Longrightarrow d'(f(x), f(x_0)) < \epsilon \Longrightarrow f(x) \in O$. If we write $G = \{x : d(x, x_0) < \delta\}$ then G is an open set containing x_0 and $f(G \cap D) \subset O$. Hence f is continuous according to Def. 7.2.1(c).

To complete the equivalence of the definitions we shall show that Def. 7.2.1(c) $\Longrightarrow$ Def. 7.2.1(a).

Let $\epsilon > 0$. $\{y : d'(y, f(x_0)) < \epsilon\}$ is open and contains $f(x_0)$. By Def. 7.2.1(c) there exists an open set G in X containing x_0 such that $f(G \cap D) \subset \{y : d'(y, f(x_0)) < \epsilon\}$. This can be written as $d'(f(x), f(x_0)) < \epsilon$ for every $x \in G \cap D$. Surely G contains a δ-neighborhood of x_0 and with this δ, Def. 7.2.1(a) holds.

The idea of continuity can be extended to a set as follows.

Definition 7.2.2. *A function f is said to be continuous on a set S, S being a subset of the domain of f, if the restriction of f to S is continuous at every point of S.*

It is very interesting to note that a function f with a domain D may be discontinuous at some points of $S(S \subset D)$, yet its restriction on S may still be continuous on S. We illustrate this with the following two examples.

EXAMPLE 7.2.1. Let $f(x) = \dfrac{1}{x}$ for $x \in (0,1]$ and $f(0) = 0$.

It is obvious that f is discontinuous at 0. Now if we let $S = \left[\dfrac{1}{2}, 1\right] \cup \{0\}$, then f is continuous on S, since 0 is only an isolated point of S. Notice that in the original domain $[0,1]$, 0 was not an isolated point.

EXAMPLE 7.2.2. Let $f(x) = x^2$ for $x \in [0,1]$ and $f(x) = 3 - x$ for $x \in (1,2]$.

It is a simple exercise in elementary calculus to show that f is discontinuous at $x = 1$. However, if we let $S = [0,1]$, then f is continuous on S. (Obviously, $f(x) = x^2$ is continuous on $[0,1]$.)

When we consider the continuity of a function $f : D \longrightarrow Y$ on a set $S \subset D$, we do not worry about the behavior of the function outside S.

So far we have given discussion of continuous functions in a general metric space. In the remaining part of this chapter, we shall deal mostly with those functions whose domains and ranges are subsets of real numbers, though many of the theorems can be generalized for a complete metric space.

Let us now consider some properties of continuous functions.

Theorem 7.2.1. Let f and g be two continuous functions on a domain D. Then (i) $f + g$ is continuous on D; (ii) $f \times g$ is continuous on D where $(f \times g)(x) = f(x) \cdot g(x)$, and (iii) f/g is continuous on D, if $g(x) \neq 0$ anywhere on D.

The proofs are not difficult and are left as exercises for the reader.

The following two theorems are very useful.

Theorem 7.2.2. If f is continuous on a compact set, then it is bounded on that set.

Proof. Suppose C is the compact set on which f is continuous. Let $x_0 \in C$. Since f is continuous on C, for $\epsilon(=K) > 0$, $\exists$ a neighborhood N_{x_0} of x_0 such that $|f(x) - f(x_0)| < K$ for $x \in N_{x_0} \cap C$ (Def. 7.2.1(a)). But $|f(x)| = |f(x) - f(x_0) + f(x_0)| \leqslant |f(x) - f(x_0)| + |f(x_0)| \leqslant |f(x_0)| + K$ for $x \in N_{x_0} \cap C$.

If we now let x_0 vary over C then the class of all sets $\{N_{x_0} \cap C : x_0 \in C\}$ is an open covering of C and by the Heine-Borel theorem there exists a finite number of sets, say $N_{x_1} \cap C$, $N_{x_2} \cap C, \ldots, N_{x_n} \cap C$ such that $C \subset \bigcup\limits_{i=1}^{n} (N_{x_i} \cap C)$. Let $M = \max \{|f(x_1)| + K, |f(x_2)| + K, \ldots, |f(x_n)| + K\}$. Then $|f(x)| < M$ for every $x \in C$ which completes the proof. ∎

Theorem 7.2.3. If f is continuous on a compact set (of a complete metric space) then it assumes its least upper bound and its greatest lower bound on that set.

Proof. Let f be continuous on a compact set C. By the last theorem f is bounded on C. Let m and M be the glb and the lub, respectively, of f on C. Then $m \leqslant f(x) \leqslant M$ for every $x \in C$. Suppose there does not exist any $x \in C$ such that $f(x) = m$. In that case, $f(x) > m$ for every $x \in C \Longrightarrow f(x) - m > 0$ for $x \in C$. Let g be a function defined as follows: $g(x) = \dfrac{1}{f(x) - m}, x \in C$. From Theorem 7.2.1, it follows that g is also continuous on C; moreover, $g(x) > 0$ for every $x \in C$. Again using the last theorem, g is bounded on C. Let $u(> 0)$ be the lub of g on C. Then $\dfrac{1}{f(x) - m} \leqslant u$ for $x \in C \Longrightarrow \dfrac{1}{u} \leqslant f(x) - m$ for $x \in C \Longrightarrow f(x) \geqslant m + \dfrac{1}{u}$ for every $x \in C$, which contradicts the fact that m is the glb of f on C. Therefore, $\exists$ a point $x_1 \in C$ such that $f(x_1) = m$. Similarly, we can show that $\exists$ a point $x_2 \in C$ such that $f(x_2) = M$. The proof is complete. ∎

The following examples illustrate these theorems.

EXAMPLE 7.2.3. $f(x) = x^2$ is continuous on the entire set R of real numbers but is not bounded on R. (R is not compact.) However, x^2 is bounded on any closed interval.

EXAMPLE 7.2.4. $f(x) = \dfrac{1}{x}$ is continuous on $(0,1)$ but is not bounded on $(0,1)$. ($(0,1)$ is not compact.)

EXAMPLE 7.2.5. $f(x) = x^2$ is continuous (and bounded) on $(0,1)$ but it does not assume its glb and lub anywhere on $(0,1)$. (glb = 0 and lub = 1).

The next result and Theorem 7.2.5 are true for any metric space.

Theorem 7.2.4. Let $f: D \longrightarrow R$ be a function which is continuous on D. If D is connected then $f(D)$ is connected. (Continuity preserves connectedness.)

Proof. Assume by way of contradiction that $f(D)$ is disconnected. Then there exist two open sets O_1 and O_2 such that $O_1 \cap O_2 \cap f(D) = \emptyset, O_1 \cup O_2 \supset f(D)$, and $O_1 \cap f(D), O_2 \cap f(D)$ are nonempty.

Let $y_1 \in O_1 \cap f(D)$. Then $\exists \, x_1 \in D$ such that $f(x_1) = y_1$. Since f is continuous on D, there is an open set G_1 containing x_1 such that $f(G_1 \cap D) \subset O_1$. Now let y_1 vary over $O_1 \cap f(D)$. The collection of all sets $f(G_1 \cap D)$ will cover $O_1 \cap f(D)$ (why?). Furthermore, $\cup (G_1 \cap D)$ for all G_1 will be a set S_1, which will be relatively open in D. Similarly, we can construct a set $S_2 = \cup (G_2 \cap D)$, such that $f(G_2 \cap D) \subset O_2$, and that S_2 will be relatively open in D. It is easy to see that S_1 and S_2 form a partition for D, which means that D is disconnected. Thus we get the contradiction and the proof is complete.

We are about to prove a generalization of the famous Weierstrass Intermediate Value Theorem (abbreviated as WIVT) which says, "If f is continuous on a closed interval $[a,b]$, then f assumes all the values between $f(a)$ and $f(b)$ on $[a,b]$." Intuitively, this theorem implies that one can draw the graph of a continuous function on an interval without lifting his pencil. This theorem may be generalized for any connected set as shown in Theorem 7.2.5. *The brunt of the proof of Theorem 7.2.5 is absorbed in Theorem 7.2.4.*

Theorem 7.2.5. Let D be a connected subset of a metric space and $f: D \longrightarrow R$ be defined and continuous on D. If m and M are glb and lub, respectively, of f on D, then for every y_0 lying between m and M, there is $x_0 \in D$ such that $f(x_0) = y_0$.

Proof. From Theorem 7.2.4 it follows that $f(D)$ is connected. By way of contradiction, we assume that for some y_0 between m and M there is *no* x_0 in D such that $f(x_0) = y_0$. This implies $y_0 \notin f(D)$. But that would mean $f(D)$ is not path-connected, and hence disconnected. This gives us a contradiction, and the theorem is proved.

The converse of WIVT is not necessarily true as shown by the following example.

EXAMPLE 7.2.6. Let $f(x) = \sin \dfrac{1}{x}$ if $x \neq 0$ and $x \in (0,1]$, and $f(0) = 0$. f takes on all the values between $f(0)$ and $f(1)$ on $[0,1]$ but f is not continuous on $[0,1]$. (It is discontinuous at $x = 0$.)

The following result is true for any metric space.

Theorem 7.2.6. Let $f: D \longrightarrow M$ be a function defined on an open subset D of a metric space $\{M,d\}$. f is continuous on D if and only if for every open set O in $M, f^{-1}(O)$ is open.

Proof. We first prove that if f is continuous on D, then for every open set O in $R, f^{-1}(O)$ is open. If $f^{-1}(O)$ is empty, then it is open. If $f^{-1}(O)$ is not empty, then for $x \in f^{-1}(O)$ there is an open set G_x such that $f(G_x \cap D) \subset O$. Also, D is open, and therefore $G_x \cap D$ is open. This implies that there exists a neighborhood N_x of x such that $N_x \subset G_x \cap D \Longrightarrow f(N_x) \subset O \Longrightarrow f^{-1} f(N_x) \subset f^{-1}(O)$. But $N_x \subset f^{-1} f(N_x)$ (Cf. Exercise 1.2(1)). Thus, $N_x \subset f^{-1}(O)$, which shows that x is an interior point of $f^{-1}(O)$, and hence $f^{-1}(O)$ is open.

Conversely, if $f^{-1}(O)$ is open for every open set O in R, then we shall show that f is continuous on D. Let $x_0 \in D$. If O is an open set containing $f(x_0)$ then $f^{-1}(O)$ is open. Let $G = f^{-1}(O)$. Obviously, $x_0 \in G$. Furthermore, $ff^{-1}(O) \subset O$ (Cf. Exercise 1.2(2)), which means $f(G) \subset O$. Hence, f is continuous at x_0 according to Def. 7.2.1(c). The proof is now complete. ∎

In some parts of this chapter and in Chapters 12 and 13 we will be dealing with many sets of the form $\{x : f(x) < k\}$, $\{x : f(x) > k\}$, $\{x : f(x) \leqslant k\}$ and $\{x : f(x) \geqslant k\}$.

The set $\{x : f(x) < k\}$ is the set of all points x in the domain of f such that $f(x) < k$ and is sometimes written as $f^{-1}\{f(x) < k\}$. This set could be the void set or even the entire domain.

Theorem 7.2.7. Let f be a function defined on an open set D. f is continuous on $D \Longleftrightarrow$ for every number k the sets $\{x : f(x) < k\}$ and $\{x : f(x) > k\}$ are open.

The proof follows directly from Theorem 7.2.6 and is left as an exercise.

We conclude this section with a brief introduction of "homeomorphism."

Definition 7.2.3. Let $\{M_1, d_1\}$, $\{M_2, d_2\}$ be two metric spaces. A function $f : M_1 \longrightarrow M_2$ is called a homeomorphism if f is bijective and f and f^{-1} are both continuous. The metric spaces M_1 and M_2 are said to be homeomorphic to each other or topologically equivalent.

We know that the inverse image of an open set $f : M_1 \longrightarrow M_2$ is open if and only if f is continuous. It will follow that under a homeomorphism, every open set would have an open set as its image. (Why?) The converse is also true. (Why?) A similar statement can be made about the closed sets.

Thus we have the following result.

Theorem 7.2.8. A function $f : M_1 \longrightarrow M_2$ is a homeomorphism $\Longleftrightarrow$ every subset G of M_1 is open if and only if $f(G)$ is open in $M_2 \Longleftrightarrow$ every subset F of M_1 is closed if and only if $f(F)$ is closed in M_2.

Homeomorphism is an equivalence relation. It can be easily shown that all closed intervals are homeomorphic to each other and so are all open intervals. However, a closed interval is not homeomorphic to an open interval nor to the real line, although an open interval is homeomorphic to the real line by virtue of the fact that the function $f : \left(-\dfrac{\pi}{2}, \dfrac{\pi}{2}\right) \longrightarrow R$ with $f(x) = \tan x$ is a homeomorphism.

Exercise 7.2

1. Let f be a real-valued function with domain D (a subset of reals). Prove that f is continuous on $D \Longleftrightarrow$ for every open set $G, f^{-1}(G)$ is relatively open in $D \Longleftrightarrow$ for every closed set $F, f^{-1}(F)$ is relatively closed in D.
2. Prove the Weierstrass Intermediate Value Theorem by using the theorem of nested intervals.
3. Let f be a function defined on a *closed* set. Prove that f is continuous on that set if and only if for every real number k, the sets $\{x : f(x) \leqslant k\}$ and $\{x : f(x) \geqslant k\}$ are closed.
4. Let f be a function defined on $[a,b]$ where a and b are rationals as follows:

$$f(x) = x, \text{ if } x \text{ is rational and } x \in [a,b]$$

$$f(x) = a + b - x, \text{ if } x \text{ is irrational and } x \in [a,b]$$

At what points of $[a,b]$ is f continuous?
5. Give the complete proof of Theorem 7.2.8.
6. Prove that it is impossible to have a homeomorphism between the open interval (a,b) and the closed interval $[a,b]$.
7. Show that the set of all real numbers is not homeomorphic to $[a,b]$.

7.3. Discontinuities

It is important for the student to grasp as fully as possible the nature of the concept of continuity, at least with respect to real valued functions. Many intriguing questions can arise. For example: Can a function fail to be continuous everywhere? If a function is continuous at a point, need it be continuous in some neighborhood of that point, or is it possible that a function be continuous at merely one point. While the reader should investigate these questions exhaustively and certainly raise others, the following examples may be helpful.

EXAMPLE 7.3.1. Let $f(x) = 0$ if x is rational, and $f(x) = 1$ if x is irrational. It can be easily shown that f is discontinuous everywhere on R. When f is defined in this manner it is referred to as the *characteristic function of the set of all irrationals*. This concept will be encountered later in the study of measure theory.

EXAMPLE 7.3.2. Let $g(x) = x$ if x is rational, and $g(x) = 1 - x$ if x is irrational. This function is continuous only at one point. (Which one?)

Discontinuities of a function may arise in more than one way. In calculus texts one encounters the term "removable discontinuity" which occurs when $\lim_{x \to x_0} f(x)$ exists but is not equal to $f(x_0)$. For example:

$$f(x) = x \sin \frac{1}{x}, \ x \neq 0.$$

$$f(0) = 1$$

Obviously, $\lim\limits_{x \to 0} f(x) = 0 \neq f(0)$. We say f has a removable discontinuity at $x = 0$, for it can be removed in the sense that redefining f at $x = 0$ can make the function continuous at $x = 0$. If, on the other hand, $\lim\limits_{x \to a} f(x)$ does not exist then f is said to have a *nonremovable* or *essential discontinuity* at $x = a$.

At this point we introduce upper and lower limit of a function.

Definition 7.3.1. *Let $f : D \longrightarrow R$ be a function, and $x_0 \in D$. Let f be defined on at least a deleted neighborhood of x_0. We say L is the upper limit of f as $x \longrightarrow x_0$, if for $\epsilon > 0 \ \exists \ \delta > 0$ such that $f(x) < L + \epsilon$ for $x \in (x_0 - \delta, x_0 + \delta) - \{x_0\}$ and there is a sequence $\{x_n\}$ with limit x_0 such that $f(x_n) > L - \epsilon$ for every n. We write $\overline{\lim\limits_{x \to x_0}} \ f(x) \ (or \ \lim\limits_{x \to x_0} \ \sup f(x)) = L$.*

The lower limit of a function can be defined in the same way, and is denoted by $\underline{\lim} f$ or $\lim \inf f$.

With slight modifications we may include the possibility of any of these limits being $+\infty$ or $-\infty$ (cf. Def. 4.2.2.).

As an example, $\overline{\lim\limits_{x \to 0}} \sin \frac{1}{x} = 1$ and $\underline{\lim\limits_{x \to 0}} \sin \frac{1}{x} = -1$.

Definition 7.3.2. *A function f is said to be upper-semicontinuous at x_0, if $\overline{\lim\limits_{x \to x_0}} \ f(x) \leqslant f(x_0)$; equivalently, f is upper semicontinuous at x_0 if for $\epsilon > 0 \ \exists$ a $\delta > 0$ such that $f(x) < f(x_0) + \epsilon$ for $|x - x_0| < \delta$.*

Definition 7.3.3. *A function f is said to be lower semicontinuous at x_0 if $\underline{\lim\limits_{x \to x_0}} \ f(x) \geqslant f(x_0)$.*

Clearly, f is upper semicontinuous $\Longleftrightarrow -f$ is lower semicontinuous.

EXAMPLE 7.3.3. Let $f(x) = \sin \frac{1}{x}$, if $x \neq 0$ and $f(0) = 1$.

Here $\overline{\lim\limits_{x \to 0}} \sin \frac{1}{x} = 1 = f(0)$, but $\underline{\lim\limits_{x \to 0}} f(x) = -1 < f(0)$,

Thus, f is upper semicontinuous at 0, but not lower semicontinuous.

In the above example, if we define $f(0) = -1$ (instead of $+1$) then f becomes lower semicontinuous at $x = 0$, yet no longer upper semicontinuous at that point.

As another example, consider the characteristic function of irrationals (Example 7.3.1) which is upper semicontinuous at every irrational and lower semicontinuous at every rational.

It is easy to observe that a function is continuous at a point if and only if it is upper semicontinuous and lower semicontinuous at that point.

If the restriction of a function to S is upper semicontinuous at every point of S then we say it is upper semicontinuous on S.

Theorem 7.3.1. Let f be a function defined on a closed set. f is upper semicontinuous $\Longleftrightarrow$ the set $\{x : f(x) \geqslant k\}$ is closed for every real number k.

The proof of this theorem is left to the reader (cf. Ex. 7.2(3)).

A similar result can be stated for lower semicontinuous functions.

Definition 7.3.4. *For a bounded function f defined on D, the oscillation of f on D is defined to be* lub $\{f(x) : x \in D\}$ - glb $\{f(x) : x \in D\}$. *This is denoted by $O_s(f,D)$.*

Definition 7.3.5. *Let f be a bounded function defined on D, and let $x_0 \in D$. The oscillation of f at x_0 is defined to be* glb $\{O_s(f,N_{x_0} \cap D): N_{x_0}$ is a neighborhood of $x_0\}$ and is written as $O_s(f,x_0)$.

Another name for oscillation is "saltus."

It follows from this definition that a function f is continuous at x_0 if and only if $O_s(f,x_0) = 0$, and discontinuous if and only if $O_s(f,x_0) > 0$ or undefined. For any bounded function f defined on D we will have oscillation of f defined for every point of D (it will be zero at points where f is continuous); therefore, whenever f is bounded, $O_s(f,x)$ is a nonnegative, real-valued function with domain identical to that of f. This function may be called the *oscillation function of f*, and may be denoted by $O_f(x)$.

The following theorem is quite significant since it gives the structure of the set of points of discontinuity of a function whose domain is a closed set.

Theorem 7.3.2. Let f be a bounded, real-valued function with domain D. Then $O_f(x)$ is upper semicontinuous on D.

The proof is left for the reader.

Theorem 7.3.3. Let f be a bounded function defined on the set of all reals (or on a closed set). Then the set of points of discontinuity of f is of the type F_σ.

Proof. Let S be the set of points of discontinuity of f and let $O_f(x)$ be the oscillation function of f. If $x_0 \in S$, then $O_f(x_0) > 0 \Longrightarrow \exists$ a natural number n such that $O_f(x_0) \geqslant \dfrac{1}{n}$; consequently, $x_0 \in \bigcup_{m=1}^{\alpha} \left\{ x : O_f(x) \geqslant \dfrac{1}{m} \right\}$.

Conversely, if $x_0 \in \bigcup_{m=1}^{\infty} \left\{ x : O_f(x) \geqslant \dfrac{1}{m} \right\}$ then $O_f(x_0) \geqslant \dfrac{1}{n}$ for some $n \Longrightarrow x_0 \in S$.

Therefore $S = \bigcup\limits_{m=1}^{\infty} \left\{ x : O_f(x) \geqslant \dfrac{1}{m} \right\}$. But $O_f(x)$ is upper semicontinuous,

and so $\left\{ x : O_f(x) \geqslant \dfrac{1}{m} \right\}$ is closed for every m, which implies S is of type F_σ. That completes the proof.

This theorem has the following interesting (but obvious) corollary.

Corollary. It is impossible to construct a function which is continuous at all rational points and discontinuous at all irrational points.

On the other hand, if we are given a set of type F_σ there exists a function which is discontinuous at every point of that set and continuous everywhere else. (The proof of this may be found in B. Gelbaum and J. Olmsted, *Counterexamples in Analysis*, p. 30). In the following example we shall discuss a very special case when such a set is denumerable.

EXAMPLE 7.3.4. Let $S = \{x_1, x_2, \ldots, x_n, \ldots\}$ be a denumerable set. Let $f(x_n) = \dfrac{1}{n}$, and $f(x) = 0$ for $x \notin S$.

To show that it is discontinuous at x_n, we just have to choose $\epsilon \leqslant \dfrac{1}{n}$. For every neighborhood of x_n will contain some points of cS, and then for such points $|f(x) - f(x_n)| = \left| 0 - \dfrac{1}{n} \right| = \dfrac{1}{n} \geqslant \epsilon$.

However, if $x_0 \in cS$, we would have $f(x_0) = 0$, and then for $\epsilon > 0$ we can choose a natural number $m > \dfrac{1}{\epsilon}$. Then we select a neighborhood N_{x_0} such that it excludes $x_1, x_2, \ldots, x_m$. Now if $x_n \in N_{x_0}$, then $n > m$ and $f(x_n) = \dfrac{1}{n} <$ $\dfrac{1}{m} < \epsilon$, and at other points of N_{x_0}, $f(x) = 0$. Thus $|f(x) - f(x_0)| < \epsilon$ for $x \in N_{x_0}$. Hence f is continuous at every point of cS.

Exercise 7.3

1. Construct a function which is upper semicontinuous at a point but not continuous from the right at that point.
2. Construct a function which is continuous from the left at a point but not lower semicontinuous at that point.
3. Prove that if f is upper semicontinuous on an open set, then the set $\{x : f(x) > k\}$ is open for every real number k.
4. State and prove a similar result for a lower semicontinuous function.

5. Prove the corollary of Theorem 7.3.3.
6. Construct a function which is continuous at every irrational point, and discontinuous at every rational point.
7. Prove that the "characteristic function of the set of all irrationals" (cf. Example 7.3.1) is discontinuous everywhere.
8. Find the point of discontinuity of the function g of Example 7.3.2.

7.4. Uniform Continuity

Although we have defined continuity on a set it was merely an extension of the concept of continuity at a point. In other words, continuity is basically a "local" concept. That is, it describes what is happening to a function in the neighborhood of a point.

In contrast to this we are going to introduce the *concept of "uniform continuity" which is not a "local" but a "global" concept.* It must be noted in the definition of continuity that δ depends not only upon ϵ but also upon the point. We now define "uniform continuity" by removing this dependence on the particular point. In this section we confine ourselves to the real valued function of real variables, though the ideas may be generalized to any complete metric space.

Definition 7.4.1. *A function f defined on D is said to be uniformly continuous on a set $S \subset D$ if for $\epsilon > 0$ $\exists$ a $\delta(\epsilon) > 0$ such that $|f(x) - f(y)| < \epsilon$ for all x and y $\in S$ and $|x - y| < \delta$.*

It must be observed here that δ depends upon ϵ, but is independent of the point chosen. The above definition can be restated as follows:

"A function f is uniformly continuous on S if for $\epsilon > 0$ $\exists$ a $\delta > 0$ such that the image of any point y, which is in a δ-neighborhood of any other point x, lies in the ϵ-neighborhood of $f(x)$."

From the definition it follows that the uniform continuity always implies continuity; however, continuity does not necessarily imply uniform continuity. This will be shown by some examples later on. *A sufficient condition for the "continuity" to imply "uniform continuity" is the compactness of the domain.*

Theorem 7.4.1. If a function f is continuous on a compact set C then it is uniformly continuous on C.

Proof. Since f is continuous on C, for $\epsilon > 0$ and for every $x \in C$, $\exists \delta_x > 0$ such that $|f(y) - f(x)| < \dfrac{\epsilon}{2}$ for $|y - x| < \delta_x$. Let I_x be the open interval $\left(x - \dfrac{\delta_x}{2}, x + \dfrac{\delta_x}{2}\right)$. The class of open intervals $\{I_x : x \in C\}$ is an open covering of C, and since C is compact, there exists a finite subclass $J = \{I_{x_1}, I_{x_2}, \ldots I_{x_n}\}$ of these intervals which cover C (Heine-Borel theorem).

Let $\delta = \min\left\{\dfrac{\delta_{x_1}}{2}, \dfrac{\delta_{x_2}}{2}, \ldots \dfrac{\delta_{x_n}}{2}\right\}$ (notice that this δ is independent of x).

Now if x and y are any two arbitrary points such that $|x - y| < \delta$ then $x \in I_{x_m}$
for at least one $I_{x_m} \in J$, which implies $|x - x_m| < \dfrac{\delta_{x_m}}{2}$.

Now $|y - x_m| = |y - x + x - x_m| \leqslant |y - x| + |x - x_m| < \delta + \dfrac{\delta_{x_m}}{2} \leqslant \delta_{x_m}$. But

$|y - x_m| < \delta_{x_m} \implies |f(y) - f(x_m)| < \dfrac{\epsilon}{2}$. Thus,

$$|f(y) - f(x)| = |f(y) - f(x_m) + f(x_m) - f(x)|$$

$$\leqslant |f(y) - f(x_m)| + |f(x) - f(x_m)| < \frac{\epsilon}{2} + \frac{\epsilon}{2} = \epsilon$$

Or $|f(y) - f(x)| < \epsilon$ for $|y - x| < \delta$, and this is exactly the condition for uniform continuity. ▮

The usefulness of this theorem in establishing uniform continuity is obvious, since for compact sets the problem becomes one of establishing ordinary continuity.

We now discuss some examples to illustrate the difference between continuity and uniform continuity.

EXAMPLE 7.4.1. Let $f(x) = \dfrac{1}{x}, x \in (0,2)$.

It is an easy matter to show that f is continuous on $(0,2)$; however, f is not uniformly continuous. If $\epsilon = \dfrac{1}{2}$ and δ is any positive number, then for $n > \dfrac{1}{\delta}$ we
have $\left(\dfrac{1}{n} - \dfrac{1}{n+1}\right) < \delta$. Now we have merely to choose $x = \dfrac{1}{n}$ and $y = \dfrac{1}{n+1}$ to get
$|f(x) - f(y)| = 1 > \epsilon$ while $|x - y| < \delta$.

In this example f is uniformly continuous on (a, ∞) if $a > 0$; for if $\epsilon > 0$ and
x and y are in (a, ∞), then $x > a > 0$, and $y > a > 0$ and $\dfrac{1}{xy} < \dfrac{1}{a^2}$. Now choose

$\delta < \epsilon a^2$. Then $|x - y| < \delta \implies |f(x) - f(y)| = \left|\dfrac{1}{x} - \dfrac{1}{y}\right| = \left|\dfrac{x - y}{xy}\right| < \dfrac{\delta}{a^2} < \epsilon$.

EXAMPLE 7.4.2. Let $f(x) = x^3, x \in (0,1)$.

For $\epsilon > 0$, we choose $\delta = \dfrac{\epsilon}{5}$. Now if $x, y \in (0,1)$ such that $|x - y| < \delta$,
then we have $|x^3 - y^3| = |(x - y)(x^2 + xy + y^2)| = |x - y| \, |(x + y)^2 - xy| <$

$\delta((x+y)^2 + |xy|)$ and since $0 < x, y < 1$, it follows that $(x+y)^2 + xy < 2^2 + 1 = 5$. Thus $|x^3 - y^3| < 5\,\delta = \epsilon$ for $|x - y| < \delta$.

However, f is not uniformly continuous on R, the entire set of reals, for if we choose $\epsilon = 1$, and if δ is any positive number then $\exists$ a natural number $n > \sqrt{\dfrac{2}{3\delta}}$. Now let $x = n$, and $y = n + \dfrac{\delta}{2}$.

Then $|f(x) - f(y)| = \left| \left(n + \dfrac{\delta}{2}\right)^3 - n^3 \right| = 3n^2 \dfrac{\delta}{2} + 3n \dfrac{\delta^2}{4} + \dfrac{\delta^3}{8} > 1 +$

$3 \sqrt{\dfrac{2}{3\delta}} \cdot \dfrac{\delta^2}{4} + \dfrac{\delta^3}{8} > 1 = \epsilon.$

The following is an interesting example of a function which is continuous but not uniformly continuous.

EXAMPLE 7.4.3. Let f be defined on $\mathfrak{Q}$ (the set of all rationals) in the following manner:

$$f(x) = 1 \text{ if } x < \pi \text{ and } x \in \mathfrak{Q}$$

and

$$f(x) = 2 \text{ if } x > \pi \text{ and } x \in \mathfrak{Q}$$

Note that f is continuous on $\mathfrak{Q}$, for if $\epsilon > 0$ and if $x_0 < \pi$ then with $\delta < (\pi - x_0)$, we have $|f(x) - f(x_0)| = 1 - 1 = 0 < \epsilon$ whenever $|x - x_0| < \delta$. Thus f is continuous for $x_0 < \pi$. Similarly, we can prove that f is continuous for $x_0 > \pi$. f is not uniformly continuous on $\mathfrak{Q}$, however. Indeed, consider $0 < \epsilon < 1$. For any $\delta > 0$ we can find $x < \pi$ and $y > \pi$ such that $y - x < \delta$ (cf. Theorem 2.1.6). Now $|f(x) - f(y)| = |1 - 2| > \epsilon$.

Whenever we are given a function continuous on a set D, then for $x_0 \in D$ and $\epsilon > 0$ we can find $\delta(\epsilon, x_0) > 0$ such that $|f(x) - f(x_0)| < \epsilon$ for $x \in (x_0 - \delta, x_0 + \delta)$. The set $\{\delta(\epsilon, x_0) : x_0 \in D\}$ is always bounded below. If this set has a *positive greatest lower bound* then the function must be uniformly continuous, and not otherwise. See the last example.

Exercise 7.4

1. Prove that $f(x) = x^2$ is uniformly continuous on any bounded set, but not uniformly continuous on R.

2. Is $\cos \dfrac{1}{x}$ uniformly continuous on $(0, 1)$?

3. Is $\dfrac{1}{1 + x^3}$ uniformly continuous on R.

4. Prove that $f(x) = \sqrt{x}$ is uniformly continuous on R.

7.5. Continuity in R^n

To discuss the continuity of a function $f: D^n \longrightarrow R^m$ (D^n being a subset of R^n) we could either use "norm" or "open sets."

Definition 7.5.1. *A function $f: D^n \longrightarrow R^m$ is continuous at a point $P \in D^n$ if for $\epsilon > 0$ there is a $\delta > 0$ such that $\|f(X) - f(P)\| < \epsilon$ for $\|X - P\| < \delta$.*

From the context it is quite obvious that the norm in $\|f(X) - f(P)\|$ is that of R^m and in $\|X - P\|$ is the one that is used in R^n.

Alternatively, we can say.

Definition 7.5.1(a). *A function $f: D^n \longrightarrow R^m$ is continuous at $P \in D^n$ if for every open set G (of R^m) containing $f(P)$ there exists an open set O (of R^n) containing P such that $f(O \cap D^n) \subset G$.*

We would leave it to the reader to establish the equivalence of these two definitions.

For practical purposes the following criterion may be helpful:

Theorem 7.5.1. A function $f: D^n \longrightarrow R^m$ is continuous at a point $P \in D^n$ if and only if every coordinate function of f is continuous at P.

Proof. Let $f: D^n \longrightarrow R^m$ be described as follows

$$f(\langle p_1, p_2, \ldots, p_n \rangle) = \langle q_1, q_2, \ldots, q_m \rangle$$

where $q_i = f_i \langle p_1, p_2, \ldots, p_m \rangle$, $i = 1, 2, \ldots, n$. We have to show that each f_i is continuous. Since f is continuous at P, for $\epsilon > 0$, there is a $\delta > 0$ such that $\|\langle y_1, y_2, \ldots, y_m \rangle - \langle q_1, q_2, \ldots, q_m \rangle\| < \epsilon$ for $\|\langle x_1, x_2, \ldots, x_n \rangle - \langle p_1, p_2, \ldots, p_n \rangle\| < \delta$, where $y_i = f_i(\langle x_1, x_2, \ldots, x_n \rangle) = f_i(X)$, $X = \langle x_1, \ldots, x_n \rangle$. Now $\|f_i(X) - f_i(P)\| = |y_i - q_i| \leqslant \|\langle y_1, y_2, \ldots, y_n \rangle - \langle x_1, x_2, \ldots, x_n \rangle\| < \epsilon$ for $\|X - P\| < \delta$. Hence, f_i is continuous at P.

Conversely, if every f_i is continuous at P, then for $\epsilon > 0$ there is a $\delta > 0$ such that $\|f_i(X) - f_i(P)\| = |y_i - q_i| < \dfrac{\epsilon}{m}$ for $\|X - P\| < \delta$. In that case, we have $\|f(X) - f(P)\| = \|\langle y_1, \ldots, y_m \rangle - \langle q_1, \ldots, q_m \rangle\| \leqslant |y_1 - q_1| + |y_2 - q_2| + \cdots + |y_m - q_m| < \dfrac{\epsilon}{m} + \dfrac{\epsilon}{m} + \cdots + \dfrac{\epsilon}{m} = \epsilon$ for $\|X - P\| < \delta$, which shows that f is continuous at P and that completes the proof.

Because of the preceding theorem the discussion of continuity of any function $f: D^n \longrightarrow R^m$ simply reduces to that of real-valued coordinate functions $f_i: D^n \longrightarrow R$ which is easier to handle.

Most of the results discussed in Sec. 7.2 are valid for the functions $f: D^n \longrightarrow R^m$. For examples, "A continuous image of a connected set is connected," and "a continuous image of a compact set is compact."

Exercise 7.5

1. Let $f: D^n \longrightarrow R^m$ be a function. Prove that (a) if D^n is connected then $f(D^n)$ is connected, and (b) if D^n is compact then $f(D^n)$ is compact.

2. Define uniform continuity of $f: D^n \longrightarrow R^m$, and prove that if D^n is compact then the continuity of f implies the uniform continuity.

3. Prove that the function $f: D^n \longrightarrow R^m$ is continuous at $P_0 \in D^n$ if and only if for every sequence of points $\{P_k\}$ in D^n which converges to P_0 the sequence $\{f(P_k)\}$ in R^m converges to $f(P_0)$.

4. Show that the function $f: R^n \longrightarrow R$ described by $f(X) = \|X\|$ is continuous at every point of R^n.

5. Let $f: D^n \longrightarrow S^m$ ($S^m \subset R^m$) be a bijective and continuous function. Prove that if D^n is compact then $f^{-1}: S^m \longrightarrow D^n$ is continuous.

6. Let $f: I \longrightarrow E$ be a function where $I = [0, 2\pi)$ and E the standard ellipse in the plane R^2 with $f(x) = \langle a \cos x, b \sin x \rangle$. Show that in spite of the fact that f is continuous, bijective, f^{-1} is not continuous. (Compare it with the result in Exercise 5.)

7.6. Connectedness in R^n

In Sec. 5.4 we postponed the discussion of connectedness of subsets of R^n. We now introduce it here. The definition of "connectedness" need not be repeated since it was given in Sec. 5.4 for a metric space. We start with the definition of paths in R^n.

Definition 7.6.1. *Let A and B be two points of R^n. A path in R^n from A to B is the image of a continuous function $f: [a, b] \longrightarrow R^n$ such that $f(a) = A$ and B. (Indeed, $[a, b]$ is the closed interval in R.) The point A is called the initial point of the path and the point B is called the terminal point.*

Since the closed interval $[a, b]$ is homeomorphic to $[0, 1]$, we could conveniently replace $[a, b]$ by $[0, 1]$. Furthermore, this definition of a path does not have to be restricted to R^n. It applies to any metric space.

A special case of a path is a directed line segment defined as follows.

Definition 7.6.2. *A directed line segment in R^n with initial point at A and the terminal point at B is a path from A to B described by the function $f: [0, 1] \longrightarrow R^n$ with $f(x) = A + x(B - A)$.*

It follows from the definition that for two given points there is a unique directed line segment.

Next we define a polygonal path.

Definition 7.6.3. *A polygonal path in R^n from a point A to a point B is a well-ordered set of a finite number of line segments $\{L_1, L_2, \ldots, L_m\}$ such that*

(i) *A is the initial point of L_1*, (ii) *B is the terminal point of L_m, and* (iii) *the terminal point of L_k is the initial point of L_{k+1}.*

It is obvious from the definition that a polygonal path is a path, and a directed line segment is a polygonal path. Furthermore, if $\{L_1, L_2, \ldots, L_m\}$ is a polygonal path from A to B and $\{M_1, M_2, \ldots, M_k\}$ a polygonal path from B to C, then $\{L_1, L_2, \ldots, L_m; M_1, M_2, \ldots, M_k\}$ is a polygonal path from A to C.

In the system of real numbers a path is simply a closed interval and there is one and only one path from one point to another, and the path connectedness of a set S was defined as having the property of containing the entire path (the closed interval) if the end points belonged to S. It was then proved that a subset of real numbers is path-connected if and only if it is connected. (Cf. Theorem 5.4.3.)

The situation in R^n for $n \geqslant 2$ becomes much more complicated since the paths from one point to another are not unique. Let us first define path-connectedness in R^n.

Definition 7.6.4. *A set S in R^n is said to be path-connected if for every two points A and B in S there is a path between them which lies entirely in S. In case there exists a polygonal path between A and B that lies in S, then S is said to be polygonally path-connected.*

It is obvious that a polygonally path-connected set is path-connected but the converse is not necessarily true (Example 7.6.1). In the real number system the "path-connectedness" is equivalent to "connectedness" (cf. Theorem 5.4.3), but that is not true in R^n for $n \geqslant 2$ (cf. Exercise 4, Sec. 7.6). The following result is true in any case.

Theorem 7.6.1. If a set S in R^n is path-connected then it is connected.

Proof. Suppose by way of contradiction that S is disconnected. In that case there exist two open sets G_1 and G_2 such that $G_1 \cup G_2 \supset S$; $S \cap G_1$ $S \cap G_2$ are nonempty and $S \cap G_1 \cap G_2 = \emptyset$.

Now let $P_1 \in G_1 \cap S$ and $P_2 \in G_2 \cap S$ and $I = [0,1]$. Since S is path-connected there exists a path $f: I \longrightarrow S$ joining P_1 and P_2. From Theorem 7.2.4 it is obvious that $f(I)$ is a connected subset of S.

Let $O_1 = G_1 \cap f(I) \subset G_1 \cap S$ and $O_2 = G_2 \cap f(I) \subset G_2 \cap S$. O_1 and O_2 are nonempty since they contain P_1 and P_2, respectively. Furthermore, $O_1 \cap O_2 = G_1 \cap G_2 \cap f(I) \subset G_1 \cap G_2 \cap S = \emptyset$ and $O_1 \cup O_2 = f(I)$ since $G_1 \cup G_2 \supset S \supset f(I)$.

Thus O_1 and O_2 form a partition for $f(I)$ which mean $f(I)$ is disconnected and that gives us a contradiction, and the proof is complete.

Corollary 1. If a subset of R^n is polygonally path-connected, then it is connected.

Corollary 2. A path, in particular a polygonal path, is path-connected and hence connected.

It may be pointed out that Theorem 7.6.1 is valid for every metric space. The converse of the theorem is not necessarily true in R^n for $n \geq 2$ because of a well-known example of a set in R^2 given by $S = \left\{ \langle x,y \rangle : y = \sin\dfrac{1}{x}, x \neq 0 \right\} \cup \{\langle x,y \rangle : x = 0, -1 \leq y \leq 1\}$ (cf. Exercise 4, Sec. 7.6).

In the next couple of theorems we establish connectedness of some of the sets of R^n.

Theorem 7.6.2. R^n is connected.

Proof. It is easy to show that R^n is path-connected for every two points A and B of R^n; there is a directed line segment from A to B which is a path lying in R^n. Hence by Theorem 7.6.1, R^n is connected, and that completes the proof. $\blacksquare$

Theorem 7.6.3. Each of the following subsets of R^n is path-connected and hence connected: (a) an open ball, (b) a closed ball, (c) an open box, (d) a closed box.

The proof is easy and is left as an exercise.

We claimed earlier that the connectedness of a set does not necessarily imply path-connectedness much less the polygonal path-connectedness. However, if a set of R^n is connected as well as open, then it is polygonally path-connected because of the following theorem.

Theorem 7.6.4. If G is a connected open set in R^n, then for every two points of G there exists a polygonal path between them that lies entirely in G, and therefore G is path-connected.

Proof. If G is empty then the theorem is trivially satisfied. Therefore we assume that G is nonempty. Let $P \in G$, and G_1 be the set of all those points in G which can be joined to P by a polygonal path lying entirely in G. Let X_1 be an arbitrary point of G_1, then $X_1 \in G$ and so there must be an open ball $B = \{X : \|X - X_1\| < \epsilon\}$ contained in G. Every point of the ball must be in G_1 since it can be joined to X_1 by a directed line segment lying entirely in G and X_1 in turn can be joined to P by a polygonal path. Therefore, $B \subset G_1$ which makes G_1 open. If $G - G_1 \neq \emptyset$ then we can show in similar manner that $G - G_1$ is open. In that case, G_1 and $G - G_1$ will form a partition for G making G disconnected. Thus $G - G_1$ is empty and $G = G_1$, which means any two points of G can be joined to P by polygonal paths lying entirely in G, and therefore there is a polygonal path between them lying entirely in G. The proof is complete. $\blacksquare$

An open connected set in R^n is also called an open region of R^n.

In general, we may define a region as follows.

Definition 7.6.5. *A region in R^n consists of an open connected set with some, all or none of its boundary points. It is called a closed region if it contains all its boundary points, an open region if it contains none of its boundary points and partially closed if it contains some but not all of its boundary points.*

We have the following property for a region.

Theorem 7.6.5.   A region is polygonally path-connected.

The proof of this theorem is left as an exercise.

The converse of this theorem is not necessarily true. For if we consider a polygonal path by itself, any two points of this polygonal path can be joined by a polygonal path, but it is not a region.

It is quite obvious that polygonal path-connectedness implies path-connectedness, and we said earlier that the converse of this statement is not necessarily true. This follows from the following example.

Example 7.6.1.   In R^2 consider the following set. Let $S = \{\langle x,y\rangle : 0 < x \leqslant y^2\} \cup \langle 0,0\rangle$. S is pathwise connected; we can write $S = S_1 \cup S_2 \cup \langle 0,0\rangle$ where

$$S_1 = \{\langle x,y\rangle : 0 < x \leqslant y^2 \text{ and } y > 0\}$$
$$S_2 = \{\langle x,y\rangle : 0 < x \leqslant y^2 \text{ and } y < 0\}.$$

For any two points in $S_1 \cup \{\langle 0,0\rangle\}$ there exists a path lying entirely in $S_1 \cup \{\langle 0,0\rangle\}$. The same thing can be said for $S_2 \cup \{\langle 0,0\rangle\}$.

Now if $P_1 \in S_1$ and $P_2 \in S_2$. Then there is a path from P_1 to P_2 as indicated in the diagram. Draw a line parallel to y-axis from P_1 and P_2 meeting the

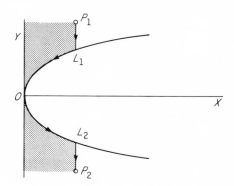

parabola in L_1 and L_2, respectively. It is easy to show that the union of $P_1 L_1, L_1 L_2$ (along the parabola), $L_2 P_2$ is a path.

However, P_1 and P_2 cannot be joined by means of a polygonal path, for such a polygonal path would have to pass through the origin and any line from a point on the parabola to the origin lies outside S.

Exercise 7.6

1. If $f: [0,1] \longrightarrow R^n$ represents a path with initial point at A and terminal point at B, then prove that $g: [0,1] \longrightarrow R^n$, where $g(x) = f(1 - x)$, represent a path whose terminal point is B and initial point is A and $f\{[0,1]\} = g\{[0,1]\}$. (The latter path is basically the same as the original path except that its orientation is reversed.)

2. Prove Theorem 7.6.3.

3. Prove Theorem 7.6.5.

4. Show that the set $S = \left\{ \langle x,y \rangle : y = \sin \dfrac{1}{x}, x \neq 0 \right\} \cup \{ \langle x,y \rangle : -1 \leqslant y \leqslant 1$ and $x = 0 \}$ is connected, but not path-connected. (*Hint*: Use Exercise 5.4(4).)

8

Infinite Series

8.1. Introduction

In this chapter we give a brief development of the theory of infinite series. Formally an infinite series of real or complex numbers a_n is written as

$$\sum_{n=1}^{\infty} a_n = a_1 + a_2 + a_3 + \cdots \qquad (8.1)$$

Let

$$S_n = a_1 + a_2 + \cdots + a_n. \qquad (8.2)$$

The sequence $\{S_n\}_1^{\infty}$ is called the *sequence of partial sums of the series* $\sum_{n=1}^{\infty} a_n$

and the number S_n is called the nth *partial sum of the series* $\sum_{n=1}^{\infty} a_n$. The number a_n is called the nth *term of the series*.

Definition 8.1.1. *If the sequence of partial sums* $\{S_n\}_1^{\infty}$ *converges to a real or complex number S, that is, if* $\lim_{n \to \infty} S_n = S$, *then the infinite series* $\sum_{n=1}^{\infty} a_n$ *is said to converge to the sum S. If* $\{S_n\}_1^{\infty}$ *does not converge, the series* $\sum_{n=1}^{\infty} a_n$ *is said to* *diverge. If $\{S_n\}_{n=1}^{\infty}$ is bounded and the series $\sum_{n=1}^{\infty} a_n$ is divergent then the series $\sum_{n=1}^{\infty} a_n$ is said to be finitely oscillating.*

8.2. Tests of Convergence

In the next theorem we give Cauchy criterion of convergence for a series $\sum_{n=1}^{\infty} a_n$ of real or complex terms.

Theorem 8.2.1. The series $\sum_{n=1}^{\infty} a_n$ converges if and only if given $\epsilon > 0$ there exists a positive integer $n_0(\epsilon)$ such that $n \geqslant n_0(\epsilon)$ implies that

$$|a_{n+1} + a_{n+2} + \cdots + a_{n+p}| < \epsilon \tag{8.3}$$

for $p = 1, 2, \ldots$.

Proof. The series $\sum_{n=1}^{\infty} a_n$ converges if and only if the sequence of partial sums $\{S_n\}_1^{\infty}$ converges. By Cauchy's criterion the sequence $\{S_n\}_1^{\infty}$ converges if and only if given $\epsilon > 0$ there exists a positive integer $n_0(\epsilon)$ such that $n \geqslant n_0(\epsilon)$ and p a positive integer imply

$$|S_n - S_{n+p}| < \epsilon$$

That is,

$$\left| \sum_{k=1}^{n} a_k - \sum_{k=1}^{n+p} a_k \right| = \left| a_{n+1} + \cdots + a_{n+p} \right| < \epsilon$$

and the proof is complete. ∎

Taking $p = 1$, we deduce the following theorem:

Theorem 8.2.2. If $\sum_{n=1}^{\infty} a_n$ converges, then

$$\lim_{n \to \infty} a_n = 0. \tag{8.4}$$

The following example shows that condition (8.4) is necessary but not sufficient for the convergence of $\sum_{n=1}^{\infty} a_n$.

EXAMPLE 8.2.1. (The Harmonic Series). The series $\sum_{n=1}^{\infty} \dfrac{1}{n}$ diverges. We have

$$S_1 = 1, S_2 = \frac{3}{2}$$

$$S_4 = 1 + \frac{1}{2} + \left(\frac{1}{3} + \frac{1}{4}\right) > \frac{3}{2} + \frac{2}{4},$$

$$S_4 = 1 + \frac{1}{2} + \left(\frac{1}{3} + \frac{1}{4}\right) < 1 + \frac{1}{2} + \left(\frac{1}{2} + \frac{1}{2}\right) = \frac{3}{2} + \frac{2}{2}.$$

In general, we have for $k > 1$,

$$S_{2k} = 1 + \frac{1}{2} + \left(\frac{1}{3} + \frac{1}{4}\right) + \left(\frac{1}{5} + \frac{1}{6} + \frac{1}{7} + \frac{1}{8}\right) + \cdots + \left(\frac{1}{2^{k-1} + 1} + \cdots + \frac{1}{2^k}\right)$$

$$> \frac{3}{2} + (k - 1)\frac{1}{2} = 1 + \frac{k}{2}$$

$$S_{2k} < \frac{3}{2} + (k - 1) = \frac{1}{2} + k.$$

Hence for $2^k \leqslant n < 2^{k+1}$ $(k > 1)$ we have

$$S_n \geqslant S_{2k} > 1 + \frac{k}{2}, S_n < S_{2^{k+1}} < \frac{3}{2} + k.$$

That is,

$$1 + \frac{k}{2} < S_n < \frac{3}{2} + k.$$

Since $k \longrightarrow \infty$ as $n \longrightarrow \infty$, it follows that $\lim_{n \to \infty} S_n = \infty$. Therefore the series $\sum_{n=1}^{\infty} \dfrac{1}{n}$ is divergent. Note that $\lim_{n \to \infty} a_n = \lim_{n \to \infty} \dfrac{1}{n} = 0$.

Exercise 8.2

1. Prove that the series

$$1 - 2 + 3 - 4 + \cdots = \sum_{n=1}^{\infty} (-1)^{n+1} n$$

is divergent.

$$\begin{pmatrix} Hint: \qquad S_{2n} = 1 + 2 + 3 + \cdots + 2n - 2(2 + 4 + \cdots 2n) \\[4pt] = \frac{2n(2n+1)}{2} - 4\frac{n(n+1)}{2} = -n \\[4pt] S_{2n+1} = S_{2n} + (2n+1) = n+1 \end{pmatrix}$$

Note that $\{S_n\}$ is not bounded and hence the series is not finitely oscillating. Note also that $a_n \nrightarrow 0$ (a_n does not tend to zero), and so Theorem 8.2.2 also gives the divergence of the series.

2. Let K be a positive constant. Prove that the series $\sum_{n=1}^{\infty} a_n$ and $\sum_{n=1}^{\infty} Ka_n$ are both convergent or both divergent $\left(Hint\colon \text{ Let } \sigma_n = \sum_{k=1}^{n} Ka_k. \text{ Then } \sigma_n = \right.$

$K \sum_{k=1}^{n} a_k = KS_n. \text{ The result follows on letting } n \longrightarrow \infty. \Bigg)$

3. Let N be a given positive integer. Prove:

   (i) The series $\sum_{n=1}^{\infty} a_n$ and $\sum_{n=N+1}^{\infty} a_n$ are both convergent or both divergent.

   (ii) If $\sum_{n=1}^{\infty} a_n$ converges to S, then the series $\sum_{n=N+1}^{\infty} a_n$ converges to $S - (a_1 + a_2 + \cdots + a_N)$.

$\Bigg(Hint\colon \text{ Let } n > N \text{ and } \sigma_n = \sum_{k=N+1}^{n} a_k. \text{ Put } K = a_1 + \cdots + a_N. \text{ Then } S_n = a_1 +$

$a_2 + \cdots + a_N + \sum_{k=N+1}^{n} a_k = K + \sigma_n. \text{ Let } n \longrightarrow \infty. \Bigg)$

4. Prove that if the series $\sum_{n=1}^{\infty} a_n$ converges to A and the series $\sum_{n=1}^{\infty} b_n$ converges to B then the series $\sum_{n=1}^{\infty} (a_n + b_n)$ converges to $A + B$.

$\Bigg(Hint\colon \text{ Let } A_n = \sum_{k=1}^{n} a_k \text{ and } B_n = \sum_{k=1}^{n} b_k. \text{ Then } A_n \text{ tends to } A \text{ and } B_n \text{ tends}$

to B. Now

$$\sum_{k=1}^{n}(a_k + b_k) = (a_1 + b_1) + (a_2 + b_2) + \cdots + (a_n + b_n)$$

$$= (a_1 + a_2 + \cdots + a_n) + (b_1 + b_2 + \cdots b_n).\Bigg)$$

5. Prove that if the series

$$a_1 + a_2 + \cdots$$

is convergent to S, then the series

$$(a_1 + a_2 + \cdots + a_{n_1}) + (a_{n_1+1} + \cdots + a_{n_2}) + \cdots$$

obtained by inserting parenthesis in the first series, is also convergent to S. (*Hint*: Let $S_n = a_1 + a_2 + \cdots + a_n$. Then $S_n \longrightarrow S$. Hence $S_{n_p} \longrightarrow S$ as $n_p \longrightarrow \infty$. But the partial sum of the second series is

$$(a_1 + \cdots + a_{n_1}) + (a_{n_1+1} + \cdots + a_{n_2}) + \cdots + (a_{n_{p-1}+1} + \cdots + a_{n_p})$$

$$= a_1 + \cdots + a_{n_p} = S_{n_p}.)$$

6. Show that the series

$$(i) \quad \left(\frac{3}{2} - \frac{4}{3}\right) + \left(\frac{5}{4} - \frac{6}{5}\right) + \cdots + \left(\frac{2n+1}{2n} - \frac{2n+2}{2n+1}\right) + \cdots$$

is convergent whereas the series

$$(ii) \quad \frac{3}{2} - \frac{4}{3} + \frac{5}{4} - \frac{6}{5} + \cdots$$

obtained from the first series by omitting parenthesis is divergent.
$\Bigg($*Hint*: Write the first series as

$$\sum_{n=1}^{\infty} a_n \quad \text{where } a_n = \left(\frac{2n+1}{2n} - \frac{2n+2}{2n+1}\right)$$

Then

$$S_n = \sum_{k=1}^{n} a_k = \sum_{1}^{n}\left(1 + \frac{1}{2k} - 1 - \frac{1}{2k+1}\right)$$

Since this is a finite sum we can omit parenthesis. Hence

$$S_n = \frac{1}{2} - \frac{1}{3} + \frac{1}{4} - \frac{1}{5} + \cdots + \frac{1}{2n} - \frac{1}{2n+1}$$

and

$$S_{n+1} - S_n = \frac{1}{2n+2} - \frac{1}{2n+3} > 0.$$

Further,

$$S_n = \frac{1}{2} - \left(\frac{1}{3} - \frac{1}{4}\right) - \cdots - \frac{1}{2n + 1} < \frac{1}{2}.$$

Hence $\{S_n\}$ is an increasing bounded sequence, and this gives the convergence of (i). To prove the divergence of (ii) note that the nth term of (ii), $(-1)^{n+1}\dfrac{n + 2}{n + 1}$, does not tend to zero as $n \longrightarrow \infty$. Note also that the series (ii) is finitely oscillating for if σ_n denotes the partial sum then $\sigma_{2n} = S_n < \dfrac{1}{2}$ and

$$\sigma_{2n+1} = \sigma_{2n} + \frac{2n + 3}{2n + 2} = \sigma_{2n} + 1 + \frac{1}{2n + 2} < \frac{7}{4}\bigg)$$

8.3. Series of Nonnegative Terms

In this section we shall suppose that a_n is real and nonnegative for all n. Hence

$$S_{n+1} - S_n = a_{n+1} \geqslant 0$$

and so the sequence $\{S_n\}$ is monotonic increasing. If $\{S_n\}$ is bounded then S_n tends to a finite limit S and the series $\sum\limits_{n=1}^{\infty} a_n$ is convergent. If $S_n \leqslant K$, where K is a constant then $\lim\limits_{n\to\infty} S_n = S \leqslant K$ and $\sum\limits_{n=1}^{\infty} a_n = S \leqslant K$. If $\{S_n\}$ is unbounded then $S_n \longrightarrow +\infty$ and the series $\sum\limits_{n=1}^{\infty} a_n$ is divergent. We combine these results in the following theorem.

Theorem 8.3.1. A series $\sum\limits_{n=1}^{\infty} a_n$, $a_n \geqslant 0$, is convergent if and only if the sequence of partial sums $\{S_n\}$ is bounded.

EXAMPLE 8.3.1. The series $\sum\limits_{k=1}^{\infty} \dfrac{1}{k^2}$ is convergent for

$$S_n = \sum_{1}^{n} \frac{1}{k^2} < 1 + \sum_{2}^{n} \frac{1}{(k - 1) k}$$

$$= 1 + \sum_{2}^{n} \left(\frac{1}{k-1} - \frac{1}{k} \right)$$

$$= 1 + \left(1 - \frac{1}{n} \right) < 2.$$

This inequality shows that $\sum_{k=1}^{\infty} \frac{1}{k^2} \leqslant 2$.   (Actually the sum of this series is $\pi^2/6$.)

Theorem 8.3.2. (Comparison Test).   Let $\sum_{n=1}^{\infty} c_n$ be a convergent series of positive terms and let $\sum_{n=1}^{\infty} d_n$ be a divergent series of positive terms.

(i)   If $0 \leqslant a_n \leqslant c_n$ for all $n \geqslant 1$, then $\sum_{n=1}^{\infty} a_n$ is convergent.

(ii)   If $a_n \geqslant d_n$ for all $n \geqslant 1$ then $\sum_{n=1}^{\infty} a_n$ is divergent.

Proof.   (i)   Since $S_n = \sum_{k=1}^{n} a_k \leqslant \sum_{k=1}^{n} c_k < \sum_{k=1}^{\infty} c_k = C$ (say), $\sum_{n=1}^{\infty} a_n$ is convergent.

(ii)   Since $S_n = \sum_{k=1}^{n} a_k \geqslant \sum_{k=1}^{n} d_k$, and $\sum_{k=1}^{n} d_k \longrightarrow \infty$ as $n \longrightarrow \infty$, the sequence $\{S_n\}$ is unbounded and $\sum_{n=1}^{\infty} a_n$ is divergent.   ∎

Corollary.   (i)   If there exist a positive integer N_0 and a constant $K > 0$, such that $0 \leqslant a_n \leqslant Kc_n$ for all $n > N_0$, then $\sum_{n=1}^{\infty} a_n$ is convergent.

(ii)   If there exist a positive integer N_1 and a constant $k > 0$, such that $a_n \geqslant kd_n$ for all $n > N_1$, then $\sum_{n=1}^{\infty} a_n$ is divergent.

Theorem 8.3.3. (Geometric Series). Let $r \geqslant 0$. The series $\sum_{n=0}^{\infty} r^n$ is convergent if $r < 1$ and divergent if $r \geqslant 1$.

Proof. Let $r < 1$. Then

$$S_n = 1 + r + \cdots + r^{n-1} = \frac{1 - r^n}{1 - r} < \frac{1}{1 - r}$$

Hence, by Theorem 8.3.1, $\sum_{n=0}^{\infty} r^n$ is convergent.

If $r \geqslant 1$ then $S_n \geqslant n$ and so $\sum_{n=0}^{\infty} r^n$ is divergent. $\blacksquare$

Theorem 8.3.4. (Ratio Test). Let $a_n > 0$ and let

$$\limsup_{n \to \infty} \frac{a_{n+1}}{a_n} = L, \quad \liminf_{n \to \infty} \frac{a_{n+1}}{a_n} = l.$$

Then $\sum_{n=1}^{\infty} a_n$ is convergent if $L < 1$ and divergent if $l > 1$.

Proof. (i) Let $L < 1$ and $\lambda = \frac{1 + L}{2}$. Then $\lambda < 1$ and we can find m such that for all $n \geqslant m$,

$$a_{n+1} < \lambda a_n$$

In particular we have

$$a_{m+1} < \lambda a_m, \quad a_{m+2} < \lambda a_{m+1} < \lambda^2 a_m, \ldots$$
$$a_{m+p} < \lambda a_{m+p-1} < \cdots < \lambda^p a_m$$

Since $0 < \lambda < 1$, the series $\sum_{p=1}^{\infty} \lambda^p$ and $\sum_{p=1}^{\infty} \lambda^p a_m$ are convergent. It follows by comprison test that the series $\sum_{p=1}^{\infty} a_{m+p}$ is convergent. Hence $\sum_{n=1}^{\infty} a_n$ is convergent.

(ii) Suppose next that $l > 1$. Let $\mu = \frac{1 + l}{2} > 1$. Then there exists M such that for all $n \geqslant M$:

$$a_{n+1} > \mu a_n$$

As in (i), we obtain that $a_{M+p} > \mu^p a_M$. This implies that $\lim\limits_{p \to \infty} a_{M+p} = \infty$. Hence by Theorem 8.2.2 the series $\sum\limits_{n=1}^{\infty} a_n$ is divergent. $\blacksquare$

Remark. If $l = 1$ and $\dfrac{a_{n+1}}{a_n} \geqslant 1$ for all $n \geqslant n_0$, where n_0 is some fixed integer, then we get $a_{n+p} \geqslant a_{n_0}$ and so $\lim\limits_{p \to \infty} \sup a_{n+p} \neq 0$. Theorem 8.2.2 gives the divergence of $\sum\limits_{n=1}^{\infty} a_n$.

EXAMPLE 8.3.2. The series $\sum\limits_{n=1}^{\infty} \dfrac{1}{n}$ is divergent and the series $\sum\limits_{n=1}^{\infty} \dfrac{1}{n^2}$ is convergent. For both series we have $L = l = \lim\limits_{n \to \infty} \dfrac{a_{n+1}}{a_n} = 1$. The ratio test does not give any information when $\lim\limits_{n \to \infty} \dfrac{a_{n+1}}{a_n} = 1$. It fails to give information when $l \leqslant 1 \leqslant L$. (See Example 8.3.3.)

Theorem 8.3.5. (Root Test). If $\lim\limits_{n \to \infty} \sup a_n^{1/n} = L$, then the series $\sum\limits_{n=1}^{\infty} a_n$ is convergent if $L < 1$ and divergent if $L > 1$.

Proof. Suppose first that $L < 1$ and let $\lambda = \dfrac{1 + L}{2} < 1$. There exists N such that for all $n \geqslant N$

$$a_n^{1/n} < \lambda$$

That is,

$$a_n < \lambda^n$$

Since $\lambda < 1$, the geometric series $\sum\limits_{n=N+1}^{\infty} \lambda^n$ is convergent and so by comparison test the series $\sum\limits_{n=N+1}^{\infty} a_n$ is convergent. Hence $\sum\limits_{n=1}^{\infty} a_n$ is convergent.

Suppose next that $L > 1$ and let $\mu = \dfrac{1 + L}{2} > 1.$ Then

$$a_n^{1/n} > \mu$$

for infinitely many values of n. Hence $a_n > \mu^n > 1$ for these indices and so $a_n \not\to 0$ as $n \longrightarrow \infty$. This shows that $\displaystyle\sum_{n=1}^{\infty} a_n$ is divergent. ∎

The root test does not give any information when $L = 1$. See Example 8.3.4.

In the next two theorems we consider series of positive nonincreasing terms. We write $a_n \downarrow$ if $a_{n+1} \leqslant a_n$ for each $n \geqslant 1$ and $a_n \uparrow$ if $a_{n+1} \geqslant a_n$ for each $n \geqslant 1$.

Theorem 8.3.6. (Abel-Pringsheim). If $\displaystyle\sum_{n=1}^{\infty} a_n$ is a convergent series of positive and nonincreasing terms, then $\displaystyle\lim_{n\to\infty} na_n = 0$.

Proof. Given $\epsilon > 0$, there exists $n_0(\epsilon)$ such that for $n \geqslant n_0(\epsilon)$ and $p = 1, 2, \ldots$ we have

$$|a_{n+1} + a_{n+2} + \cdots + a_{n+p}| < \epsilon/4.$$

Since $a_{n+1} \geqslant a_{n+2} \geqslant \cdots,$

$$|a_{n+1} + \cdots + a_{n+p}| = a_{n+1} + \cdots + a_{n+p} \geqslant p\, a_{n+p}.$$

Taking $p = n$ we get

(i) $\qquad\qquad\qquad\qquad 2n\, a_{2n} < \epsilon/2.$

Further

(ii) $\qquad\qquad\qquad (2n + 1)\, a_{2n+1} < 4n\, a_{2n} < \epsilon.$

These two inequalities prove the theorem. ∎

EXAMPLE 8.3.3. The series $\displaystyle\sum_{n=1}^{\infty} \dfrac{1}{an + b}, a > 0, b \geqslant 0$ is divergent.

For $a_n > 0, \downarrow$. If the series be convergent, then $\lim na_n = 0$, but $\displaystyle\lim_{n\to\infty}$

$\dfrac{n}{an + b} = \dfrac{1}{a}.$

Theorem 8.3.7. (Cauchy's Condensation Test). If $a_n > 0, \downarrow$, then the two series

$$\sum_1^{\infty} a_n, \quad \sum_1^{\infty} 2^n a_{2^n}$$

converge or diverge together.

Proof. We have

$$a_3 + a_4 \geq 2a_4$$
$$\cdots$$
$$a_{2^{n-1}+1} + a_{2^{n-1}+2} + \cdots + a_{2^n} \geq 2^{n-1} a_{2^n}.$$

Adding these inequalities, we get

$$\sum_{k=1}^{2^n} a_k > \sum_{k=3}^{2^n} a_k > \sum_{k=2}^{n} 2^{k-1} a_{2^k}.$$

If $\sum_{k=1}^{\infty} 2^k a_{2^k}$ diverges, then so does $\sum_{k=1}^{\infty} 2^{k-1} a_{2^k}$, and the above inequality shows

that $\sum_{k=1}^{\infty} a_k$ diverges. To complete the proof we note that

$$a_2 + a_3 \leq 2a_2$$
$$\cdots$$
$$a_{2^{n-1}} + a_{2^{n-1}+1} + \cdots + a_{2^n-1} \leq 2^{n-1} a_{2^{n-1}}.$$

Hence

$$a_2 + a_3 + \cdots + a_{2^n-1} \leq \sum_{k=1}^{n-1} 2^k a_{2^k}.$$

Suppose that $\sum_{1}^{\infty} 2^k a_{2^k}$ is convergent and has the sum S. Then for every n

$$a_2 + a_3 + a_4 + \cdots + a_{2^n-1} \leq \sum_{k=1}^{n} 2^k a_{2^k} < S.$$

Since for any integer $p \geq 1$ we can choose n so large that $2^n - 1 \geq p$, we have

$$a_1 + a_2 + \cdots + a_p < a_1 + \cdots + a_{2^n-1} < a_1 + S.$$

This proves the convergence of $\sum_{k=1}^{\infty} a_k.$ ∎

EXAMPLE 8.3.4. Consider the series $\sum_{n=1}^{\infty} a_n = 1 + \dfrac{1}{2^p} + \dfrac{1}{3^q} + \dfrac{1}{4^p} + \dfrac{1}{5^q} + \cdots$

where $1 \leq p < q$.

Suppose first that $p > 1$ and let $c_n = 1/n^p$. Then $a_n \leq c_n$ for every n and $\sum_{n=1}^{\infty} c_n$ is convergent. By the comparison test, $\sum_{n=1}^{\infty} a_n$ is convergent.

If $p = 1$, then

$$\sum_{k=1}^{2n} a_k > 1 + \frac{1}{2} + \frac{1}{4} + \frac{1}{6} + \cdots + \frac{1}{2n}$$

$$> \frac{1}{2} \left(1 + \frac{1}{2} + \frac{1}{3} + \cdots + \frac{1}{n}\right).$$

Since the harmonic series is divergent, the last expression diverges to ∞. This proves the divergence of $\sum_{n=1}^{\infty} a_n$ when $p = 1$.

Note here that *in either case $p > 1$ or $p = 1$*, we have

$$\limsup_{n \to \infty} \frac{a_{n+1}}{a_n} = \lim_{n \to \infty} \frac{n^q}{(n+1)^p} = \infty$$

$$\liminf_{n \to \infty} \frac{a_{n+1}}{a_n} = \lim_{n \to \infty} \frac{n^p}{(n+1)^q} = 0$$

$$L = \limsup_{n \to \infty} a_n^{1/n} = \liminf_{n \to \infty} a_n^{1/n} = 1 \text{ (See Exercise 11.7(2)).}$$

We now consider the rearrangement of the terms of a series Σa_n of nonnegative terms.

Definition 8.3.1. *A series $\sum_{n=1}^{\infty} b_n$ is said to be a rearrangement of $\sum_{n=1}^{\infty} a_n$ if every a_n occurs exactly once in $\sum_{n=1}^{\infty} b_n$ and if every b_n occurs exactly once in $\sum_{n=1}^{\infty} a_n$.*

Theorem 8.3.8. *If $\sum_{n=1}^{\infty} a_n$ is a series of nonnegative terms and $\sum_{n=1}^{\infty} b_n$ is a rearrangement of $\sum_{n=1}^{\infty} a_n$, then*

$$\sum_{n=1}^{\infty} a_n = \sum_{n=1}^{\infty} b_n$$

in the sense that if one of the two series is convergent then the second series is also convergent and both have the same sum; and if one of the two series is divergent then the second series is also divergent.

Proof. Let $\sum_{k=1}^{n} a_k = A_n$, $\sum_{k=1}^{n} b_k = B_n$. Suppose first that $\sum_{1}^{\infty} a_k$ is convergent and has sum A. Consider B_n and let $b_1 = a_{m_1}$, $b_2 = a_{m_2}, \ldots$. If m be the largest of the integers $m_1, m_2, \ldots, m_n$, then

$$B_n \leqslant A_m \leqslant A.$$

Since $B_n \uparrow$, B_n tends to a finite limit B. Hence $\sum_{n=1}^{\infty} b_n$ is convergent and $B \leqslant A$. By considering A_n we have $A \leqslant B$, and so $A = B$.

Suppose now that $\sum_{1}^{\infty} a_k$ is divergent. Then for all sufficiently large m,

$$B_m > A_n.$$

Since $\lim_{n \to \infty} A_n = \infty$, $\lim_{m \to \infty} B_n = \infty$, and this proves the divergence of $\sum_{k=1}^{\infty} b_k$. The proof is complete. ▌

Exercise 8.3

1. Prove that if $a_n > 0, b_n > 0$ and $\dfrac{a_n}{b_n} \longrightarrow c, c \neq 0$, as $n \longrightarrow \infty$, the series $\sum_{n=1}^{\infty} a_n$ and $\sum_{n=1}^{\infty} b_n$ are both convergent or both divergent.

$\left[\text{Hint:} \ 0 \leqslant a_n \leqslant (c + \epsilon) b_n, \epsilon > 0 \text{ and } 0 \leqslant b_n \leqslant \dfrac{a_n}{c - \epsilon} \text{ for } n > n_0(\epsilon). \right]$

2. (Ratio Comparison Test). Let $a_n > 0$, $b_n > 0$, $(a_{n+1}/a_n) \leqslant (b_{n+1})/b_n$ for all $n \geqslant N$. Prove that (i) $\sum_{n=1}^{\infty} a_n$ converges if $\sum_{n=1}^{\infty} b_n$ converges, and (ii) $\sum_{n=1}^{\infty} b_n$

diverges if $\sum_{n=1}^{\infty} a_n$ diverges. $\left(Hint: a_n = a_N \dfrac{a_{N+1}}{a_N} \cdots \dfrac{a_n}{a_{n-1}} \leqslant a_N \dfrac{b_{N+1}}{b_N} \cdots\right.$

$\dfrac{b_n}{b_{n-1}} = \dfrac{a_N}{b_N} b_n.$ Now use Corollary Theorem 8.3.2.$\Big)$

3. Prove that $\sum_{n=1}^{\infty} \dfrac{1}{n^p}$ is convergent when $p > 1$ and divergent when $p \leqslant 1$.

$\Big(Hint:$ Let $a_n = \dfrac{1}{n^p}$ $(p > 0)$ in the condensation test and note that the

geometric series $\sum_{n=1}^{\infty} \left\{\dfrac{1}{2^{(p-1)}}\right\}^n$ is convergent when $p > 1$ and divergent when

$p \leqslant 1.\Big)$

4. Deduce the Abel-Pringsheim theorem from the Cauchy condensation test. (*Hint:* If Σa_n is convergent and $a_n \downarrow$, then $\Sigma 2^n a_{2^n}$ is convergent and so $\lim_{n \to \infty} 2^n a_{2^n} = 0$. Given any integer p, we can choose n such that $2^n \leqslant p < 2^{n+1}$ and then $p a_p < 2^{n+1} a_{2^n} = 2\{2^n a_{2^n}\} \longrightarrow 0$ as $p \longrightarrow \infty$.)

5. Examine for convergence the series

   (i) $\Sigma \dfrac{1}{2 + 3n^2}$

   (ii) $\Sigma \dfrac{1 + n}{2 + 3n^2}$

   (iii) $\Sigma \dfrac{n^3}{2^n + 5}$

   (iv) $\Sigma \dfrac{(n + 1)^2}{n!}$

   (v) $\Sigma \dfrac{(n + 2)^n}{n^{2n}}$

8.4. Absolute Convergence

We consider in this section series whose terms may be real or complex.

Definition 8.4.1. *If a series* $\sum_{n=1}^{\infty} a_n$ *of real or complex terms* a_n *is such that*

$\sum_{n=1}^{\infty} |a_n|$ *is convergent then the series* $\sum_{n=1}^{\infty} a_n$ *is said to be absolutely convergent.*

Definition 8.4.2. *If* $\sum\limits_{n=1}^{\infty} a_n$ *is convergent and* $\sum\limits_{n=1}^{\infty} |a_n|$ *is divergent then the*

series $\sum\limits_{n=1}^{\infty} a_n$ *is said to be conditionally convergent.*

A convergent series with real nonnegative terms is *absolutely convergent.*

Theorem 8.4.1. *If* $\sum\limits_{n=1}^{\infty} a_n$ *is absolutely convergent then it is convergent.*

Proof. Since $\Sigma |a_n|$ is convergent, given $\epsilon > 0$, there exists an $n_0(\epsilon)$ such that for $n \geqslant n_0(\epsilon)$ and all values of p,

$$|a_{n+1}| + |a_{n+2}| + \cdots + |a_{n+p}| < \epsilon$$

But for $n \geqslant n_0, p = 1, 2, \ldots$

$$|a_{n+1} + a_{n+2} + \cdots + a_{n+p}| \leqslant |a_{n+1}| + \cdots + |a_{n+p}| < \epsilon$$

Hence Σa_n is convergent. ∎

EXAMPLE 8.6. The series

$$\sum_{n=1}^{\infty} \frac{\sin n\theta}{n^p}, \quad \sum_{n=1}^{\infty} \frac{\cos n\theta}{n^p}, \quad 0 \leqslant \theta \leqslant 2\pi, \quad p > 1$$

are absolutely convergent.

8.5. Series of Positive and Negative Terms

We shall now consider in Secs. 8.5–8.6 series whose terms are all real but not necessarily of the same sign.

Theorem 8.5.1. (Alternating Series). The series

$$a_1 - a_2 + a_3 - a_4 + \cdots + = \sum_{n=1}^{\infty} (-1)^{n+1} a_n \qquad (8.5)$$

is convergent if (i) $a_n \downarrow$, and (ii) $\lim\limits_{n \to \infty} a_n = 0$.

Proof. We note that the hypothesis implies $a_n \geqslant a_{n+1} \geqslant 0$ for every n. Consider first the sequence $\{S_{2n}\}_1^{\infty}$. Since

$$S_{2n+2} - S_{2n} = a_{2n+1} - a_{2n+2} \geqslant 0$$

the sequence $\{S_{2n}\}_1^\infty$ is monotonic increasing. It is bounded above, for

$$S_{2n} = a_1 - (a_2 - a_3) - \cdots - (a_{2n-2} - a_{2n-1}) - a_{2n} \leqslant a_1$$

Hence $\{S_{2n}\}$ converges and $\lim\limits_{n \to \infty} S_{2n} = L \leqslant a_1$. Next consider the sequence $\{S_{2n+1}\}_0^\infty$. This sequence is monotonic decreasing, for

$$S_{2n+1} - S_{2n-1} = -a_{2n} + a_{2n+1} \leqslant 0.$$

It is bounded below, for

$$S_{2n+1} = (a_1 - a_2) + (a_3 - a_4) + \cdots + a_{2n+1} \geqslant a_1 - a_2.$$

Hence $\{S_{2n+1}\}$ converges and $\lim\limits_{n \to \infty} S_{2n+1} = l \geqslant a_1 - a_2$. Now $a_{2n+1} = S_{2n+1} - S_{2n} \longrightarrow l - L$. But by hypothesis (ii), $\lim\limits_{n \to \infty} a_{2n+1} = 0$. Hence $l = L$. This proves the convergence of $\{S_n\}_1^\infty$ and hence that of the series. ∎

Note that the sum l of the series (8.5) lies in the interval (S_{2n}, S_{2n+1}), $n = 1, 2, \ldots$, and hence in the interval $(a_1 - a_2, a_1)$.

EXAMPLE 8.5.1. The series

$$1 - \frac{1}{2} + \frac{1}{3} - \frac{1}{4} + \cdots$$

is conditionally convergent. Note that

$$S_4 = \frac{1}{3} + \frac{1}{4} < S < S_5 = \frac{1}{3} + \frac{1}{4} + \frac{1}{5}.$$

EXAMPLE 8.5.2. The series

$$\sum_{n=1}^{\infty} a_n = 1 + \frac{1}{3} - \frac{1}{2} + \frac{1}{5} + \frac{1}{7} - \frac{1}{4} + \frac{1}{9} + \frac{1}{11} - \frac{1}{6} + \cdots$$

is conditionally convergent.

For $\sum\limits_{n=1}^{\infty} \dfrac{1}{n}$ is divergent. Also $\{S_{3n}\}_1^\infty$, where

$$S_{3n} = \left(1 + \frac{1}{3} - \frac{1}{2}\right) + \left(\frac{1}{5} + \frac{1}{7} - \frac{1}{4}\right) + \cdots + \left(\frac{1}{4n-3} + \frac{1}{4n-1} - \frac{1}{2n}\right)$$

is monotone increasing for

$$S_{3n+3} - S_{3n} = \frac{1}{4n+1} + \frac{1}{4n+3} - \frac{1}{2(n+1)} > 0.$$

Further,

$$S_{3n} = 1 + \frac{1}{3} + \left(-\frac{1}{2} + \frac{1}{5} + \frac{1}{7}\right) + \left(-\frac{1}{4} + \frac{1}{9} + \frac{1}{11}\right)$$

$$+ \cdots + \left(-\frac{1}{2(n-1)} + \frac{1}{4n-3} + \frac{1}{4n-1}\right) - \frac{1}{2n} < 1 + \frac{1}{3}.$$

Hence $\{S_{3n}\}_1^\infty$ converges. Let $\lim_{n\to\infty} S_{3n} = S$. Then

$$S_{3n+1} = S_{3n} + \frac{1}{4n+1} \longrightarrow S$$

and

$$S_{3n+2} = S_{3n} + \frac{1}{4n+1} + \frac{1}{4n+3} \longrightarrow S$$

Hence $\{S_n\}_1^\infty$ converges to S. Note that $\frac{4}{3} > S > \frac{5}{6} > \frac{1}{3} + \frac{1}{4} + \frac{1}{5}$.

Now we show that $\sum_{n=1}^\infty a_n$ is not absolutely convergent. Let Σ_n denote the

nth partial sum of the series $\sum_{n=1}^\infty |a_n|$. Then

$$\sum_{3n} = 1 + \frac{1}{3} + \frac{1}{2} + \frac{1}{5} + \frac{1}{7} + \frac{1}{4} + \cdots + \frac{1}{4n-3} + \frac{1}{4n-1} + \frac{1}{2n}$$

$$> \frac{1}{2} + \frac{1}{4} + \cdots + \frac{1}{2n}$$

$$= \frac{1}{2}\left(1 + \frac{1}{2} + \cdots + \frac{1}{n}\right)$$

which $\longrightarrow \infty$ as $n \longrightarrow \infty$ (see Example 8.2.1).

EXAMPLE 8.5.3. The series

$$1 - \frac{1}{5} + \frac{1}{9} - \frac{1}{13} + \frac{1}{17} - \cdots$$

is conditionally convergent.

For Theorem 8.5.1 shows that the series is convergent. If it were absolutely convergent then by Abel-Pringsheim theorem we must have $\lim_{n\to\infty} n|a_n| = 0$. But

$\lim_{n\to\infty} n|a_n| = \lim_{n\to\infty} \frac{n}{4n-3} = \frac{1}{4}$. Hence the series is not absolutely convergent.

Theorem 8.5.2. (Rearrangement of Terms). If $\sum_{n=1}^{\infty} a_n$ is an absolutely convergent series and $\sum_{n=1}^{\infty} b_n$ is a rearrangement of $\sum_{n=1}^{\infty} a_n$, then $\sum_{n=1}^{\infty} b_n$ is convergent and the sums of the two series are the same.

Proof. Let $|a_n| = \alpha_n$ and

$$a_n^+ = \frac{1}{2}(\alpha_n + a_n), \quad a_n^- = \frac{1}{2}(\alpha_n - a_n).$$

Then $a_n^+ \geqslant 0$, $a_n^- \geqslant 0$ and $a_n^+ + a_n^- = \alpha_n$, $a_n^+ - a_n^- = a_n$. Hence $a_n^+ \leqslant \alpha_n$, $a_n^- \leqslant \alpha_n$. Since $\Sigma \alpha_n$ is convergent both series Σa_n^+ and Σa_n^- are convergent by comparison test. If we define similarly b_n^+ and b_n^- then Σb_n^+ and Σb_n^- are also convergent. Now the series Σb_n^+ is a rearrangement of a series Σa_n^+ of nonnegative terms and so, by Theorem 8.3.8, it is convergent and $\sum_{n=1}^{\infty} b_n^+ = \sum_{n=1}^{\infty} a_n^+$. Similarly, Σb_n^- is convergent and $\sum_{n=1}^{\infty} b_n^- = \sum_{n=1}^{\infty} a_n^-$. Hence

$$\sum_{n=1}^{\infty} b_n = \sum_{n=1}^{\infty}(b_n^+ - b_n^-) = \sum_{n=1}^{\infty} b_n^+ - \sum_{n=1}^{\infty} b_n^- \qquad \text{(See Exercise 8.2(4))}.$$

$$= \sum_{n=1}^{\infty} a_n^+ - \sum_{n=1}^{\infty} a_n^-$$

$$= \sum_{n=1}^{\infty}(a_n^+ - a_n^-) = \sum_{n=1}^{\infty} a_n \quad \blacksquare$$

Note that Σa_n^+, Σa_n^- are both convergent. So in an absolutely convergent series, the series formed by its positive terms alone is convergent and the series formed by its negative terms alone is convergent.

Theorem 8.5.3. If a series $\sum_{n=1}^{\infty} a_n$ is conditionally convergent, then the series of its positive terms and the series of its negative terms are both divergent.

Proof. Let $S_n = \sum_{k=1}^{n} a_k$. Denote the sum of the positive terms in this partial sum by α_n and the sum of the negative terms in this partial sum by $-\beta_n$. Then $\alpha_n \geqslant 0$, $\beta_n \geqslant 0$, and

$$\sigma_n = \sum_{k=1}^{n} |a_k| = \alpha_n + \beta_n, \quad S_n = \alpha_n - \beta_n$$

Furthermore, the sequences $\{\alpha_n\}$ and $\{\beta_n\}$ are monotone increasing. By hypothesis $\sigma_n \longrightarrow \infty$, $S_n \longrightarrow S$ as $n \longrightarrow \infty$. Since

$$\alpha_n = \frac{\sigma_n + S_n}{2}, \quad \beta_n = \frac{\sigma_n - S_n}{2}; \quad \lim_{n \to \infty} \alpha_n = \infty, \quad \lim_{n \to \infty} \beta_n = \infty.$$

This completes the proof. $\blacksquare$

Note that if the series Σa_n is absolutely convergent then $\{\sigma_n\}$ is convergent.

Exercise 8.5

1. Let $a_n \neq 0$ for each $n = 1, 2, \ldots$ and let $A = \limsup\limits_{n \to \infty} \left| \dfrac{a_{n+1}}{a_n} \right|$ and $B = \liminf\limits_{n \to \infty} \left| \dfrac{a_{n+1}}{a_n} \right|$. Prove the following:
   (i) If $A < 1$ then the series Σa_n converges absolutely, and in particular converges.
   (ii) If $B > 1$ then the series Σa_n diverges.
   (iii) If $\left| \dfrac{a_{n+1}}{a_n} \right| \geqslant 1$ for all $n \geqslant n_0$, where n_0 is some fixed integer then the series Σa_n diverges.
   (iv) If $B \leqslant 1 \leqslant A$ then the test fails to give information.
   In this and the next exercise, a_n may be real or complex.
2. Let $L = \limsup\limits_{n \to \infty} |a_n|^{1/n}$. Prove the following:

   (a) If $L < 1$ then $\displaystyle\sum_{n=1}^{\infty} a_n$ converges absolutely.

   (b) If $L > 1$ then $\displaystyle\sum_{n=1}^{\infty} a_n$ diverges.

   (c) If $L = 1$ then the test fails to give information. (*Hint*: See Theorem 8.3.5 and Example 8.3.4.)

8.6. Abel's and Dirichlet's Tests

We consider now the convergence of the series $\displaystyle\sum_{k=1}^{\infty} a_k b_k$ and begin with Abel's (N. H. Abel, 1802–1829) method of partial summation. Let $A(m,k) =$

$a_m + a_{m+1} + \cdots + a_k$, $1 \leqslant m \leqslant k$.   Then $A(m,j) - A(m,j - 1) = a_j$, $j = m + 1$, $m + 2, \ldots k$; and

$$\sum_{k=m}^{n} a_k b_k = A(m,m)b_m + (A(m,m + 1) - A(m,m))b_{m+1}$$

$$+ \cdots + (A(m,n - 1) - A(m,n - 2))b_{n-1} + (A(m,n) - A(m,n - 1))b_n$$

$$= \sum_{k=m}^{n-1} A(m,k)(b_k - b_{k+1}) + A(m,n)b_n \tag{8.6}$$

$$= \sum_{k=m}^{n-1} (A(1,k) - A(1,m - 1))(b_k - b_{k+1}) + (A(1,n) - A(1,m - 1))b_n$$

$$= \sum_{k=m}^{n-1} A(1,k)(b_k - b_{k+1}) - A(1,m - 1)(b_m - b_n)$$

$$+ A(1,n)b_n - A(1,m - 1)b_n$$

$$= \sum_{k=m}^{n-1} A(1,k)(b_k - b_{k+1}) - A(1,m - 1)b_m + A(1,n)b_n \tag{8.7}$$

Here $A(1,m - 1) = 0$ if $m = 1$. The identities (8.6) and (8.7), known as Abel's partial summation, lead us to the following Lemma.

Lemma (Abel's lemma). If $\{b_k\}_1^\infty$ is a positive monotonic decreasing sequence and if $A_n = \sum_{k=1}^{n} a_k$, $n \geqslant 1$, $A_0 = 0$,

$$H(m,n) = \max \{|a_m|, |a_m + a_{m+1}|, \ldots |a_m + a_{m+1} + \cdots + a_n|\}$$
$$M = \max \{|A_{m-1}|, |A_m|, \ldots |A_n|\}$$

then

$$\left| \sum_{k=m}^{n} a_k b_k \right| \leqslant H(m,n)b_m \leqslant 2\, Mb_m. \tag{8.8}$$

Proof. Since $b_m - b_{m+1}, \ldots b_{n-1} - b_n, b_n$ are all nonnegative we have, from (8.6),

$$\left| \sum_{k=m}^{n} a_k b_k \right| \leqslant H(m,n) \left\{ \sum_{k=m}^{n-1} (b_k - b_{k-1}) + b_n \right\}$$

$$= H(m,n)b_m$$

Now

$$|a_m| = |A_m - A_{m-1}| \leqslant |A_m| + |A_{m-1}| \leqslant 2M$$

and

$$|a_m + \cdots + a_n| = |A_n - A_{m-1}| \leqslant |A_n| + |A_{m-1}| \leqslant 2M;$$

hence $H(m,n) \leqslant 2M$. The proof of (8.8) is complete. ∎

Theorem 8.6.1. (Dirichlet's Test). The series $\displaystyle\sum_{k=1}^{\infty} a_k b_k$ is convergent if $A_n = \displaystyle\sum_{k=1}^{n} a_k$ is bounded and $\{b_k\}_1^{\infty}$ decreases to 0.

Proof. Suppose $|A_n| \leqslant M$. Since $\displaystyle\lim_{m \to \infty} M b_m = 0$, the inequality (8.8) shows that Cauchy criterion for convergence is satisfied by $\Sigma\, a_k b_k$. ∎

Remark. The case $a_k = (-1)^{k+1}$ gives Theorem 8.5.1. Note that in this and the next test b_k is real.

Theorem 8.6.2. (Abel's Test). The series $\displaystyle\sum_{k=1}^{\infty} a_k b_k$ is convergent if $\displaystyle\sum_{k=1}^{\infty} a_k$ converges and $\{b_k\}_1^{\infty}$ is monotone and bounded.

Proof. Let $\displaystyle\lim_{k \to \infty} b_k = b$. Write

$$v_k = \begin{cases} b - b_k & \text{if} & b_k \uparrow \\ b_k - b & \text{if} & b_k \downarrow \end{cases}$$

In either case $v_k \downarrow$, $\displaystyle\lim_{k \to \infty} v_k = 0$, and we can write $b_k = b \pm v_k$. Let $M = \sup\,(|A_{m-1}|, |A_m|, \ldots, |A_n|)$. Then

$$\left| \sum_{k=m}^{n} a_k b_k \right| = \left| \sum_{k=m}^{n} a_k (b \pm v_k) \right|$$

$$= \left| b \sum_{k=m}^{n} a_k \pm \sum_{k=m}^{n} a_k v_k \right|$$

$$\leqslant |b| \sum_{k=m}^{n} |a_k| + 2 v_m M$$

$$< |b|\epsilon + \epsilon = (|b| + 1)\epsilon$$

if $m \geqslant m_0(\epsilon), n \geqslant m$. This proves the convergence of $\Sigma a_k b_k$. ∎

EXAMPLE 8.6.1. If (i) $\left| \sum_{k=1}^{p} a_k \right| \leqslant H$, $p = 1, 2, \ldots, n$, and

(ii) $\sum_{k=1}^{\infty} |b_k - b_{k+1}| = V$, then $\lim_{n \to \infty} |b_n| = b$ exists and

$$\left| \sum_{k=1}^{n} a_k b_k \right| \leqslant H(V + b).$$

Since the series $\Sigma |b_k - b_{k+1}|$ is convergent we have

$$|b_n - b_{n+p}| \leqslant |b_n - b_{n+1}| + |b_{n+1} - b_{n+2}| + \cdots + |b_{n+p-1} - b_{n+p}| < \epsilon$$

for $n \geqslant n_0(\epsilon)$, $p = 1, 2, , \ldots$ This proves that the sequence $\{b_n\}$ is convergent and $\lim_{n \to \infty} b_n$, and hence $\lim_{n \to \infty} |b_n| = b$ exist. The identity (8.6) now gives

$$\left| \sum_{k=1}^{n} a_k b_k \right| \leqslant H \left\{ \sum_{k=1}^{n-1} |b_k - b_{k+1}| + |b_n| \right\}$$

Since $|b_n| \leqslant |b_n - b_{n+1}| + |b_{n+1}|$, the expression in parenthesis increases with n and tends to $V + b$.

EXAMPLE 8.6.2. If $\sum_{k=1}^{\infty} a_k$ and $\sum_{k=1}^{\infty} |b_k - b_{k+1}|$ are convergent, then

$\sum_{k=1}^{\infty} a_k b_k$ converges.

As in Example 8.6.1, $\lim_{n \to \infty} |b_n| = b$ exists. Since $\Sigma_1^{\infty} a_k$ is convergent, given $\epsilon > 0$, we can choose m such that for $k \geqslant m$,

$$|A(m,k)| = |a_m + a_{m+1} + \cdots + a_k| < \epsilon.$$

Hence we have from identity (8.6),

$$\left| \sum_{k=m}^{n} a_k b_k \right| < \epsilon \left\{ \sum_{k=m}^{n-1} |b_k - b_{k+1}| + |b_n| \right\}$$

$$\leqslant \epsilon \left\{ \sum_{k=m}^{n} |b_k - b_{k+1}| + b \right\}$$

This shows that Cauchy criterion for convergence is satisfied by $\Sigma a_k b_k$.

EXAMPLE 8.6.3. Let $b_n \downarrow$, $\lim_{n \to \infty} b_n = 0$. Then $\sum_{n=1}^{\infty} b_n \sin n\theta$ is convergent

for all (real) θ and $\sum_{n=1}^{\infty} b_n \cos n\theta$ is convergent for all (real) θ other than multi-

ples of 2π.

Let $\sin \frac{\theta}{2} \neq 0$. Then since $2 \sin \alpha \sin \beta = \cos (\alpha - \beta) - \cos (\alpha + \beta)$ we have

$$2 \sin \frac{\theta}{2} \left(\sin \theta + \sin 2\theta + \cdots + \sin n\theta \right)$$

$$= \left(\cos \frac{\theta}{2} - \cos \frac{3\theta}{2} \right) + \left(\cos \frac{3\theta}{2} - \cos \frac{5\theta}{2} \right) + \cdots + \left(\cos \left(n - \frac{1}{2} \right) \theta - \cos \left(n + \frac{1}{2} \right) \theta \right)$$

$$= \cos \frac{\theta}{2} - \cos \left(n + \frac{1}{2} \right) \theta$$

Hence for $\sin \frac{\theta}{2} \neq 0$,

$$\sum_{k=1}^{n} \sin k\theta = \frac{\cos \frac{\theta}{2} - \cos \left(n + \frac{1}{2} \right) \theta}{2 \sin \frac{\theta}{2}}$$

and

$$\left| \sum_{k=1}^{n} \sin k\theta \right| \leq \frac{1}{\left| \sin \frac{\theta}{2} \right|}.$$

Similarly we use the formula $2 \sin \alpha \cos \beta = \sin (\alpha + \beta) + \sin (\alpha - \beta)$ and obtain

$$2 \sin \frac{\theta}{2} (\cos \theta + \cos 2\theta + \cdots + \cos n\theta)$$

$$= \left(\sin \frac{3\theta}{2} - \sin \frac{\theta}{2} \right) + \cdots + \left(\sin \left(n + \frac{1}{2} \right) \theta - \sin \left(n - \frac{1}{2} \right) \theta \right)$$

$$= \sin \left(n + \frac{1}{2} \right) \theta - \sin \frac{\theta}{2}$$

and for $\sin \frac{\theta}{2} \neq 0$,

$$\left| \sum_{k=1}^{n} \cos k\theta \right| = \left| \frac{\sin\left(n + \frac{1}{2}\right)\theta - \sin\frac{\theta}{2}}{2\sin\frac{\theta}{2}} \right| \leqslant \frac{1}{\left|\sin\frac{\theta}{2}\right|}.$$

The convergence of both series follows from Dirichlet's test when $\sin\theta/2 \neq 0$.

If $\sin\theta/2 = 0$, that is, $\theta = 2p\pi$, then the cosine series is divergent unless $\sum_{n=1}^{\infty} b_n$ is convergent and the sine series becomes $0 + 0 + \cdots$ and so converges trivially to 0.

Remarks. (*i*) The cosine series converges absolutely if and only if $\sum_{n=1}^{\infty} b_n$ is convergent. For if $\sum_{n=1}^{\infty} b_n$ is convergent,

$$|b_n \cos n\theta| \leqslant b_n$$

and so $\Sigma b_n \cos n\theta$ is absolutely convergent. On the other hand,

$$|b_n \cos n\theta| \geqslant b_n \cos^2 n\theta = \frac{1}{2} b_n (1 + \cos 2n\theta).$$

We have proved above that $\sum_{n=1}^{\infty} b_n \cos 2n\theta$ is convergent when $\theta \neq p\pi$. Hence $\sum_{n=1}^{\infty} |b_n \cos n\theta|$ cannot be convergent when $\theta \neq p\pi$ unless Σb_n converges. When $\theta = p\pi$ then $|b_n \cos n\theta| = b_n$.

Similarly the sine series cannot converge absolutely when $\theta \neq p\pi$ unless Σb_n converges.

(*ii*) The series $\sum_{n=1}^{\infty} \cos n\theta$ and $\sum_{n=1}^{\infty} \sin n\theta$ are both finitely oscillating, for when $\theta \neq 2p\pi$,

$$\left| \sum_{k=m+1}^{n} \cos k\theta \right| \leqslant \frac{1}{\left|\sin\frac{\theta}{2}\right|}, \quad \left| \sum_{k=m+1}^{n} \sin k\theta \right| \leqslant \frac{1}{\left|\sin\frac{\theta}{2}\right|}. \qquad (8.9)$$

8.7. Power Series

In this section we consider the convergence of a power series, that is, a series of the form

$$\sum_{n=0}^{\infty} a_n z^n = a_0 + a_1 z + \cdots + a_n z^n + \cdots \qquad (8.10)$$

Here z and a_n are complex numbers and the coefficients a_n depend on n but not on z. (In a special case a_n and z could be real.)

EXAMPLE 8.7.1. The series $\sum_{n=0}^{\infty} z^n$ is absolutely convergent if $|z| < 1$ and divergent if $|z| \geqslant 1$.

By the root test the series $\sum_{n=0}^{\infty} |z|^n$, is convergent if $\limsup_{n \to \infty} |z^n|^{1/n} = |z| < 1$ and divergent if $|z| > 1$. If $|z| \geqslant 1$ then $|z|^n \geqslant 1$ and so by Theorem 8.2.2, the series $\sum_{n=0}^{\infty} z^n$ is divergent if $|z| \geqslant 1$.

Note that if $z \neq 1$,

$$\sum_{k=0}^{n} z^k = \frac{1 - z^{n+1}}{1 - z}$$

and $z^{n+1} \longrightarrow 0$ as $n \longrightarrow \infty$ if $|z| < 1$. Hence $\sum_{k=0}^{\infty} z^k$ converges to $\frac{1}{1 - z}$ if $|z| < 1$.

EXAMPLE 8.7.2. The series $\sum_{n=0}^{\infty} \frac{z^n}{n!}$ is absolutely convergent for every z.

It is more convenient here to use the ratio test.

We now show that if the series (8.10) converges at a point z_0 then it is absolutely convergent if $|z| < |z_0|$.

Theorem 8.7.1. (*i*) If the series $\sum_{n=0}^{\infty} a_n z^n$ is convergent for $z = z_0 (\neq 0)$ then it is absolutely convergent if $|z| < |z_0|$.

(*ii*) If $\sum_{n=0}^{\infty} a_n z^n$ is divergent for $z = z_1$, then it is divergent if $|z| > |z_1|$.

Proof. (*i*) Since $|a_n z_0^n| < K$, we have

$$|a_n z^n| = |a_n z_0^n| \left| \frac{z}{z_0} \right|^n < K \left| \frac{z}{z_0} \right|^n$$

The series $\Sigma \left| \dfrac{z}{z_0} \right|^n$ is convergent if $|z| < |z_0|$ and so $\displaystyle\sum_0^\infty a_n z^n$ is absolutely convergent if $|z| < |z_0|$.

(*ii*) Let $|z_2| > |z_1|$ and suppose that $\Sigma a_n z_2^n$ is convergent. Then by (*i*), $\Sigma a_n z_1^n$ is absolutely convergent, contradicting our hypothesis. ∎

Theorem 8.7.2. Given a power series $\displaystyle\sum_{n=0}^\infty a_n z^n$, let

$$\limsup_{n \to \infty} |a_n|^{1/n} = L, \mathfrak{R} = \frac{1}{L} \tag{8.11}$$

where $\mathfrak{R} = 0$ if $L = \infty$ and $\mathfrak{R} = \infty$ if $L = 0$. Then the series converges absolutely if $|z| < \mathfrak{R}$ and diverges if $|z| > \mathfrak{R}$.

Proof. We apply the root test to the series (8.10):

$$\limsup_{n \to \infty} |a_n z^n|^{1/n} = |z| \limsup_{n \to \infty} |a_n|^{1/n} = \frac{|z|}{\mathfrak{R}}. \tag{8.12}$$

Hence the series (8.10) is absolutely convergent if $|z| < \mathfrak{R}$. If $|z| > \mathfrak{R}$ then (8.12) shows that $\lim_{n \to \infty} |a_n z^n| \neq 0$ as $n \longrightarrow \infty$. Hence the series is divergent if $|z| > \mathfrak{R}$. ∎

Note. $\mathfrak{R}$ is called the *radius of convergence* of the power series (8.10) and $|z| = \mathfrak{R}$ is called the *circle of convergence*. At every point z inside this circle (and in particular on the open interval $(-\mathfrak{R}, \mathfrak{R})$ on the real line) the series is absolutely convergent and at every point this circle the series is divergent. At points on this circle the series may be convergent at all points or at some points or at no point.

EXAMPLE 8.7.3. The radius of convergence of the series $\displaystyle\sum_{n=1}^\infty \frac{z^n}{n^2}$ is 1.

This series is absolutely convergent at every point on $|z| = 1$ for $\left| \dfrac{z^n}{n^2} \right| = \dfrac{1}{n^2}$ and

$\displaystyle\sum_{n=1}^\infty \frac{1}{n^2}$ is convergent. The radius of convergence of $\displaystyle\sum_{n=1}^\infty z^n$ is 1 and the series is divergent at every point on $|z| = 1$. The radius of convergence of $\sum\limits_{n=1}^{\infty} \dfrac{z^n}{n}$ is 1, and this series is divergent for $z = 1$ and convergent for $z = -1$. (The reader can prove, with the help of inequalities (8.5) and Abel's lemma that this series is convergent for every z such that $|z| = 1$ and $z \neq 1$.)

Exercise 8.7

1. Determine the radius of convergence of the following series.

(i) $\displaystyle\sum_{n=1}^{\infty} \dfrac{(-1)^{n+1} z^n}{n}$

(ii) $\displaystyle\sum_{n=1}^{\infty} n z^n$

(iii) $1 + \displaystyle\sum_{n=1}^{\infty} \dfrac{a(a+1)\cdots(a+n-1)b(b+1)\cdots(b+n-1)}{n!\,c(c+1)\cdots(c+n-1)} z^n$

  where a, b, c are complex numbers and $c \neq 0, -1, -2, \ldots$.

2. Prove that

$$\liminf_{n \to \infty} \left| \frac{a_n}{a_{n+1}} \right| \leq 1/\limsup_{n \to \infty} |a_n|^{1/n} \leq 1/\liminf_{n \to \infty} |a_n|^{1/n}$$

$$\leq \limsup_{n \to \infty} \left| \frac{a_n}{a_{n+1}} \right|$$

Deduce that if $\Re$ is the radius of convergence of $\sum\limits_{n=0}^{\infty} a_n z^n$, then

$$\liminf_{n \to \infty} \left| \frac{a_n}{a_{n+1}} \right| \leq \Re \leq \limsup_{n \to \infty} \left| \frac{a_n}{a_{n+1}} \right|$$

$\left(\text{Hint: Write } M = \limsup\limits_{n \to \infty} \left| \dfrac{a_n}{a_{n+1}} \right|.\right.$ If $M = \infty$ then there is nothing to prove. Let $M < \infty$. Then

$$\left| \frac{a_n}{a_{n+1}} \right| < M + \epsilon, n > n_0(\epsilon)$$

Multiplying these inequalities we get for $n > n_0$

$$\left| \frac{a_{n_0}}{a_{n_\cdot}} \right| < (M + \epsilon)^{n - n_0}$$

and hence

$$\frac{1}{M + \epsilon} \leqslant \liminf_{n \to \infty} |a_n|^{1/n}$$

This implies the extreme right-hand inequality of the first part. The extreme left-hand inequality can be proved similarly.

8.8. Multiplication of Series

Suppose that $\sum_{n=0}^{\infty} a_n$ and $\sum_{n=0}^{\infty} b_n$ are two series of real or complex numbers.

If we multiply formally the two power series $\sum_{n=0}^{\infty} a_n z^n$ and $\sum_{n=0}^{\infty} b_n z^n$ we obtain

the series $\sum_{n=0}^{\infty} c_n z^n$ where

$$c_n = a_0 b_n + a_1 b_{n-1} + \cdots + a_n b_0. \tag{8.13}$$

The series $\sum_{n=0}^{\infty} c_n$ is called the *Cauchy product* of the two series $\sum_{n=0}^{\infty} a_n$ and

$\sum_{n=0}^{\infty} b_n$. If the two series $\sum_{n=0}^{\infty} a_n$ and $\sum_{n=0}^{\infty} b_n$ are both conditionally convergent

then $\sum_{n=0}^{\infty} c_n$ may not converge.

EXAMPLE 8.8.1. Let

$$a_n = b_n = \frac{(-1)^{n+1}}{\sqrt{n}}, \quad n \geqslant 1; \ a_0 = b_0 = 0$$

Then

$$c_n = (-1)^{n+1} \left\{ \frac{1}{\sqrt{1}} \frac{1}{\sqrt{n-1}} + \frac{1}{\sqrt{2}\sqrt{n-2}} + \cdots + \frac{1}{\sqrt{n-1}} \frac{1}{\sqrt{1}} \right\}$$

Since

$$\sqrt{p(n-p)} \leqslant \frac{p+n-p}{2} = \frac{n}{2} \quad \text{for } p = 1, 2, \ldots, n-1$$

we have

$$|c_n| > \frac{2(n-1)}{n} \longrightarrow 2$$

Hence $\displaystyle\sum_{n=0}^{\infty} c_n$ is divergent.

Theorem 8.8.1. Let $\displaystyle\sum_{k=0}^{\infty} a_k = A$ and $\displaystyle\sum_{k=0}^{\infty} b_k = B$. If one of these two series converges absolutely then their Cauchy product $\displaystyle\sum_{k=0}^{\infty} c_k$ converges to AB.

Proof. Without loss of generality we may assume that $\displaystyle\sum_{k=0}^{\infty} |a_k| = M < \infty$.

Let

$$A_n = \sum_{k=0}^{n} a_k, B_n = \sum_{k=0}^{n} b_k, B_n - B = \beta_n, \text{ and } C_n = \sum_{k=0}^{n} c_k.$$

Thus

$$\begin{aligned} C_n &= c_0 + c_1 + \cdots + c_n \\ &= (a_0 b_0) + (a_0 b_1 + a_1 b_0) + \cdots + (a_0 b_n + \cdots + a_n b_0) \end{aligned}$$

Since this is a finite sum we can rearrange the terms without affecting the sum and obtain

$$\begin{aligned} C_n &= a_0(b_0 + \cdots + b_n) + a_1(b_0 + \cdots + b_{n-1}) + \cdots + a_n b_0 \\ &= a_0(B + \beta_n) + a_1(B + \beta_{n-1}) + \cdots + a_n(B + \beta_0) \end{aligned}$$

(i)
$$= BA_n + \sum_{k=0}^{n} a_k \beta_{n-k}$$

Write $\displaystyle\gamma_n = \sum_{k=0}^{n} a_k \beta_{n-k}$. Then $\displaystyle\lim_{n \to \infty} \beta_n = 0$ and $|\beta_n| \leqslant M_1$ for some M_1 and $n = 0, 1, \ldots$. Given $\epsilon > 0$ choose $n_0(\epsilon)$ such that $|\beta_n| < \dfrac{\epsilon}{2M}$ for $n \geqslant n_0$. Then

$$|\gamma_n| \leqslant |a_0\beta_n + \cdots + a_{n-n_0-1}\beta_{n_0+1}| + |a_{n-n_0}\beta_{n_0} + \cdots + a_n\beta_0|$$

$$< \frac{\epsilon}{2M}(|a_0| + |a_1| + \cdots + |a_{n-n_0-1}|) + \left|\sum_{k=n-n_0}^{n} a_k\beta_{n-k}\right|$$

Fix n_0 and choose n_1 such that for $n > n_1$:

$$\left|\sum_{k=n-n_0}^{n} a_k\beta_{n-k}\right| < (n_0+1)M_1 \max_{n-n_0 \leqslant k \leqslant n} (|a_k|)$$

$$< \epsilon/2$$

Then for $n > \max(n_0, n_1)$,

$$|\gamma_n| < \frac{\epsilon}{2M}M + \frac{\epsilon}{2} = \epsilon$$

Since $A_n \longrightarrow A$ we have from (i) that $\lim\limits_{n \to \infty} C_n = AB$.

Theorem 8.8.2. If the series $\sum\limits_0^{\infty} a_k = A$ and $\sum\limits_0^{\infty} b_k = B$ converge absolutely,

then their Cauchy product $\sum\limits_0^{\infty} c_k = C$ converges absolutely, and $AB = C$.

Proof. By Theorem 8.8.1, the Cauchy product of the series $\sum\limits_0^{\infty} |a_k|$ and

$\sum\limits_{k=0}^{\infty} |b_k|$ is convergent; that is, $\sum\limits_0^{\infty} (|a_0| \, |b_n| + |a_1| \, |b_{n-1}| + \cdots + |a_n| \, |b_0|)$ is convergent. Hence the Cauchy product of the original series converges absolutely.

The second part $C = AB$ follows also from Theorem 8.8.1. $\blacksquare$

Miscellaneous Exercises for Chapter 8

1. Examine for convergence the series

(i) $\sum\limits_{n=0}^{\infty} \frac{1}{(n+1)^2 - n^2}$

(ii) $\displaystyle\sum_{n=1}^{\infty} \frac{\sqrt{n+1} - \sqrt{n}}{n}$

(iii) $\displaystyle\sum_{n=1}^{\infty} \frac{1}{n^{1+1/n}}$

(iv) $\displaystyle\sum_{n=1}^{\infty} \frac{n^n}{n!}$

(v) $\displaystyle\sum_{n=2}^{\infty} \frac{(-1)^n (\log n)}{\sqrt{n}}$

2. Prove that if $\Sigma \, |a_n|^2$ and $\Sigma \, |b_n|^2$ are convergent then $\Sigma \, a_n b_n$ is absolutely convergent.

3. Suppose $D_n > 0, a_n > 0$ and let

$$t(n) = D_n \frac{a_n}{a_{n+1}} - D_{n+1}$$

Prove the following

(i) $\displaystyle\sum_{n=1}^{\infty} a_n$ is convergent if $\displaystyle\liminf_{n \to \infty} t(n) > 0$

(ii) $\displaystyle\sum_{n=1}^{\infty} a_n$ is divergent if $\displaystyle\limsup_{n \to \infty} t(n) < 0$ and $\displaystyle\sum_{1}^{\infty} \frac{1}{D_n} = \infty$

$\Bigg($ *Hint* (i): Let

$$\liminf t(n) = l > 0.$$

Then

$$t(n) > \frac{l}{2} \quad \text{for } n > m$$

Hence

$$D_n a_n - D_{n+1} a_{n+1} > \frac{l}{2} a_{n+1}, n > m.$$

Adding these inequalities for $n = m + 1, m + 2, \ldots, m + p$ we have

$$D_{m+1}a_{m+1} - D_{m+p+1}a_{m+p+1} > \frac{l}{2} \sum_{k=m+2}^{m+p+1} a_k$$

Hence

$$D_{m+1}a_{m+1} > \frac{l}{2} \sum_{k=m+2}^{m+p+1} a_k$$

and this implies that

$$S_n = \sum_{k=1}^{n} a_k$$

is bounded.

Hint (ii): If

$$\lim \sup t\,(n) = L < 0,$$

then

$$t\,(n) < 0 \text{ for } n > N.$$

Hence

$$D_{N+1}\,a_{N+1} < D_{N+2}\,a_{N+2} < \cdots < D_n a_n$$

and this proves the divergence of $\Sigma\,a_n$, for

$$a_n > \frac{K}{D_n} \text{ when } n > N \text{ and } K = D_{N+1}\,a_{N+1} \cdot \Big)$$

4. Prove that the series

$$\sum_{n=1}^{\infty} \frac{1.3.5. \ldots (2n-1)}{2.4.6. \ldots (2n)}$$

is divergent.

5. Suppose

$$\sum_{n=1}^{\infty} c_n \text{ converges, } c_n > 0$$

and

$$r_n = \sum_{k=n+1}^{\infty} c_k.$$

Prove the following:

(i) $\displaystyle\sum_{n=1}^{\infty} \frac{c_n}{\sqrt{r_{n-1}}}$ converges

(ii) $\displaystyle\sum_{n=1}^{\infty} \frac{c_n}{r_{n-1}}$ diverges

$\left(\text{\textit{Hint}}:\right.$

$$\frac{c_n}{\sqrt{r_{n-1}}} = \frac{r_{n-1} - r_n}{\sqrt{r_{n-1}}}$$

$$= \frac{\sqrt{r_{n-1}} - \sqrt{r_n}}{1} \cdot \frac{\sqrt{r_{n+1}} + \sqrt{r_n}}{\sqrt{r_{n-1}}}$$

$$< 2(\sqrt{r_{n-1}} - \sqrt{r_n}) \qquad \frac{c_n}{r_{n-1}} + \frac{c_{n+1}}{r_n} + \cdots + \frac{c_{n+p}}{r_{n+p-1}}$$

$$= \frac{r_{n-1} - r_{n+p}}{r_{n-1}}$$

$$= 1 - \frac{r_{n+p}}{r_{n-1}}$$

$\left.\text{Note that } r_n \downarrow 0.\right)$

9

Derivatives and Integrals

9.1. Derivatives

One of the most fundamental operations in calculus is that of differentiation. It is safe to assume that the reader has had some experience with it. However, we shall approach this concept from a broader base and generalize it. We do not intend to discuss the applications of "derivatives." Again, only the functions of real variables will be used.

Definition 9.1.1. *Let $f: D \longrightarrow R$ be a function, where D is a subset of reals. Let x_0 be a limit point of D which is in D. Then a real number d is called the derivative of f at x_0, if $d = \lim\limits_{x \to x_0} \dfrac{f(x) - f(x_0)}{x - x_0}$.*

We also write $d = f'(x_0)$ or $Df(x_0)$ or $(df/dx)_{x=x_0}$.

Derivatives at isolated points are not defined. It may be recalled, however, that a function is always continuous at an isolated point of the domain. The domain of the derivative of $f: D \longrightarrow R$ is contained in $D \cap D'$, where D' is the set of all limit points of D. In most cases D is an interval.

A function which has a derivative at a point is called *differentiable* at that point.

Definition 9.1.2. *A function $f: D \longrightarrow R$ is said to be differentiable on a set $S \subset D$, if the restriction of f to S is differentiable at every point of S.*

Many functions fail to have derivatives at certain points. For such functions the idea of Dini derivates is found to be useful. In discussing these derivates we use the notation $x \longrightarrow x_0+$ which means x approaches x_0 through values greater than x_0. A similar meaning is attached to the symbol $x \longrightarrow x_0-$.

Definition 9.1.3. *Let $f: D \longrightarrow R$ be a function and let $x_0 \in D \cap D'$. Consider the following limits:*

$$D^+f(x_0) = \limsup_{x \to x_0^+} \frac{f(x) - f(x_0)}{x - x_0}$$

$$D_+f(x_0) = \liminf_{x \to x_0^+} \frac{f(x) - f(x_0)}{x - x_0}$$

$$D^-f(x_0) = \limsup_{x \to x_0^-} \frac{f(x) - f(x_0)}{x - x_0}$$

$$D_-f(x_0) = \liminf_{x \to x_0^-} \frac{f(x) - f(x_0)}{x - x_0}$$

These limits are called *upper right derivate*, *lower right derivate*, *upper left derivate*, and *lower left derivate* respectively. They are also called *Dini derivates* of f.

If we include $+\infty$ and $-\infty$ in these limits then we can say that Dini derivates of all functions exist at every interior point of $D \cap D'$.

The following two theorems follow easily from the Def. 9.1.3, and as such their proofs are left as exercises.

Theorem 9.1.1. $D_-f(x) \leqslant D^-f(x)$ and $D_+f(x) \leqslant D^+f(x)$.

Theorem 9.1.2. f' exists at an interior point x_0 if and only if all the four Dini derivates exist (being different from $-\infty$ or ∞) and are equal to each other.
Now we prove the following result:

Theorem 9.1.3. A necessary and sufficient condition for the existence of a derivative of f at an interior point x_0 is that $D^+f(x_0) \leqslant D_-f(x_0)$ and $D^-f(x_0) \leqslant D_+f(x_0)$.

Proof. The condition is obviously necessary, for if f' exists then all the four Dini derivates are equal.

Conversely, if $D^+f(x_0) \leqslant D_-f(x_0)$ and $D^-f(x_0) \leqslant D_+f(x_0)$, then using Theorem 9.1.1 we would have $D^+f(x_0) \leqslant D_-f(x_0) \leqslant D^-f(x_0) \leqslant D_+f(x_0) \leqslant D^+f(x_0)$. This would necessitate the equality of all the four Dini derivates. Hence $f'(x_0)$ exists and the theorem is proved. ∎

It may happen that two of the four Dini derivates may be equal to each other. Therefore, we introduce the following definitions.

Definition 9.1.4. f *is said to have a right derivative at* x_0 *if* $D^+f(x_0) = D_+f(x_0)$. *We may denote the right derivative by* $D_Rf(x_0)$.

A similar definition may be given for a left derivative which may be denoted by $D_Lf(x_0)$.

Definition 9.1.5. *If* $D^+f(x_0) = D^-f(x_0)$, *then* f *is said to have an upper derivative and it is denoted by* $\overline{D}f(x_0)$.

A similar definition may be given for lower derivative $\underline{D}f(x_0)$.

We now give some examples of functions which do not have derivatives at certain points but possess Dini derivates.

EXAMPLE 9.1.1. Let

$$f(x) = x \sin \frac{1}{x} \text{ for } x > 0$$

$$f(0) = 0$$

$$f(x) = x \sin \frac{1}{x} + 3x \text{ for } x < 0$$

Here,

$$D^+f(0) = \lim_{x \to 0+} \sup \left[\frac{x \sin \frac{1}{x} - 0}{x} \right] = 1$$

$$D_+f(0) = \lim_{x \to 0+} \inf \left[\frac{x \sin \frac{1}{x} - 0}{x} \right] = -1$$

$$D^-f(0) = \lim_{x \to 0-} \sup \left[\frac{\left(x \sin \frac{1}{x} + 3x \right) - 0}{x} \right] = 4$$

$$D_-f(0) = \lim_{x \to 0-} \inf \left[\frac{\left(x \sin \frac{1}{x} + 3x \right) - 0}{x} \right] = 2$$

All the four Dini derivates are unequal, and therefore none of the upper derivative, lower derivative, right derivative, left derivative exists.

EXAMPLE 9.1.2. Let

$$f(x) = |x|$$

In this case,

$$D^+f(0) = 1 = D_+f(0)$$

and

$$D^-f(0) = -1 = D_-f(0)$$

Thus the right derivative of f at 0 exists and so does the left derivative, but they are not equal.

Upper and lower derivatives do not exist at $x = 0$.

EXAMPLE 9.1.3. Let

$$f(x) = x \sin \frac{1}{x} \ (x \neq 0)$$

$$f(0) = 0$$

The upper derivative of f at $x = 0$ is 1, and the lower derivative is -1.

The right and left derivatives do not exist at $x = 0$.

The following example shows that

$D^+ [f(x) + g(x)]$ is not necessarily equal to $D^+ f(x) + D^+ g(x)$

EXAMPLE 9.1.4. Let

$$f(x) = x \sin \frac{1}{x} \ (x \neq 0)$$

and

$$f(0) = 0$$

Let

$$g(x) = -x \sin \frac{1}{x} \ (x \neq 0)$$

and

$$g(0) = 0$$

$$f(x) + g(x) = 0 \qquad \text{for all } x$$

Hence

$$D^+ [f(0) + g(0)] = 0$$

But

$$D^+ f(0) = 1 \quad \text{and} \quad D^+ g(0) = 1$$

By some similar examples, it can be shown that

$D^- [f(x) + g(x)]$ is not necessarily equal to $D^- f(x) + D^- g(x)$,

and

$D_+ [f(x) + g(x)]$ is not necessarily equal to $D_+ f(x) + D_+ g(x)$,

Also,

$D_- [f(x) + g(x)]$ is not necessarily equal to $D_- f(x) + D_- g(x)$.

It is well known in elementary calculus that every function which has a derivative at a point must be continuous at that point. *The converse, of course, is not true.* There are many examples to illustrate this point. The function $f(x) = |x|$ (Example 9.1.2) is continuous at $x = 0$, yet it does not have a derivative at that point. Example 9.1.3 is another illustration, since it can be shown to be continuous at $x = 0$, and is not differentiable there. Of course, if the domain of a function has an isolated point then the function would be continuous there without being differentiable.

If a function is continuous on a connected set and does not have a derivative at one of its points, then, geometrically, it means that there is a sharp corner in the graph of that function at that particular point; (since the existence of the derivative at that point would have guaranteed a unique tangent at that point thereby making the curve smooth). Normally one may expect a continuous function to have only a few sharp corners (if any); but the great analyst Karl T. Weierstrass surprised the mathematical world by constructing a function which was continuous everywhere but did not have a derivative at any point. Later on some more functions like that of Weierstrass were constructed by Van der Waerdan and others. One such function is as follows:

EXAMPLE 9.1.5

$$f(x) = \sum_{n=1}^{\infty} \frac{1}{2^n} \cos (3^n x)$$

is continuous everywhere, since the series is uniformly convergent (cf. Sec. 11.4) and $\frac{1}{2^n} \cos 3^n x$ is continuous. However, it can be shown that f does not have a derivative anywhere.

The following theorem is a generalization of a well known theorem in calculus.

Theorem 9.1.4. If the maximum of a function $f : D \longrightarrow R$ occurs at x_0 then $D_+ f(x_0) \leqslant 0$ (if it exists at x_0), and $D^- f(x_0) \geqslant 0$ (if it exists at x_0).

The proof of this theorem follows directly from the Def. 9.1.3.

The following is an obvious but significant corollary of the last theorem.

Corollary. If the maximum of a function $f : D \longrightarrow R$ occurs at an interior point x_0 of D and f' exists at x_0, then $f'(x_0) = 0$.

A similar result holds for the minimum of a function.

We now come to Rolle's theorem which is given in every Calculus book.

Theorem 9.1.5. (Rolle's Theorem). If a function f is continuous on a closed interval $[a,b]$, and if it is differentiable on the open interval (a,b) then $f(a) = f(b) \Longrightarrow$ there is a point x_0 in (a,b) such that $f'(x_0) = 0$.

The proof of this theorem is left as an exercise. (Use Theorem 9.1.4 and Theorem 7.2.3.)

It may be remarked here that the conditions of "the existence of the derivative of f on (a,b)" and that of "continuity of f on $[a,b]$" are independent of each other.

For example, $|x|$ is continuous on $[-1,1]$, yet its derivative does not exist on $(-1,1)$ and thus the conclusion of the Rolle's theorem is not satisfied;

Whereas if we consider

$$f(x) = \sin \frac{1}{x} \qquad x \neq 0$$

$$f(0) = 0$$

then f has a derivative at every point $\left(0,\frac{1}{\pi}\right)$, but is not continuous on $\left[0,\frac{1}{\pi}\right]$. In spite of this, the conclusion of Rolle's theorem is valid.

The function $\sqrt{c^2 - x^2}$ $(c > 0)$ is continuous on $[-c,c]$, but is differentiable only on $(-c,c)$ (not at $x = \pm c$). Since its values at $-c$ and c are equal, the conclusion of the Rolle's theorem holds, indeed.

Rolle's theorem may be generalized as follows.

Theorem 9.1.6 (Cauchy Mean Value Theorem). Let f and g be two functions which are continuous on $[a,b]$ and have derivatives on (a,b). Then, there is a point x_0 in (a,b) such that

$$[g(b) - g(a)] f'(x_0) = [f(b) - f(a)] g'(x_0)$$

Proof. If $f(b) = f(a)$ then by the use of Rolle's theorem a point x_0 can be found in (a,b) such that $f'(x_0) = 0$, and the conclusion of the theorem is true since both sides will be equal.

Let us now assume that $f(b) \neq f(a)$. In that case let $h(x) = \lambda f(x) - g(x)$, where λ is so determined that $h(a) = h(b)$ or $\lambda [f(b) - f(a)] = g(b) - g(a)$. The rest of the proof may be easily completed by the application of Rolle's theorem and is left as an exercise.

It is interesting to note that the Rolle's theorem may be easily obtained from the last theorem by letting $g(x) = x$ and $f(b) = f(a)$. However, we could not easily prove the last theorem without using the Rolle's theorem. This is one of those cases where we have to use a special case to prove the general result.

For a function of a real variable the existence of the derivative does not guarantee the continuity of the derivative, as illustrated by the following example.

EXAMPLE 9.1.7. Let

$$f(x) = x^2 \cos \frac{1}{x} \qquad (x \neq 0)$$

and

$$f(0) = 0$$

In this case

$$f'(0) = \lim_{x \to 0} \frac{x^2 \cos \dfrac{1}{x} - 0}{x}$$

exists and is equal to zero, whereas,

$$f'(x) = 2x \cos \frac{1}{x} + \sin \frac{1}{x}$$

and therefore $\lim_{x \to 0} f'(x)$ does not exist.

In the last example, the existence of f' on $[a,b]$ did not make it (the derivative) continuous. However, the image of f' is always connected, and as such f' satisfies the intermediate value property. We prove it in the next theorem. Compare it with Theorem 7.3.5.

Theorem 9.1.7 (Intermediate Value Theorem for a Derivative). If f' exists on $[a,b]$ then the image of $f' : [a,b] \longrightarrow R$ is connected.

Proof. In view of Theorem 5.4.3, we are supposed to show that if $f'(c)$ and $f'(d)$ are two values of the derivative and if $f'(c) < f'(d)$ (c,d are in $[a,b]$), then for every y_0 for which $f'(c) < y_0 < f'(d)$ there exists x_0 between c and d such that $f'(x_0) = y_0$.

Let us assume $c < d$. Consider the function

$$g(x) = y_0 x - f(x)$$

Now g' exists on $[c,d] \subset [a,b]$ and as such is continuous on $[c,d]$. By Theorem 7.2.3 it attains its maximum on $[c,d]$. If c is the maximum then $g'(c) = D_+ g(c) \leqslant 0$ (Theorem 9.1.4). But $g'(c) = y_0 - f'(c) > 0$. Therefore c cannot be the maximum of g. Similarly, we can show that d is not the maximum of g, either. Thus the maximum of g must occur at an interior point x_0 of $[c,d]$ (which would also be the interior point of $[a,b]$). By the corollary of Theorem 9.1.4, $g'(x_0) = 0$, which means that $f'(x_0) = y_0$.

The reasoning would be similar for the case when $d < c$, and that completes the proof. ∎

Exercise 9.1

1. Prove that if $f : D \longrightarrow R$ is a function, then the Dini derivates exist at every interior point of D or are $\pm \infty$.

2. Let $f : \left(-\dfrac{\pi}{2}, \dfrac{\pi}{2}\right) \longrightarrow R$ be a function defined as follows:

$$f(x) = x, \quad x \text{ rational in } \left(-\frac{\pi}{2}, \frac{\pi}{2}\right)$$

$$f(x) = \tan x, \quad x \text{ irrational in } \left(-\frac{\pi}{2}, \frac{\pi}{2}\right)$$

Show that all the four Dini derivates are equal at $x = 0$. At what other points (if any) does the derivative exist?

3. Let $f : [a,b] \longrightarrow R$ be a function. Prove that if f has a left derivative at $c \in [a,b]$ then the *restriction* of f to $[a,c]$ is continuous at c.

4. Without using Cauchy mean value theorem (Theorem 9.1.6) prove that if f is continuous on $[a,b]$ and f' exists on (a,b) then there is a point $x_0 \in (a,b)$ such that

$$f'(x_0) = \frac{f(b) - f(a)}{b - a}$$

5. Let $f : [0,2] \longrightarrow R$ be defined as follows.

$$f(x) = 0 \text{ for } x \in [0,1)$$
$$f(x) = 2 \text{ for } x \in [1,2]$$

Does there exist a function $g : [0,2] \longrightarrow R$ such that $g'(x) = f(x)$? (*Hint:* use the mean value theorem of Exercise 4.)

6. For the function $f : [a,b] \longrightarrow R$ what can you say about the left derivative at a or the right derivative at b?

7. Prove that if $f'(x_0) < 0$, then there exists a neighborhood of x_0 in which f in monotonically decreasing. Show by means of an example that the converse is not necessarily true.

8. Prove Theorems 9.1.1, 9.1.2, and 9.1.5.

9. Complete the proof of Theorem 9.1.6.

10. Let

$$f(x) = x^{3/2} \quad \text{if } x \text{ is rational}$$
$$f(x) = 0 \quad \text{if } x \text{ is irrational}$$

Find $f'(0)$ if it exists.

9.2. Riemann Integration

Definite integration of functions over a closed interval as discussed in elementary calculus texts is one of many types, and is called Riemann integration. This idea is due to Bernhard Riemann (1826–66). Riemann was one of the three main contributors to analytic function theory mentioned earlier in this book, the other two being Cauchy and Weierstrass. Riemann is even better known for his very significant contributions to geometry.

In this section we are going to give a rigorous discussion of the theory of Riemann integration. We start with some definitions.

Definition 9.2.1. *A partition or a subdivision of a closed interval* $[a,b]$ *is a finite suborder of the linear order of the closed interval such that a is always the first and b is always the last element of this suborder.*

We may describe a partition of $[a,b]$ as follows:

$$a = x_0 < x_1 < x_2 < \cdots < x_n = b$$

A partition like this may be denoted by P, and the points $x_0, \ldots, x_n$ are called the points of the subdivision.

Definition 9.2.2. *A refinement P' of a partition P of $[a,b]$ is a partition of the same closed interval such that P is a suborder of P'.*

In other words, a refinement of a partition P is itself a partition containing all the points of the subdivision of P and may have some other points of $[a,b]$ in it.

We now define Darboux sums for a bounded function on $[a,b]$. In this section we shall deal with bounded functions only unless otherwise specified.

Definition 9.2.3. *Let f be a bounded function defined over a closed interval* $[a,b]$ *and let* $P = \{a = x_0 < x_1 < x_2 < \cdots < x_n = b_n\}$ *be a partition of* $[a,b]$. *Let* $m_k = \text{glb} \{f(x) : x \in [x_{k-1}, x_k]\}$,

$$m_n = \text{glb} \{f(x) : x \in [x_{n-1}, x_n]\}$$

and

$$M_k = \text{lub} \{f(x) : x \in [x_{k-1}, x_k]\}.$$
$$M_n = \text{lub} \{f(x) : x \in [x_{n-1}, x_n]\}.$$

Then $\sum_{k=1}^{n} m_k (x_k - x_{k-1})$ *and* $\sum_{k=1}^{n} M_k (x_k - x_{k-1})$ *are called lower and upper* Darboux sums, respectively, of f for the partition P; and they are denoted by s(f,P) and S(f,P) respectively.

One may notice easily that these sums depend upon the function and the partition. It may also be observed that we have not assumed the continuity of the function. These sums do exist, however, for any bounded function.

It follows trivially from the definition that $s(f,P) \leqslant S(f,P)$.

In the next theorem we prove that any refinement of P does not lower the lower sum and does not raise the upper sum.

Theorem 9.2.1. If P' is a refinement of a partition P of $[a,b]$ and f is a bounded function defined over this closed interval, then $s(f,P) \leqslant s'(f,P')$ and $S(f,P) \geqslant S'(f,P')$.

Proof. If $P' = P$ then the equalities will hold. We, therefore, assume that P' has at least one more point than P has. Suppose such a point is x'_j in jth sub-interval $[x_{j-1}, x_j)$.

Let $m'_j = \text{glb} \ \{f(x) : x \in [x_{j-1}, x'_j)\}$ and $m''_j = \text{glb} \ \{f(x) : x \in [x'_j, x_j)\}$. Then it is obvious that $m_j \leqslant m'_j$ and also $m_j \leqslant m''_j$; and thus

$$m_j(x_j - x_{j-1}) \leqslant m'_j(x'_j - x_{j-1}) + m''_j(x_j - x'_j).$$

Taking into consideration all such additional points it follows that $s(f,P) \leqslant s'(f,P')$. Similarly, we can show that $S(f,P) \geqslant S'(f,P')$ and the proof is complete. ∎

Since the closed interval $[a,b]$ may be regarded a partition of itself $\{a = x_0 < x_1 = b\}$, we have the following corollary.

Corollary. For any partition P of $[a,b]$ and for any function $f: [a,b] \longrightarrow R$, we have $m(b - a) \leqslant s(f,P) \leqslant S(f,P) \leqslant M(b - a)$, where m and M are the greatest lower bound and the least upper bound, respectively, of f on $[a,b]$.

In the next theorem we prove that the lower sum for a partition is always less than or equal to the upper sum for any partition.

Theorem 9.2.2. If f is bounded function defined over a closed interval, and if P_1 and P_2 are two partitions of this interval, then $s_1(f,P_1) \leqslant S_2(f,P_2)$ and $s_2(f,P_2) \leqslant S_1(f,P_1)$.

Proof. Let P_3 be a partition which is a refinement of P_1 and P_2 (Inclusion of all the points of subdivision of P_1 and P_2 would assure the existence of such a partition).

By the last theorem $s_1(f,P_1) \leqslant s_3(f,P_3)$, and $S_3(f,P_3) \leqslant S_2(f,P_2)$. But $s_3(f,P_3) \leqslant S_3(f,P_3)$. Therefore, $s_1(f,P_1) \leqslant s_3(f,P_3) \leqslant S_3(f,P_3) \leqslant S_2(f,P_2)$; which implies that $s_1(f,P_1) \leqslant S_2(f,P_2)$.

Similarly, we show that $s_2(f,P_2) \leqslant S_1(f,P_1)$, and the proof is complete.

Summing up the last two theorems we observe that for a bounded function defined on a closed interval, every lower Darboux sum is less than or equal to every upper Darboux sum; and increasing the number of points of a subdivision does not decrease (raise) the lower (upper) sum. Therefore, every upper sum is an upper bound of the class of all the lower sums for a given function on a fixed interval; and, of course, every lower sum is a lower bound of the class of all the upper sums.

Denoting the class of all partitions of $[a,b]$ by $\mathcal{P}$ we introduce the following definitions:

Definition 9.2.4. *Let*

$$R \underline{\int}_a^b f = \text{lub } \{s(f,P): P \in \mathcal{P}\}$$

and

$$R \overline{\int}_a^b f = \text{glb } \{S(f,P): P \in \mathcal{P}\}$$

$R \underline{\int}_a^b f$ *and* $R \overline{\int}_a^b f$ *are called lower and upper Riemann integrals, respectively.* *

From the corollary of Theorem 9.2.2 we have the following result:

Theorem 9.2.3. *If* f *is a bounded function defined on a closed interval* $[a,b]$ *and* P, *a partition of this interval, then* $m(b-a) \leqslant s(f,P) \leqslant R \underline{\int}_a^b f \leqslant R \overline{\int}_a^b f \leqslant S(f,P) \leqslant M(b-a).$

The proof is a simple exercise.

We now define Riemann integral of a function.

Definition 9.2.5. *A bounded function* f *is said to be Riemann-integrable over* $[a,b]$ *if* $\underline{\int}_a^b f = \overline{\int}_a^b f.$

The common value is called the Riemann integral of f, and is denoted by $\int_a^b f$. It may be remarked here that we use the symbol $\int_a^b f$ instead of $\int_a^b f(x)dx$.

Not all bounded functions are Riemann integrable.

EXAMPLE 9.2.1. Let

$$f(x) = 1 \quad \text{if} \quad x \in [0,1] \text{ and } x \text{ is rational}$$

and

$$f(x) = 0 \quad \text{if} \quad x \in [0,1] \text{ and } x \text{ is irrational}$$

Then for every partition of $[0,1]$, $m_k = 0$ and $M_k = 1$ in all the subintervals.

---

*We may drop R from $R \underline{\int}_a^b$ and $R \overline{\int}_a^b$ in this chapter.

This implies $s(f,P) = 0$ and $S(f,P) = 1$ for every partition P. Therefore, $\displaystyle\int_{\underline{a}}^{b} f =$

0 and $\displaystyle\int_{a}^{\overline{b}} f = 1$; hence f is not Riemann integrable over $[0,1]$.

The function discussed here is called the characteristic function of the set of all rationals in $[0,1]$.

The following result gives a necessary and sufficient condition for the existence of Riemann integral of a function.

Theorem 9.2.4. A bounded function f is R-integrable on $[a,b]$ $\Longleftrightarrow$ for every $\epsilon > 0$ there is a partition P such that $S(f,P) - s(f,P) < \epsilon$.

Proof. If f is R-integrable on $[a,b]$ then $\displaystyle\int_{\underline{a}}^{b} f = \int_{a}^{\overline{b}} f$. Using the properties of the greatest lower bound and the least upper bound, respectively; we have for $\epsilon > 0$, a partition P_1 and a partition P_2 such that

$$s_1(f,P_1) + \frac{\epsilon}{2} > \int_{\underline{a}}^{b} f$$

and

$$S_2(f,P_2) < \int_{a}^{\overline{b}} f + \frac{\epsilon}{2}\,.$$

Now, let P be a refinement of both P_1 and P_2. Then $S(f,P) \leqslant S_2(f,P_2) <$

$$\int_{a}^{\overline{b}} f + \frac{\epsilon}{2} = \int_{\underline{a}}^{b} f + \frac{\epsilon}{2} < s_1(f,P_1) + \frac{\epsilon}{2} + \frac{\epsilon}{2} \leqslant s(f,P) + \epsilon.$$

Therefore, $S(f,P) - s(f,P) < \epsilon$.

Conversely, if for every $\epsilon > 0$ there is a partition P such that $S(f,P) - s(f,P) < \epsilon$, then from Theorem 9.2.3 it follows that $\displaystyle\int_{a}^{\overline{b}} f - \int_{\underline{a}}^{b} f < \epsilon$ for every ϵ,

which means that $\displaystyle\int_{a}^{\overline{b}} f \leqslant \int_{\underline{a}}^{b} f$. But $\displaystyle\int_{\underline{a}}^{b} f \leqslant \int_{a}^{\overline{b}} f$, and thus $\displaystyle\int_{a}^{\overline{b}} f = \int_{\underline{a}}^{b} f$, implying that f is R-integrable. The proof is complete. ∎

This theorem is useful in proving various other results, but one disadvantage of this theorem is that in practice it is not always convenient to use the criterion of this theorem to determine the Riemann integrability of an arbitrary function.

The following theorem follows easily from the definition and its proof is left as a simple exercise.

Theorem 9.2.5. If $a < c < b$, then for any bounded function f

$$\underline{\int_a^b} f = \underline{\int_a^c} f + \underline{\int_c^b} f$$

$$\overline{\int_a^b} f = \overline{\int_a^c} f + \overline{\int_c^b} f$$

and if f is R-integrable, then

$$\int_a^b f = \int_a^c f + \int_c^b f$$

Next we prove that the continuity of a function is a sufficient condition for its integrability, (in the sense of Riemann).

Theorem 9.2.6. If a function f is continuous on $[a,b]$ then it is Riemann integrable on $[a,b]$.

Proof. By Theorem 7.2.2 f is bounded on $[a,b]$ and attains its glb and lub on $[a,b]$ and on every closed subinterval of it. Furthermore, f is uniformly continuous on $[a,b]$ because of Theorem 7.4.1. Thus for $\epsilon > 0$ there is a $\delta > 0$ such that $|f(x') - f(x'')| < \dfrac{\epsilon}{b - a}$ for $|x' - x''| < \delta$. Now let $n > \dfrac{b - a}{\delta}$, and divide the interval $[a,b]$ into n equal parts, which would mean that the length of each subinterval would be less than δ. Let this partition be denoted by P.

Using the usual notation, m_k and M_k would represent the glb and lub of f on $[x_k, x_{k+1}]$. Let $f(x_k') = m_k$ and $f(x_k'') = M_k$. But $|x_k' - x_k''| < \delta$, and therefore

$$M_k - m_k < \frac{\epsilon}{b - a}. \text{ Now, } S(f,P) - s(f,P) = \sum_{k=1}^n (M_k - m_k) \Delta x_k$$

$$< \sum_{k=1}^n \left(\frac{\epsilon}{b - a} \right) \Delta x_k = \frac{\epsilon}{b - a} \sum_k^n \Delta x_k = \frac{\epsilon}{b - a} (b - a)$$

Hence $S(f,P) - s(f,P) < \epsilon$, and so f is integrable on $[a,b]$ and that completes the proof. ∎

The converse of this theorem is not necessarily true. As an example, we have

EXAMPLE 9.2.2

Let $f(x) = 2$ for $x \in [0,1)$
and $f(x) = 3$ for $x \in [1,2]$
It is then easy to show that for $\epsilon > 0$ there is a partition P such that it con-

tains a subinterval containing 1 and whose length is less than ϵ. Then it would follow that $S(f,P) - s(f,P) < \epsilon$. Hence f is Riemann-integrable.

A function can afford to be discontinuous at infinitely many points as long as the length (measure) of the set of points of discontinuities is zero. For example, a function can be discontinuous at every point of the Cantor set and still it may be Riemann-integrable on $[0,1]$.

Definition 9.2.6. *A function f is said to be monotonically nondecreasing (nonincreasing) if $x < y \Longrightarrow f(x) \leqslant f(y)$ $(f(x) \geqslant f(y))$. In either case it is called monotonic.*

The next theorem is useful too.

Theorem 9.2.7. Every bounded monotonic function $f: [a,b] \longrightarrow R$ is Riemann integrable on $[a,b]$.

Proof. It suffices to prove the result for the case of a monotonically non-decreasing function since the argument would be similar for a nonincreasing function.

In case $f(a) = f(b)$ then f would be constant, and therefore integrable.

Let us now assume that $f(a) < f(b)$. For $\epsilon > 0$ construct a partition P of $[a,b]$ such that the length of the each subinterval is less than $\dfrac{\epsilon}{f(b) - f(a)}$.

Now $S(f,P) - s(f,P) = \displaystyle\sum_{k=1}^{n} (M_k - m_k)\Delta x_k \ (\Delta x_k = x_k - x_{k-1})$

$$= \sum [f(x_k) - f(x_{k-1})] \Delta x_k$$

But $\Delta x_k < \dfrac{\epsilon}{f(b) - f(a)}$.

Thus $S(f,P) - s(f,P) < \dfrac{\epsilon}{f(b) - f(a)} \displaystyle\sum_{k=1}^{n} [f(x_k) - f(x_{k-1})]$

$$= \frac{\epsilon}{f(b) - f(a)} [f(b) - f(a)] = \epsilon$$

Hence f is Riemann integrable over $[a,b]$ and the proof is complete. ∎

Another useful approach to this type of integral is via Riemann approximating sum (as is done in the elementary calculus). We define it as follows.

Definition 9.2.7. *Let $\{P : a = x_0 < x_1 < x_2 < \cdots x_n = b\}$ be a partition of $[a,b]$, and let $\xi_k \in [x_{k-1}, x_k)$. Then $\displaystyle\sum_{k=1}^{n} f(\xi_k) \Delta x_k$ is called a Riemann approximating sum of f for the partition P. We may denote it by $R(f,P)$.*

It is obvious that

$$s(f,P) \leqslant R(f,P) \leqslant S(f,P)$$

This set of inequalities has some geometrical interpretation (what?).

One may now define the Riemann integral of f as $\lim\limits_{\substack{n \to \infty \\ \Delta \to 0}} \sum\limits_{k=1}^{n} f(\xi_k)\Delta x_k$ *if the limit exists, and this would be the same as given in* Def. 9.2.5; *where* Δ *stands for* max $\{\Delta x_k : k = 1,2, \ldots, n\}$.

If $b < a$, then it is convenient to write $\int_a^b f = -\int_b^a f$ and we can define $\int_a^a f = 0$. Thus $\int_a^b f$ has a meaning for even $b \leqslant a$.

We now recall an important result of elementary calculus known as the "fundamental theorem of calculus."

Theorem 9.2.8 (Fundamental Theorem of Calculus).   First form: If f is continuous on $[a,b]$ and if

$$F(x) = \int_a^x f(t)\,dt \qquad \text{for } x \in [a,b]$$

then

$$F' = f \qquad \text{on } [a,b]$$

Second form: If f is continuous on $[a,b]$ and $F' = f$ on $[a,b]$, then

$$\int_a^b f = F(b) - F(a)$$

The proofs are left as exercises.

We shall conclude this section by discussing Taylor expansion.

Earlier in this chapter we gave an example (Example 9.1.7) of a function which is differentiable at a point ($x = 0$), yet its derivative is not continuous at that point, let alone being differentiable there. However, many functions have derivatives of high orders and some of them have derivatives of all orders. For such functions Taylor's expansion is very useful. First a terminology.

If a function f has derivative of order n on a set S and $f^{(n)}$ is continuous on S (indeed, $f', f'', \ldots, f^{(n-1)}$ would be continuous), then we write $f \in C^n$ on S. If f is continuous on S we simply write $f \in C$ on S.

Theorem 9.2.9 (Taylor's Theorem).   Let $f \in C^n$ in a neighborhood of a point x_0, then for every point in that neighborhood we may write

$$f(x) = f(x_0) + (x - x_0)f'(x_0) + \cdots + \frac{(x - x_0)^{n-1}}{(n - 1)!} f^{(n-1)}(x_0) + R_n(x)$$

where

$$R_n(x) = \frac{1}{(n - 1)!} \int_{x_0}^{x} f^{(n)}(u)(x - u)^{n-1} \, du$$

$R_n(x)$ is called the remainder after the nth term.

Proof. From the hypothesis it follows that all the derivatives of f up to the order n continuous in the given neighborhood of x_0. Now, using elementary calculus (integration by parts), we can easily show that

$$R_n(x) - R_{n-1}(x) = -\frac{(x - x_0)^{n-1}}{(n - 1)!} f^{(n-1)}(x_0)$$

Similarly,

$$R_{n-1}(x) - R_{n-2}(x) = -\frac{(x - x_0)^{n-2}}{(n - 2)!} f^{(n-2)}(x_0)$$

$$\cdot$$
$$\cdot$$
$$\cdot$$

$$R_2(x) - R_1(x) = -(x - x_0)f'(x_0)$$

But

$$R_1(x) = \int_{x_0}^{x} f'(u) \, du = f(x) - f(x_0)$$

Adding,

$$R_n(x) = -\frac{(x - x_0)^{n-1}}{(n - 1)!} - \cdots - (x - x_0)f'(x_0) + f(x) - f(x_0)$$

Thus,

$$f(x) = f(x_0) + (x - x_0)f'(x_0) + \cdots + \frac{(x - x_0)^{n-1}}{(n - 1)!} f^{(n-1)}(x_0) + R_n(x)$$

and that completes the proof.

The following is a useful corollary of this theorem.

Corollary. The remainder after nth term in Taylor's expansion may be expressed as:

$$R_n(x) = \frac{1}{n!} f^{(n)}(x_1)(x - x_0)^n$$

where x_1 lies between x_0 and x.

Exercise 9.2

1. Let C be the Cantor set, and

$$f(x) = \frac{1}{2n} \text{ for } x \text{ in those removed intervals}$$

$$\text{whose length is } \frac{1}{3^n}$$

$$f(x) = 1 \text{ for } x \in C$$

   Show that f is Riemann-integrable over $[0,1]$.
2. Prove Theorem 9.2.8.
3. Prove Corollary of Theorem 9.2.9.
4. Construct a function which is Riemann-integrable over $[a,b]$ and has denumberably many discontinuities.
5. Show that if the limit of $R(f,P)$ (defined in 9.2.6) as $n \longrightarrow \infty$ and the length of the largest subinterval of P approaches zero, exists, then $\int_{\underline{a}}^{b} f = \overline{\int}_{a}^{b} f$ and each one of them is equal to this limit.
6. Prove that if f is Riemann-integrable on $[a,b]$ then so is $|f|$. Show by an example that the converse is not necessarily true (Hint: Modify Example 9.2.1)
7. Show that if $f(x) = g(x)$ except at most a finite number of points of $[a,b]$ then either both are Riemann-integrable or neither is.
8. If $f : [a,b] \longrightarrow R$ is continuous on $[a,b]$ and $f(x) \leqslant 0$ on $[a,b]$, then show that $\int_{a}^{b} f = 0$ implies $f(x) = 0$ for all $x \in [a,b]$. Prove a similar result for $f(x) \geqslant 0$.
9. Prove the fundamental theorem of calculus. "If $f : [a,b] \longrightarrow R$ is continuous and if $F(x) = \int_{a}^{x} f$ for $x \in [a,b]$, then $F'(x) = f(x)$."

10. Find $R_n(x)$ in the Taylor expansion of (a) $\sin x$, (b) e^x, (c) $\dfrac{1}{x-3}$, about the point 0.

11. Show by means of an example that the conclusion of Exercise 7 is not necessarily true if we substitute "denumerably many" for "a finite number." (*Hint*: Use Example 9.2.1.)

10

Inequalities and Improper Integrals

In this chapter we prove some important inequalities and introduce O (capital order), o (small order) and $\sim$ (asymptotic) symbols to compare the growth of two sequences or more generally of two real valued functions $f(x)$ and $F(x)$ as x tends to a limit a. In Sec. 10.4 we consider improper integrals.

The first inequality (Theorem 10.1.1 combined with Exercise 10.2(1)) expresses the geometrical fact that the sum of two sides of a triangle is greater than the third. The inequalities of Hölder and Minkowski are generalizations of the triangle inequality. The inequalities for integrals will be given in Chapter 13.

10.1. Inequalities

Theorem 10.1.1. (Triangle Inequality). If z_1 and z_2 are any two complex numbers, then

$$\||z_1| - |z_2|\| \leqslant |z_1 + z_2| \leqslant |z_1| + |z_2|. \tag{10.1}$$

Proof. If $z = x + iy$, where x and y are real and $i = \sqrt{-1}$, then we write $\bar{z} = x - iy$. Obviously $|z\bar{z}| = x^2 + y^2 = |z|^2$, $|\bar{z}| = |z|$, $\overline{z_1 + z_2} = \bar{z}_1 + \bar{z}_2$. Also $|x| \leqslant |z|$. Hence

$$
\begin{aligned}
|z_1 + z_2|^2 &= (z_1 + z_2)(\bar{z}_1 + \bar{z}_2) \\
&= |z_1|^2 + |z_2|^2 + z_2\bar{z}_1 + z_1\bar{z}_2 \\
&= |z_1|^2 + |z_2|^2 + 2Re(z_1\bar{z}_2) \\
&\leqslant |z_1|^2 + |z_2|^2 + 2|z_1\bar{z}_2| \\
&= |z_1|^2 + 2|z_1| |z_2| + |z_2|^2 = (|z_1| + |z_2|)^2.
\end{aligned}
$$

Since $|z_1 + z_2|$ and $|z_1| + |z_2|$ are nonnegative numbers, we have

$$|z_1 + z_2| \leqslant |z_1| + |z_2|,$$

and the right-hand inequality is proved.

Replace in this z_2 by $-z_1 + z_2$ and we get

$$|z_1 + (-z_1 + z_2)| \leqslant |z_1| + |-z_1 + z_2|.$$

That is,

$$|z_2| - |z_1| \leqslant |z_1 - z_2|.$$

If we interchange z_1 and z_2 we have

$$|z_1| - |z_2| \leqslant |z_2 - z_1| = |z_1 - z_2|.$$

These two inequalities imply

$$\|z_1| - |z_2\| \leqslant |z_1 - z_2|.$$

If we now replace z_2 by $-z_2$, we obtain

$$\|z_1| - |z_2\| \leqslant |z_1 + z_2|.$$

which completes the proof.

The following two theorems will be required in the proof of Hölder's inequality.

Theorem 10.1.2. If $0 < P < Q, a > 1$ then

$$\frac{a^P - 1}{P} < \frac{a^Q - 1}{Q}. \tag{10.2}$$

Proof. Consider

$$f(x) = \frac{x^P - 1}{P} - \frac{x^Q - 1}{Q}, \qquad x \geqslant 1.$$

Then $f'(x) < 0, x > 1$ and $f'(1) = 0$. Since $f(1) = 0$ it follows that $f(x)$ is negative for $x > 1$ and the inequality is proved.

Remark. If $0 < P < Q, a < 1$ then we have

$$\frac{a^P - 1}{P} > \frac{a^Q - 1}{Q}.$$

Theorem 10.1.3. If $p > 1$ and q is defined by $1/p + 1/q = 1$, then

$$a^{1/p} b^{1/q} \leqslant \frac{a}{p} + \frac{b}{q}, \qquad a \geqslant 0, b \geqslant 0. \tag{10.3}$$

Proof. The inequality is obvious for $b = 0$ and also for $b = a$. Suppose $b > 0$ and write $x = a/b$. Then we have to show that

$$x^{1/p} - 1 \leqslant \frac{1}{p}(x - 1).$$

This is the inequality (10.2) if $x = a/b > 1$. [Take in (10.2), $Q = 1, P = \frac{1}{p}$ and $a = x$.] If $a < b$, we write $x = b/a$ and again use (10.2). This completes the proof.

Theorem 10.1.4. (Hölder's Inequality). Let $p > 1$ and let q be defined by $\frac{1}{p} + \frac{1}{q} = 1$. Then

$$\sum_{1}^{n} |z_i w_i| \leqslant \left[\sum_{i=1}^{n} |z_i|^p\right]^{1/p} \left[\sum_{i=1}^{n} |w_i|^q\right]^{1/q} \tag{10.4}$$

for any complex numbers $z_1, \ldots, z_n; \quad w_1, \ldots, w_n$.

Proof. Write $A = \left[\sum_{1}^{n} |z_i|^p\right]^{1/p}$, $B = \left[\sum_{1}^{n} |w_i|^q\right]^{1/q}$. The inequality is obvious if either $A = 0$ or $B = 0$. So we suppose $A > 0, B > 0$. Write

$$a_i = \frac{|z_i|^p}{A^p}, \quad b_i = \frac{|w_i|^q}{B^q}.$$

Then we get, from (10.3),

$$\frac{|z_i| \, |w_i|}{AB} \leqslant \frac{a_i}{p} + \frac{b_i}{q}, \quad i = 1, 2, \ldots n.$$

Adding these n inequalities, we obtain

$$\sum_{i=1}^{n} |z_i w_i| \leqslant AB \left\{\frac{1}{p} \sum_{i=1}^{n} a_i + \frac{1}{q} \sum_{i=1}^{n} b_i\right\}$$

$$= AB\left(\frac{1}{p} + \frac{1}{q}\right) = AB. \quad \blacksquare$$

Theorem 10.1.5. [Minkowski's Inequality]. If $p \geqslant 1$, then

$$\left[\sum_{i=1}^{n} |z_i + w_i|^p\right]^{1/p} \leqslant \left[\sum_{i=1}^{n} |z_i|^p\right]^{1/p} + \left[\sum_{i=1}^{n} |w_i|^p\right]^{1/p} \tag{10.5}$$

for any complex numbers $z_1, \ldots, z_n, \quad w_1, \ldots, w_n$.

Proof. The inequality is obvious for $p = 1$. (See Theorem 10.1.1.). So we suppose that $p > 1$. Then

$$\sum_{i=1}^{n} |z_i + w_i|^p = \sum_{1}^{n} |z_i + w_i| \, |z_i + w_i|^{p-1}$$

$$\leq \sum_{1}^{n} |z_i| \, |z_i + w_i|^{p-1} + \sum_{1}^{n} |w_i| \, |z_i + w_i|^{p-1}.$$

We now use Holder's inequality for each sum on the right and get

$$\sum_{1}^{n} |z_i + w_i|^p \leq \left(\sum_{1}^{n} |z_i|^p \right)^{1/p} \left(\sum_{1}^{n} |z_i + w_i|^{(p-1)q} \right)^{1/q}$$

$$+ \left(\sum_{1}^{n} |w_i|^p \right)^{1/p} \left(\sum_{1}^{n} |z_i + w_i|^{(p-1)q} \right)^{1/q}$$

$$= \left[\left\{ \sum_{1}^{n} |z_i|^p \right\}^{1/p} + \left\{ \sum_{1}^{n} |w_i|^p \right\}^{1/p} \right] \left[\sum_{1}^{n} |z_i + w_i|^p \right]^{1/q}.$$

Since $1 - \dfrac{1}{q} = \dfrac{1}{p}$, Minkowski's inequality follows. ∎

10.2. Convex Functions

We proved in Theorem 10.1.3 that the geometric mean of two nonnegative numbers does not exceed their arithmetic mean. In this section we give a generalization of this inequality.

Definition 10.2.1. *A real-valued function f defined in an open or closed interval I is said to be convex on I, provided*

$$x, y \in I \ \text{ and } \ 0 \leq \lambda \leq 1 \implies f((1 - \lambda) x + \lambda y) \leq (1 - \lambda) f(x) + \lambda f(y).$$

Geometrically this means that any chord of the curve $y = f(x)$ lies above or on the curve.

Theorem 10.2.1. If $f: [a,b] \longrightarrow R$ is monotone increasing on $[a,b]$, then

$$F(x) = \int_{a}^{x} f(t) \, dt$$

is convex on the interval.

Proof. Let $a \leqslant x < y \leqslant b$. Then

$$(1 - \lambda) F(x) + \lambda F(y) - F((1 - \lambda)x + \lambda y)$$

$$= (1 - \lambda) \int_a^x f(t)dt + \lambda \int_a^y f(t)dt - \int_a^{(1-\lambda)x+\lambda y} f(t)\,dt$$

$$= -\lambda \int_a^x f(t)\,dt + \lambda \int_a^y f(t)\,dt - \int_x^{(1-\lambda)x+\lambda y} f(t)\,dt$$

$$= \lambda \int_x^y f(t)\,dt - \int_x^{(1-\lambda)x+\lambda y} f(t)\,dt = \int_x^{(1-\lambda)x+\lambda y} (\lambda - 1)f(t)\,dt$$

$$+ \int_{(1-\lambda)x+\lambda y}^y \lambda f(t)\,dt \geqslant (\lambda - 1)f((1 - \lambda)x + \lambda y)((1 - \lambda)x + \lambda y - x)$$

$$+ \lambda f((1 - \lambda)x + \lambda y)(y - \lambda y - (1 - \lambda)x) = 0.$$

Hence $F(x)$ is convex on $[a,b]$. ∎

Theorem 10.2.2. If $f''(x)$ exists on the interval (a,b) and is nonnegative, then $f(x)$ is convex on the interval.

Proof. Let $a < x < y < b$. Then by Taylor's theorem with remainder, when $n = 2$, we have

(i) $f(x) = f((1 - \lambda)x + \lambda y) + (x - (1 - \lambda)x - \lambda y)f'((1 - \lambda)x + \lambda y)$

$$+ \frac{(x - (1 - \lambda)x - \lambda y)^2}{2!} f''(\xi),$$

(ii) $f(y) = f((1 - \lambda)x + \lambda y) + (y - (1 - \lambda)x - \lambda y)f'((1 - \lambda)x + \lambda y)$

$$+ \frac{(y - (1 - \lambda)x - \lambda y)^2}{2!} f''(\eta),$$

where $x \leqslant \xi \leqslant (1 - \lambda)x + \lambda y \leqslant \eta \leqslant y$. From (i) and (ii) we get $(1 - \lambda)f(x) + \lambda f(y) \geqslant f((1 - \lambda)x + \lambda y)$. ∎

Theorem 10.2.3. (Jensen's Inequality). Let $a_1, \ldots, a_n$ be arbitrary positive numbers. If $f''(x)$ exists and is nonnegative on (a,b), then

$$f\left(\frac{a_1 x_1 + \cdots + a_n x_n}{a_1 + \cdots + a_n}\right) \leqslant \frac{a_1 f(x_1) + \cdots + a_n f(x_n)}{a_1 + \cdots + a_n} \tag{10.6}$$

where $a < x_1 \leqslant x_2 \leqslant \cdots \leqslant x_n < b$.

Proof. By Theorem 10.2.2, f is convex on (a,b). Write $A = a_1 + \cdots + a_n$,

$B = \dfrac{a_1 x_1 + \cdots + a_n x_n}{A}$. It is easily seen that $a < B < b$. By Taylor's theorem with remainder when $n = 2$, we have

$$f(x_p) = f(B) + (x_p - B) f'(B) + \frac{(x_p - B)^2}{2} f''(X_p)$$

where $a < X_p < b; p = 1, 2, \ldots n$

Multiply this equation by a_p. Taking $p = 1, 2, \ldots n$ and adding, we get

$$a_1 f(x_1) + \cdots + a_n f(x_n) = Af(B) + \frac{1}{2} \sum_1^n a_p (x_p - B)^2 f''(X_p) \geqslant Af(B)$$

and the inequality follows. **|**

Exercise 10.2.

1. Prove that

$$|z_1 + z_2 + \cdots + z_n| \leqslant |z_1| + |z_2| + \cdots + |z_n|.$$

Prove also that there is equality sign, if and only if $\arg z_1 = \arg z_2 = \cdots = \arg z_n$.

(*Hint*: Consider $|z_1 + \cdots + z_n|^2$ and observe that

$$\mathrm{Re}\,(z_p \bar{z}_q) = |z_p z_q| \text{ if and only if } \arg z_p = \arg z_q.)$$

2. Prove that if the numbers $a_1, \ldots, a_n, x_1, \ldots, x_n$ are all positive, then

(i) $\dfrac{a_1 + a_2 + \cdots + a_n}{\dfrac{a_1}{x_1} + \dfrac{a_2}{x_2} + \cdots + \dfrac{a_n}{x_n}} \leqslant \exp\left(\dfrac{a_1 \log x_1 + \cdots + a_n \log x_n}{a_1 + \cdots + a_n}\right)$

$$\leqslant \dfrac{a_1 x_1 + \cdots + a_n x_n}{a_1 + \cdots + a_n}.$$

Deduce that

(ii) $\dfrac{n}{\dfrac{1}{x_1} + \cdots + \dfrac{1}{x_n}} \leqslant (x_1 x_2 \ldots x_n)^{1/n} \leqslant \dfrac{x_1 + \cdots + x_n}{n}.$

[*Hint*: Take $f(x) = -\log x, a = 0$, in Theorem 10.2.3 and observe that $f''(x) = 1/x^2 > 0$. The right-half of the inequality (i) follows immediately. To prove the left-half of (i), replace x_p by $1/x_p$. The expression on the left, in (ii), is the harmonic mean of $x_1, \ldots, x_n$, the expression in the middle is the geometric mean and the expression on the right is the arithmetic mean.]

3. Prove that if n be any positive integer, then

$$\left(\frac{n}{\log ne}\right)^n \leqslant n! \leqslant \left(\frac{n+1}{2}\right)^n.$$

$\left(Hint: \text{Take } x_p = p \text{ in inequality (ii) above and note that } 1 + \frac{1}{2} + \cdots + \frac{1}{n} < 1 + \int_1^n \frac{dx}{x}.\right)$

4. Deduce Hölder's inequality from Jensen's inequality.

(*Hint:* Take $f(x) = x^q, q > 1, x > 0$ and write

$$a_r = \alpha_r^p, \ x_r = \beta_r/\alpha_r^{p/q}$$

in Theorem 10.2.3 to obtain

$$\sum_{r=1}^{n} \alpha_r \beta_r \leqslant \left(\sum_{r=1}^{n} \alpha_r^p\right)^{1/p} \left(\sum_{r=1}^{n} \beta_r^q\right)^{1/q}$$

where $\alpha_1, \ldots, \alpha_n, \ \beta_1, \ldots, \beta_n$ are positive numbers. If we add now $\alpha_{n+1}\beta_{n+1} + \cdots + \alpha_N \beta_N$ where $\alpha_i \beta_i = 0, \ \alpha_i \geqslant 0, \ \beta_i \geqslant 0, \ i = n+1, \ldots, N$ we get

$$\sum_{r=1}^{N} \alpha_r \beta_r = \sum_{r=1}^{n} \alpha_r \beta_r \leqslant \left(\sum_{r=1}^{n} \alpha_r^p\right)^{1/p} \left(\sum_{r=1}^{n} \beta_r^q\right)^{1/q} \leqslant \left(\sum_{r=1}^{N} \alpha_r^p\right)^{1/p} \left(\sum_{r=1}^{N} \beta_r^q\right)^{1/q}.$$

5. Deduce from Hölder's inequality that

$$\frac{1}{N}\sum_{r=1}^{N} \alpha_r \beta_r \leqslant \left(\frac{1}{N}\sum_{r=1}^{N} \alpha_r^p\right)^{1/p} \left(\frac{1}{N}\sum_{r=1}^{N} \beta_r^q\right)^{1/q}$$

$\left(Hint: \text{See Exercise 10.2 (4). Note that } \frac{1}{p} + \frac{1}{q} = 1.\right)$

6. By considering the identity

$$(\Sigma a_r b_r)^2 = \Sigma a_r^2 \Sigma b_r^2 - \Sigma(a_r b_s - a_s b_r)^2$$

prove that the equality sign can hold in Cauchy's inequality

$$\sum_{1}^{n} |z_i w_i| \leqslant \left(\sum_{1}^{n} |z_i|^2\right)^{1/2} \left(\sum_{1}^{n} |w_i|^2\right)^{1/2}$$

if and only if there exist real numbers c and d such that

$$c|z_i| + d|w_i| = 0, i = 1, 2, \ldots n, c^2 + d^2 > 0$$

7. Consider for convexity the function $f(x) = -\sin x$ and hence prove that

$$\sin x \geqslant \frac{2x}{\pi} \text{ for } 0 \leqslant x \leqslant \frac{\pi}{2}.$$

8. Prove that if $a_k \geqslant 0$ and $b_k \geqslant 0$, $k = 1, 2, \ldots n$ and $\alpha > 0, \beta > 0, \alpha + \beta = 1$ then

$$\left\{ \sum_{k=1}^{n} (a_k + b_k)^\alpha \right\}^{1/\alpha} \leqslant \left\{ \sum_{k=1}^{n} (a_k^\alpha + b_k^\alpha)^{1/\alpha} \right\}$$

$$\leqslant 2^{\beta/\alpha} \left\{ \left(\sum_{k=1}^{n} a_k^\alpha \right)^{1/\alpha} + \left(\sum_{k=1}^{n} b_k^\beta \right)^{1/\beta} \right\}.$$

$\left(Hint: \text{ if } y(x) = \dfrac{1 + x^\alpha}{(1 + x)^\alpha} \text{ then } y' > 0 \text{ in } (0,1) \text{ and so} \right.$

$$(1 + x)^\alpha \leqslant 1 + x^\alpha \leqslant 2^{1-\alpha} (1 + x)^\alpha \text{ for } 0 \leqslant x \leqslant 1.$$

Note that $y\left(\dfrac{1}{x} \right) = y(x)$. Replace x by $\dfrac{b_k}{a_k}$ and again by $(\Sigma b_k^\alpha / \Sigma a_k^\alpha)^{1/\alpha}.\Big)$

10.3. O, o and $\sim$ Symbols

Let $f(x)$ and $F(x)$ be given functions. Let a be a given point and suppose that $F(x)$ is positive and continuous in an interval I about a. (If $a = \infty$, then we replace "an interval I about a" by "for all sufficiently large x.")

If there is a constant K such that

$$|f(x)| < KF(x), x \in I.$$

then we say that $f(x) = O(F(x))$ as $x \longrightarrow a$. (If $a = \infty$, we replace "$x \in I$" by "for all sufficiently large x.")

If $\lim\limits_{x \to a} \dfrac{f(x)}{F(x)} = o$, then we say that $f(x) = o(F(x))$ as $x \longrightarrow a$.

If $\lim\limits_{x \to a} \dfrac{f(x)}{F(x)} = 1$, then we say that $f(x) \sim F(x)$ as $x \longrightarrow a$.

EXAMPLE 10.3.1.
   (i) $1 + 2x = O(x), x \longrightarrow \infty$.
   (ii) $3x^2 + 5x \sim 5x, x \longrightarrow 0$.
   (iii) $1 + \dfrac{3}{x} + \dfrac{10}{x^2} = O(1), x \longrightarrow \infty$.
In the last example $F(x) = 1$. Note that

$$O(1) + O(1) = O(1)$$

for if $|f| < K$, $|g| < K$ then $|f + g| < 2K$.

EXAMPLE 10.3.2.

   (i) If $a_n = o(1)$, $n \longrightarrow \infty$ and $b_n = o(1)$, $n \longrightarrow \infty$ then $a_n + b_n = o(1)$ $n \longrightarrow \infty$.

   (ii) If $a_n = o(1)$, $n \longrightarrow \infty$ and $b_n = O(1)$, $n \longrightarrow \infty$, then $a_n b_n = o(1)$, $n \longrightarrow \infty$.

   (iii) If $a_n = O\left(\dfrac{1}{n^p}\right)$, $p > 1$, then $\displaystyle\sum_1^\infty a_n$ is absolutely convergent.

10.4. Improper Integrals

In Riemann integration both the interval and the function are bounded. We now consider the cases when either the interval is infinite or the function is unbounded.

Definition 10.4.1. Let $f(x)$ be Riemann-integrable on every finite interval $[a,X]$. If

$$\int_a^X f = \int_a^X f(x)\,dx \longrightarrow \ell \qquad (\ell \text{ finite})$$

as $X \longrightarrow \infty$, we say that $\displaystyle\int_a^\infty f(x)\,dx$ converges to the value ℓ. If $\displaystyle\lim_{X \to \infty} \int_a^X f(x)\,dx$

does not exist we say that $\displaystyle\int_a^\infty f(x)\,dx$ diverges.

We define similarly

$$\int_{-\infty}^a f(x)\,dx = \lim_{X \to -\infty} \int_X^a f(x)\,dx$$

and say that the integral is convergent, if this limit exists (and is finite).

Definition 10.4.2. *Let $f(x)$ be Riemann-integrable on interval $[a,b']$ for each $b' < b$ and unbounded on $[a,b)$. If*

$$\lim_{b' \to b-} \int_a^{b'} f(x)\,dx = m \qquad (m \text{ finite})$$

we say that $\displaystyle\int_a^b f(x)\,dx$ converges to the value m.

We define similarly $\int_a^b f(x)\, dx$ when f is unbounded on $(a,b]$.

EXAMPLE 10.4.1.   The integral $\int_1^\infty \dfrac{dx}{x^p}$ is convergent when $p > 1$ and divergent otherwise. For

$$\int_1^\infty \frac{dx}{x^p} = \lim_{X \to \infty} \int_1^X \frac{dx}{x^p}$$

$$= \lim_{X \to \infty} \left\{ \frac{X^{-p+1}}{-p+1} \right\}_1^\infty, \qquad p \neq 1.$$

Hence the integral converges to $\dfrac{1}{p-1}$ if $p > 1$ and is divergent if $p < 1$. Let $[X]$ denote the integer part of X. Then

$$\int_1^X \frac{dx}{x} > \frac{1}{2} + \frac{1}{3} + \cdots + \frac{1}{[X]}$$

and the expression on the right tends to ∞ as $X \longrightarrow \infty$. (See Example 8.2.1.) Hence the integral is divergent if $p = 1$.

Comparison Tests. If $f(x)$ is nonnegative for $x > a$ then the integral $\int_a^X f(x)\, dx$ monotonically increases with X. Thus the integral $\int_a^\infty f(x)\, dx$ must either converge or diverge to ∞. It will converge if there is a positive number K such that $\int_a^\infty f(x)\, dx < K$. Hence we have the following test.

Theorem 10.4.1. If $f(x)$ and $F(x)$ are both Riemann-integrable when $x \geqslant a$ and $0 \leqslant f(x) \leqslant F(x)$, then

$$\int_a^\infty f(x)\, dx \text{ converges if } \int_a^\infty F(x)\, dx \text{ converges,}$$

and

$$\int_a^\infty F(x)\, dx \text{ diverges if } \int_a^\infty f(x)\, dx \text{ diverges.}$$

EXAMPLE 10.4.2.

(i) $\displaystyle\int_1^\infty \frac{dx}{(x^4+x^3+1)^{1/2}}$ converges, since $\displaystyle\frac{1}{(x^4+x^3+1)^{1/2}} < \frac{1}{x^2}$ when $x \geqslant 1$.

(ii) $\displaystyle\int_1^\infty \frac{dx}{(x^2+x+1)^{1/2}}$ diverges, since $\displaystyle\frac{1}{(x^2+x+1)^{1/2}} > \frac{1}{x}$ when $x \geqslant 1$.

Exercise 10.4

1. Prove that the integral

$$B(p,q) = \int_0^1 x^{p-1} (1-x)^{q-1} \, dx$$

is convergent if $p > 0, q > 0$.

[*Hint*: The integrand $f(x)$ is unbounded on $(0,\frac{1}{2}]$ if $0 < p < 1$ and on $[\frac{1}{2},1)$ if $0 < q < 1$. Consider the convergence of the two integrals $\displaystyle\int_0^{1/2} f(x)\,dx$ and $\displaystyle\int_{1/2}^1 f(x)\,dx$.]

2. Prove that the integral

$$\int_0^\pi \frac{\sin x}{x^p} \, dx, \qquad 0 \leqslant p \leqslant 1$$

is convergent.

$$\left(Hint: \ \lim_{x \to 0} \frac{\sin x}{x^p} = \lim_{x \to 0} \left(\frac{\sin x}{x} \right) x^{1-p} = 0 \qquad \begin{array}{l} \text{if } p < 1 \\ = 1 \qquad \text{if } p = 1. \end{array} \right)$$

3. Prove that if $f(x)$ is R-integrable over $a \leqslant x \leqslant X$ for every X, then a necessary and sufficient condition for the convergence of the integral $\displaystyle\int_a^\infty f(t)\,dt$ is that, given $\epsilon > 0$, there exists $X_0(\epsilon) \geqslant a$ such that

$$\left| \int_{X_1}^{X_2} f(x)\,dx \right| < \epsilon \qquad \text{for every } X_2 > X_1 > X_0.$$

4. Prove that if $\displaystyle\int_a^\infty f(x)\,dx$ is absolutely convergent, that is if $\displaystyle\int_a^\infty |f(x)|\,dx$ is

   convergent, then so is $\displaystyle\int_a^\infty f(x)\,dx$.

5. Prove that if $f(x) = O\left(\dfrac{1}{x^{1+\delta}}\right)$, $\delta > 0$, as $x \longrightarrow \infty$, and if $f(x)$ is R-integrable

   over $a \leqslant x \leqslant X$ for every X then $\displaystyle\int_1^\infty f(x)\,dx$ is absolutely convergent.

6. Prove

   (i)   The integral $\displaystyle\int_1^\infty \frac{\sin x}{x^p}\,dx, p > 1$ is absolutely convergent.

   (ii) The integral $\displaystyle\int_0^\infty \frac{\sin x}{x^p}\,dx, \ 0 < p \leqslant 1$ is convergent but not absolutely

         convergent.

   $\left(\text{Hint: } \quad \text{(i)} \quad \dfrac{|\sin x|}{x^p} \leqslant \dfrac{1}{x^p}.\right.$

   (iia) In the second integral the integrand is bounded and R-integrable over
          $[0,X]$. Let $0 < X_1 < X_2$ and integrate by parts

   $$\int_{X_1}^{X_2} \frac{\sin x}{x^p}\,dx = \frac{\cos X_1}{X_1^p} - \frac{\cos X_2}{X_2^p} + p\int_{X_1}^{X_2} \frac{\cos x\,dx}{x^{1+p}}$$

   Note that the expression on the right is in absolute values less than

   $$\frac{1}{X_1^p} + \frac{1}{X_2^p} + \frac{1}{X_1^p} - \frac{1}{X_2^p} = \frac{2}{X_1^p}.$$

   (iib) Consider

   $$\int_0^{n\pi} \frac{|\sin x|}{x^p}\,dx = \sum_{k=1}^{n}\int_{(k-1)\pi}^{k\pi} \frac{|\sin x|}{x^p}\,dx$$

   $$= \sum_{k=1}^{n}\int_0^{\pi} \frac{\sin t\,dt}{\{(k-1)\pi + t\}^p}$$

   $$> \sum_{k=1}^{n} \frac{2}{(k\pi)^p}$$

   and note that $\displaystyle\sum_{k=1}^\infty 1/k^p$ is divergent when $p \leqslant 1.\bigg)$

7. Let $-\infty \leqslant a < c < b \leqslant \infty$ and suppose that $f(x)$ is unbounded in $(c - \delta, c + \delta)$, $\delta > 0$, and that the integrals $\int_a^{c-\delta_1} f(x)\,dx$ and $\int_{c+\delta_2}^b f(x)\,dx$ exist for each $\delta_1 > 0, \delta_2 > 0$. We then define

$$\int_a^b f(x)\,dx = \lim_{\delta_1 \to 0} \int_a^{c-\delta_1} f(x)\,dx + \lim_{\delta_2 \to 0} \int_{c+\delta_2}^b f(x)\,dx$$

and say that the integral is convergent if each limit on the right exists.

The Cauchy principal value (CPV) of $\int_a^b f(x)\,dx$ is defined as

$$\text{CPV} \int_a^b f(x)\,dx = \lim_{\delta \to 0} \left\{ \int_a^{c-\delta} f(x)\,dx + \int_{c+\delta}^b f(x)\,dx \right\}$$

provided the limit on the right-hand side exists.
   Show that

$$\int_0^3 \frac{dx}{(x-1)^3}$$

does not exist but has the Cauchy principal value of $3/8$.

8. Establish convergence or divergence of the following integrals.

(i) $\displaystyle\int_0^\infty \frac{dx}{(1+x)x^{1/2}}$,   (ii) $\displaystyle\int_0^{\pi/4} \frac{x}{(\sin x)^2}\,dx$,

(iii) $\displaystyle\int_0^\infty \frac{\cos mx\,dx}{1+x^2}$,   (iv) $\displaystyle\int_0^2 \frac{dx}{x(1+x)}$,

(v) $\displaystyle\int_0^\infty e^{-t}\,t^{x-1}\,dt$.

$\bigg[$ *Hint* for (v). The integrand $f(t)$ say, is nonnegative. Note that

$$f(t) = O(t^{x-1})$$

near the origin and so $\displaystyle\int_0^1 f(t)\,dt$ is convergent if $x > 0$. The integral $\displaystyle\int_1^\infty f(t)\,dt$ is convergent for every x, since $e^{-t}\,t^{x-1} = O(1/t^2)$ as $t \to \infty$.
Thus the integral in (v) is convergent when $x > 0$ and is denoted by $\Gamma(x)$. $\bigg]$

11

Further Topics in the Theory of Series

In this chapter we prove in Sec. 11.1 the integral test for a series of positive terms. The remaining sections are devoted to uniform convergence of sequences and series of functions, power series, Cesàro and Abel summability methods and elementary transcendental functions.

11.1. Integral Test

Theorem 11.1.1. Let $f(x)$ be continuous positive and nonincreasing for $x \geqslant 1$. Then we have:

(a) The series $\sum_{k=1}^{\infty} f(k)$ converges or diverges according as the sequence $\{I_n\}$, where

$$I_n = \int_1^n f(t)\,dt$$

does or does not converge to a (finite) limit as $n \longrightarrow \infty$.

(b) $\lim\limits_{n\to\infty}\left(\sum_{k=1}^{n} f(k) - I_n\right) = d$ exists.

(c) $0 \leqslant d \leqslant f(1)$.

(d) $0 \leqslant \sum_{k=1}^{n} f(k) - I_n - d \leqslant f(n)$, for $n = 1, 2, \ldots$.

Proof. (a). For $k - 1 \leqslant x \leqslant k$ we have

$$f(k - 1) \geqslant f(x) \geqslant f(k) > 0.$$

Hence

$$f(k - 1) \geqslant \int_{k-1}^{k} f(x)\, dx \geqslant f(k).$$

Write $S_n = \sum_{k=1}^{n} f(k)$. Adding the above inequalities for $k = 2, 3, \ldots, n$ we obtain

$$f(1) + f(2) + \cdots + f(n - 1) \geqslant \int_{1}^{n} f(x)\, dx \geqslant f(2) + f(3) + \cdots + f(n).$$

Hence

$$S_n - f(n) \geqslant I_n \geqslant S_n - f(1)$$

and so

(i) $$f(1) \geqslant S_n - I_n \geqslant f(n) > 0.$$

This shows that the sequence $\{S_n - I_n\}$ is bounded. Further, it is non-increasing for

(ii) $$S_n - I_n - (S_{n+1} - I_{n+1}) = -f(n + 1) + \int_{n}^{n+1} f(x)\, dx \geqslant 0.$$

Hence the sequence $\{S_n - I_n\}$ converges to a limit d, say, and we have, from (i) and (ii),

(iii) $$f(1) \geqslant \lim_{n \to \infty} (S_n - I_n) = d \geqslant 0$$

(iv) $$S_n - I_n \geqslant d, \qquad n = 1, 2, 3, \ldots.$$

If $\{I_n\}$ converges to a limit l as $n \longrightarrow \infty$, then since $\{I_n\}$ is nondecreasing, we have

$$S_n \leqslant I_n + f(1) \leqslant l + f(1)$$

and this proves that the series $\sum_{n=1}^{\infty} f(n)$ is convergent.

If $\{I_n\}$ does not converge to a (finite) limit, then $I_n \longrightarrow \infty$ as $n \longrightarrow \infty$, and since $S_n \geqslant I_n, S_n \longrightarrow \infty$ as $n \longrightarrow \infty$. Therefore the series $\sum_{n=1}^{\infty} f(n)$ is divergent.

(b)-(d). The statements (b) and (c) follow from (iii).
To prove (d) we note that (ii) gives

$$S_n - I_n - (S_{n+1} - I_{n+1}) \leqslant f(n) - f(n+1).$$

Adding these inequalities for $n, n+1, \ldots, n+p$ we obtain

$$S_n - I_n - (S_{n+p+1} - I_{n+p+1}) \leqslant f(n) - f(n+p+1) \leqslant f(n).$$

Letting $p \longrightarrow \infty$, we have

(v) $$S_n - I_n - d \leqslant f(n), \qquad n = 1, 2, \ldots.$$

The inequalities (iv) and (v) prove (d). ∎

Corollary 1. $\displaystyle\sum_{n=1}^{\infty} \frac{1}{n^x}$ converges if and only if $x > 1$.

Corollary 2. The sequence $\{\gamma_n\}_1^{\infty} \equiv \left\{ \displaystyle\sum_{k=1}^{n} \frac{1}{k} - \log n \right\}_1^{\infty}$ is a monotone decreasing sequence of positive numbers. Denoting $\lim_{n \to \infty} \gamma_n$ by γ we have

$$0 \leqslant \gamma_n - \gamma \leqslant \frac{1}{n}, \qquad n = 1, 2, \ldots.$$

(The constant γ is known as Euler's constant. Its value is $0.5772 \ldots$.)

Exercise 11.1

1. Prove that when $x > 1$ the sum of the series $\displaystyle\sum_{n=1}^{\infty} \frac{1}{n^x}$ lies between $\dfrac{1}{x-1}$ and $\dfrac{x}{x-1}$.

2. Prove that $\displaystyle\sum_{n=2}^{\infty} \frac{1}{n(\log n)^x}$ is convergent when $x > 1$ and divergent when $x \leqslant 1$.
   Prove also that

$$\sum_{n=2}^{N} \frac{1}{n \log n} = \log \log N + O(1).$$

3. Prove that the series

$$1 - \frac{1}{2} + \frac{1}{3} - \frac{1}{4} + \cdots$$

converges to log 2.

[*Hint*: Use the theorem on alternating series to prove the convergence of the series. To find the sum, consider

$$S_{2n} = 1 - \frac{1}{2} + \frac{1}{3} - \frac{1}{4} + \cdots - \frac{1}{2n}$$

$$= 1 + \frac{1}{2} + \cdots + \frac{1}{2n} - 2\left(\frac{1}{2} + \frac{1}{4} + \cdots + \frac{1}{2n}\right)$$

$$= \gamma_{2n} + \log 2n - (\gamma_n + \log n) = \log 2 + o(1).$$

Note that

$$S_{2n+1} - S_{2n} = \frac{1}{2n+1}.\Big]$$

4. Prove that the series

$$1 + \frac{1}{3} - \frac{1}{2} + \frac{1}{5} + \frac{1}{7} - \frac{1}{4} + \frac{1}{9} + \frac{1}{11} - \frac{1}{6} + \cdots$$

converges to $\frac{3}{2}$ log 2.

[*Hint*: Consider

$$S_{3n} = \left(1 + \frac{1}{3} - \frac{1}{2}\right) + \left(\frac{1}{5} + \frac{1}{7} - \frac{1}{4}\right) + \cdots + \left(\frac{1}{4n-3} + \frac{1}{4n-1} - \frac{1}{2n}\right)$$

$$= 1 + \frac{1}{3} + \frac{1}{5} + \frac{1}{7} + \cdots + \frac{1}{4n-1} - \frac{1}{2}\left(1 + \frac{1}{2} + \cdots + \frac{1}{n}\right)$$

$$= 1 + \frac{1}{2} + \cdots + \frac{1}{4n} - \frac{1}{2}\left(1 + \frac{1}{2} + \cdots + \frac{1}{2n}\right) - \frac{1}{2}\left(1 + \frac{1}{2} + \cdots + \frac{1}{n}\right)$$

$$= \gamma_{4n} + \log 4n - \frac{1}{2}(\gamma_{2n} + \log 2n) - \frac{1}{2}(\gamma_n + \log n)$$

$$= \frac{3}{2}\log 2 + o(1).$$

Note that

$$S_{3n+1} = S_{3n} + \frac{1}{4n+1} = S_{3n} + o(1)$$

$$S_{3n+2} = S_{3n} + o(1)]$$

(Remark: The series in Exercise 3 is conditionally convergent and the series in this exercise is also a conditionally convergent series arrived at by re-arrangement of terms of the series in Exercise 3. The two series have different sums. A well-known theorem of Riemann states that *given any real number S, the terms of a conditionally convergent series Σa_n can be rearranged to yield a series which converges to S. The terms of Σa_n can also be rearranged to yield a divergent series.* See Knopp, *Theory and Application of Infinite Series*, pp. 318-319.)

5. Prove that the series

$$1 + \frac{1}{2} - \frac{2}{3} + \frac{1}{4} + \frac{1}{5} - \frac{2}{6} + \frac{1}{7} + \frac{1}{8} - \frac{2}{9} + \cdots$$

converges to the sum log 3.

6. Let $d_n > 0$, $\sum d_n = \infty$, $S_n = \sum_{k=1}^{n} d_k$ and let $f(x)$ be positive nonincreasing for $x \geqslant 1$. Show that

(i) $\sum_{n=1}^{\infty} f(S_n) d_n$ converges if $\int_1^x f(t)\,dt$ tends to a limit as $x \longrightarrow \infty$.

(ii) $\sum_{n=2}^{\infty} f(S_{n-1}) d_n$ diverges if $\int_1^x f(t)\,dt$ tends to ∞ as $n \longrightarrow \infty$.

$$\left[\text{Hint.} \quad \text{(i)} \quad d_n f(S_n) = \int_{S_{n-1}}^{S_n} f(S_n)\,dx \leqslant \int_{S_{n-1}}^{S_n} f(x)\,dx \right.$$

$$\sum_{k=2}^{n} d_k f(S_k) \leqslant \int_{S_1}^{S_n} f(x)\,dx = O(1)$$

$$\text{(ii)} \quad d_n f(S_{n-1}) \geqslant \int_{S_{n-1}}^{S_n} f(x)\,dx$$

$$\sum_{k=2}^{n} d_k f(S_{k-1}) \geqslant \int_{S_1}^{S_n} f(x)\,dx \longrightarrow \infty \text{ as } n \longrightarrow \infty.$$

Note that if we take $f(x) = \frac{1}{x}$ and write $b_n = \frac{d_n}{S_{n-1}}$, then $\frac{b_n}{d_n} \longrightarrow 0$ and

$\Sigma b_n = +\infty$. Hence given any divergent series Σd_n of positive terms we can find another series Σb_n which diverges more slowly than Σd_n.$\Big]$

7. If $f(x)$ is continuously differentiable on $[1,n]$, then show that

$$\sum_{k=1}^{n} f(k) = \int_{1}^{n} f(x)\, dx + \frac{1}{2}(f(1)+f(n)) + \int_{1}^{n}\left(x - [x] - \frac{1}{2}\right) f'(x)\, dx,$$

where $[x]$ denotes the integer part of x.

$\Big[$*Hint*:

$$\int_{1}^{n}\left(x - [x] - \frac{1}{2}\right) f'(x)\, dx$$

$$= \sum_{k=1}^{n-1} \int_{k}^{k+1}\left(x - k - \frac{1}{2}\right) f'(x)\, dx$$

$$= \sum_{k=1}^{n} f(k) - \frac{1}{2}(f(1)+f(n)) - \int_{1}^{n} f(x)\, dx.\Big]$$

8. Prove that

$$\sum_{k=1}^{n}\frac{1}{k} = \log n + \frac{1}{2} - \int_{1}^{\infty}\left(x - [x] - \frac{1}{2}\right)\frac{dx}{x^2} + o(1)$$

Deduce that $\gamma = \frac{1}{2} - \int_{1}^{\infty}\left(x - [x] - \frac{1}{2}\right)\frac{dx}{x^2}.$

11.2. Sequence of Functions: Pointwise Convergence and Uniform Convergence

In Chapter 4 we dealt with sequences of real numbers. This chapter generalizes this concept to sequences of *functions*, which shall be denoted by $\{f_n(x)\}$ or $\{f_n\}$. The properties of "pointwise convergence" (or briefly convergence) and "uniform convergence" of a sequence of functions will be discussed. It will be seen that there are some ideas, for example continuity, which remain invariant under uniform convergence but not under pointwise convergence; i.e., there exists a sequence of continuous functions whose limit function is discontinuous. As before, we shall restrict our discussion to functions of real variables; however, most of the results may be extended to a wider class of functions, e.g., functions defined on a metric space.

Definition 11.2.1. *A sequence of functions* $\{f_n(x)\}$, *with domain* D, *is said to be convergent at* $x_0 (x_0 \in D)$ *if the sequence* $\{f_n(x_0)\}$ *is convergent.*

Definition 11.2.2. *The sequence* $\{f_n(x)\}$ *is said to be pointwise convergent on* $S \subset D$, *if* $\{f_n(x_0)\}$ *converges for every* $x_0 \in S$.

The limit of $\{f_n(x_0)\}$ is unique, and S is then called the *domain of the convergence.* We may write

$$\lim_{n\to\infty} f_n(x) = f(x) \quad \text{or equivalently } f_n(x) \longrightarrow f(x) \text{ as } n \longrightarrow \infty \quad \text{for every } x \in S.$$

At this point one may raise the question if

$$\lim_{x\to x_0} f(x) = \lim_{x\to x_0} \lim_{n\to\infty} f_n(x) = \lim_{n\to\infty} \lim_{x\to x_0} f_n(x).$$

If $f_n(x)$ is continuous at x_0 for every n, then this may be restated as follows. "Is the limit function of a convergent sequence of continuous functions also continuous?" What we are really asking is if it is possible to interchange the order of the limit operation.

The answer is "no," as one sees clearly from the examples which follow.

EXAMPLE 11.2.1. Let

$$f_n(x) = \frac{a^2}{a^2 + n^2 x^2} \quad (a \neq 0) \quad (n = 1, 2, 3, \dots)$$

Then $f_n(x)$ is continuous for all x and for every n, (for there does not exist any x such that $a^2 + n^2 x^2 = 0$.) Let $f(x) = \lim_{n\to\infty} f_n(x)$. Then

$$f(x) = 0 \ (x \neq 0)$$
$$f(0) = 1$$

because $\lim_{n\to\infty} f_n(x) = 0 (x \neq 0)$ and $\lim_{n\to\infty} f_n(0) = 1$. We thus see that f is not continuous at $x = 0$ and

$$\lim_{x\to 0} \lim_{n\to\infty} f_n(x) \neq \lim_{n\to\infty} \lim_{x\to 0} f_n(x).$$

EXAMPLE 11.2.2. Let $f_n(x)$ be defined on $[0,1]$ as follows.

$$f_n(x) = 1 - nx^2 \quad \text{for } x \in \left[0, \frac{1}{\sqrt{n}}\right]$$

$$f_n(x) = 0 \quad\quad\quad x \in \left(\frac{1}{\sqrt{n}}, 1\right]$$

For this sequence of functions the graph of $f_n(x)$ (see p. 201) consists of a segment of

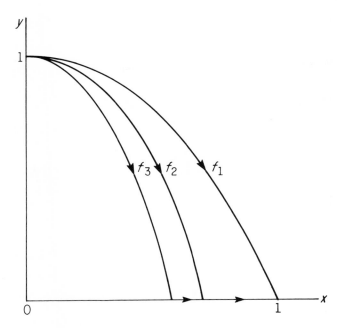

the parabola $y = 1 - nx^2$ from the point $(0,1)$ to $\left(\dfrac{1}{\sqrt{n}}, 0\right)$ and a segment of the

x-axis from $\left(\dfrac{1}{\sqrt{n}}, 0\right)$ to $(1,0)$. It is easy to see that this function is continuous.

$\left(\lim\limits_{x \to 1/\sqrt{n}} f_n(x) \text{ exists and is equal to } 0.\right)$

    If f is the limit function of this sequence of functions then $f(x) = \lim\limits_{n \to \infty} f_n(x) =$

$= 0$, for $x \neq 0$. However, $f_n(0) = 1$ for every n, and as such $f(0) = 1$. Thus

$$\lim_{x \to 0} \lim_{n \to \infty} f(x) \neq \lim_{n \to \infty} \lim_{x \to 0} f_n(x).$$

The graph of f (see p. 202) consists of the semiopen interval $(0,1]$ on the x-axis, and the point $(0,1)$. f is obviously discontinuous at $x = 0$.

    We shall now discuss the idea of uniform convergence and show how it differs from pointwise convergence.

    Definition 11.2.3. *A sequence* $\{f_n(x)\}$ *is said to be convergent to* $f(x)$ *uniformly on S if given* $\epsilon > 0 \; \exists \, n_0(\epsilon) \in \mathfrak{N} \ni |f_n(x) - f(x)| < \epsilon$ *for* $n \geqslant n_0$ *and for every* $x \in S$.

    Write $*d_n(x) = |f_n(x) - f(x)|$, $M_n = \mathrm{lub} \; \{d_n(x) : x \in S\}$.

    If the convergence is uniform, then $d_n(x) < \epsilon$ for $n \geqslant n_0(\epsilon)$ and for all $x \in$

$*d_n(x)$ is bounded if $\{f_n(x)\}$ converges.

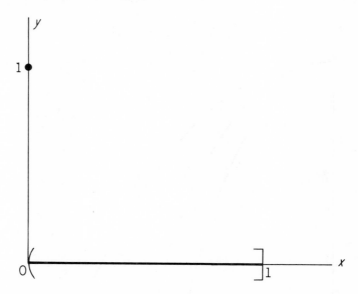

S.   Hence $M_n \leqslant \epsilon$ for $n \geqslant n_0(\epsilon)$. Since this is true for every ϵ, $\lim\limits_{n \to \infty} M_n = 0$.

Conversely, if $\lim\limits_{n \to \infty} M_n = 0$, then for a given $\epsilon > 0$, $M_n < \epsilon$ for $n \geqslant n_1(\epsilon)$ and so $d_n(x) \leqslant M_n < \epsilon$ for $n \geqslant n_1$ and for all $x \in S$, which implies uniform convergence of $\{f_n(x)\}$. Hence, we have the following:

Theorem 11.2.1.   Let $\{f_n(x)\}$ converge to $f(x)$ pointwise for every $x \in S$, and let $M_n = \mathrm{lub}\ \{|f_n(x) - f(x)|: x \in S\}$. Then $\lim\limits_{n \to \infty} M_n = 0$ is a necessary and sufficient condition for the uniform convergence of $\{f_n(x)\}$ to $f(x)$ on S.

We now show that the sequences of functions in Examples 11.2.1 and 11.2.2 are not uniformly convergent, even though they are pointwise convergent.

In Example 11.2.1,

$$f(x) = \lim_{n \to \infty} f_n(x) = 0, \text{ if } x \neq 0$$
$$f(0) = \lim_{n \to \infty} f_n(0) = 1.$$

Hence,

$$d_n(x) = \frac{a^2}{a^2 + n^2 x^2} \qquad \text{if } x{-} \neq 0$$

$$= 0 \qquad \qquad \text{if } x \ = 0.$$

Thus if S is the interval $[-1, 1]$ we see that $M_n \geqslant a^2/(a^2 + 1)$ for $n \geqslant 1$, and $\lim\limits_{n \to \infty} M_n \neq 0$. Therefore, the convergence is not uniform on $[-1, 1]$. Note, however, that if we take S to be an interval $[A, B]$ $0 < A < B$, which excludes the origin, then $M_n = a^2/(a^2 + n^2 A^2) \longrightarrow 0$ and hence the convergence is uniform on $[A, B]$.

In Example 11.2.2,

$$d_n(x) = 1 - nx^2 \qquad 0 < x < \frac{1}{\sqrt{n}}$$

$$= 0 \qquad \qquad \text{otherwise}$$

and consequently for $S = [0, 1]$, $M_n > d_n \left(\frac{1}{2\sqrt{n}} \right) = \frac{3}{4}$. Hence $\{f_n\}$ is not uniformly convergent on $[0, 1]$. |

EXAMPLE 11.2.3. Let

$$u_n(x) = nx^2, \text{ if } x \in [0, 1/n)$$
$$u_n(x) = x, \quad \text{if } x \in [1/n, 1]$$

Then

$$u(x) = \lim_{n \to \infty} u_n(x) = x \qquad \text{if } x \in (0, 1]$$

$$u_n(0) = 0$$

Hence $u_n(x)$ converges pointwise to $u(x) = x$ for $0 \leqslant x \leqslant 1$. We now show that $\{u_n(x)\}$ also converges uniformly to $u(x)$ on $[0, 1]$. Here

$$d_n(x) = |nx^2 - x|, \quad x \in [0, 1/n)$$
$$= 0 \qquad \qquad x \in [1/n, 1].$$

For a fixed n, the maximum value of $x - nx^2$ occurs at $x = 1/2n$. Thus $M_n = 1/2n \longrightarrow 0$, and the uniform convergence on $[0, 1]$ follows.

Drawing the graphs of the functions $\{u_n(x)\}$ one can observe that each $u_n(x)$ is continuous and consists of a segment of a parabola from $(0, 0)$ to $\left(\frac{1}{n}, \frac{1}{n} \right)$ and then a segment of the line $y = x$ from $\left(\frac{1}{n}, \frac{1}{n} \right)$ to $(1, 1)$. Furthermore, the limit function $u(x) = x$ is continuous at every point of $[0, 1]$.

Thus $\lim_{x \to x_0} \lim_{n \to 0} f_n(x) = \lim_{n \to \infty} \lim_{x \to x_0} f_n(x)$ for $x_0 \in [0, 1]$. (See p. 204.)

We now discuss a theorem of which Example 11.2.3 is a special case.

Theorem 11.2.2. Let $\{f_n\}$ be a sequence of functions which converges to f uniformly on S, and let each f_n be continuous on S. Then f is continuous on S.

Proof. Let $d_n(x) = |f_n(x) - f(x)|$, and $M_n = \text{lub } \{d_n(x) : x \in S\}$. Since f_n converges uniformly to f, it follows from Theorem 11.2.1 that $\lim_{n \to \infty} M_n = 0$, which means given $\epsilon > 0$ there is a natural number n_0 such that $M_n < \epsilon/3$ for $n \geqslant n_0$. This implies $d_n(x) \leqslant M_n < \epsilon/3$ for $n \geqslant n_0$ and for every $x \in S$. Let $x_0 \in S$.

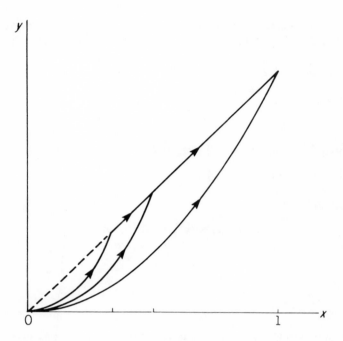

Since f_n is continuous at x_0, for $\epsilon > 0$ there is a neighborhood N_δ such that $|f_n(x) - f_n(x_0)| < \epsilon/3$ for $x \in N_\delta$. Now $|f(x) - f(x_0)| = |f(x) - f_n(x) + f_n(x) - f_n(x_0) + f_n(x_0) - f(x_0)| \leq |f(x) - f_n(x)| + |f_n(x) - f_n(x_0)| + |f_n(x_0) - f(x_0)| = d_n(x) + d_n(x_0) + |f_n(x) - f_n(x_0)| < \epsilon/3 + \epsilon/3 + \epsilon/3 = \epsilon$ for $x \in N_\delta$.

Thus f is continuous at x_0. ∎

According to this theorem, "uniform convergence" is a sufficient condition for the preservance of "continuity." It is not a necessary condition as may be seen from the following example.

EXAMPLE 11.2.4. Let $S = \{x : x > 0\}$ and $f_n(x) = \dfrac{1}{1 + nx^2}$. $\{f_n\}$ converges to f, where $f(x) = 0$ for $x \in S$, and the convergence is not uniform (why?). It is quite easy to see that each f_n is continuous on S and so is f.

Let us now investigate the following converse of Theorem 11.2.2.

"Given a sequence $\{f_n\}$ which converges pointwise to a function f on S, and that each f_n is continuous on S. Does the continuity of f imply the uniform convergence of $\{f_n\}$ on S?"

The answer is in negative as is clear from the following example.

EXAMPLE 11.2.5. Consider the sequence

$$f_n(x) = \frac{x^2}{n} \qquad \text{on the entire set of real numbers } R$$

$$f(x) = \lim_{n \to \infty} f_n(x) = 0 \qquad \text{for } x \in R.$$

Each f_n is continuous and so is f. But the convergence is not uniform. No matter what n is, we can find x such that $x^2 > n$, and that makes $|f_n(x) - f(x)| = \dfrac{x^2}{n} > 1$.

The next theorem tells us what more do we need for $\{f_n\}$ besides continuity of every f_n, and the continuity of the limit function to make the convergence uniform. This theorem is due to Ulisse Dini (1845–1918), who made significant contributions in analysis, particularly Fourier series.

Before we prove the theorem we need the following definition.

Definition 11.2.4. *A sequence $\{f_n\}$ is said to be monotone on S if either (i) for each $x \in S$ $\{f_n(x)\}$ is a nondecreasing sequence, or (ii) for each $x \in S$ $\{f_n(x)\}$ is nonincreasing.*

Theorem 11.2.3. Let $\{f_n\}$ be a sequence of functions which converges pointwise to a function f on a set S. If (i) every f_n is continuous on S, (ii) f is continuous on S, (iii) $\{f_n\}$ is monotone on S, and (iv) S is compact, then the convergence is uniform on S.

Proof. Let $\epsilon > 0$, and $y \in S$. Since $\{f_n\}$ converges to f, there exists a natural number M_y such that

$$|f_{M_y}(y) - f(y)| < \epsilon/3 \tag{11.1}$$

Since f_{M_y} and f are both continuous, there is an open set G_y containing y such that

and

$$\left.\begin{array}{c} |f_{M_y}(x) - f_{M_y}(y)| < \epsilon/3 \\[2mm] |f(x) - f(y)| < \epsilon/3 \end{array}\right\} \quad \text{for } x \in G_y \cap S \tag{11.2}$$

Now $|f_{m_y}(x) - f(x)| = |f_{m_y}(x) - f_{m_y}(y) + f_{m_y}(y) - f(y) + f(y) - f(x)| \leqslant |f_{m_y}(x) - f_{m_y}(y)| + |f_{m_y}(y) - f(y)| + |f(y) - f(x)| < \epsilon/3 + \epsilon/3 + \epsilon/3$.

Thus $|f_{m_y}(x) - f(x)| < \epsilon$ for $x \in G_y \cap S$. Since the sequence $\{f_n\}$ is monotone, for $n \geqslant m_y$, $f_n(x)$ is closer to $f(x)$ than $f_{m_y}(x)$, which means

$$|f_n(x) - f(x)| < \epsilon \qquad \text{for } n \geqslant m_y \text{ and } x \in G_y \cap S \tag{11.3}$$

The class $\{G_y : y \in S\}$ of all open sets G_y will form an open covering of S, and since S is compact there is a finite subcover $\{G_{y_1}, G_{y_2}, G_{y_3}, \ldots, G_{y_p}\}$. Let

$$M = \{\max \ m_{y_1}, m_{y_2}, \ldots, m_{y_p}\}.$$

Now, if $x_0 \in S$, then $x_0 \in G_{y_k}$, for some k $1 \leqslant k \leqslant p$; and if $n \geqslant M$

then $n \geqslant m_{y_k}$. Using (11.3) we have

$$|f_n(x_0) - f(x_0)| < \epsilon \qquad \text{for } n \geqslant M$$

x_0 is an arbitrary point of S and M is independent of x_0, and hence the convengence is uniform. ∎

We now give the Cauchy criterion for uniform convergence.

Theorem 11.2.4. The sequence of functions $\{f_n\}_{n=1}^{\infty}$ defined on S converges uniformly on S if and only if given $\epsilon > 0$, there exists $n_0(\epsilon) \in \mathfrak{N}$ such that

$$|f_m(x) - f_n(x)| \leqslant \epsilon \qquad (m, n \geqslant n_0; \ x \in S) \qquad (11.2)$$

Proof. Suppose first that $\{f_n\}_{n=1}^{\infty}$ is a uniformly convergent sequence of functions on S converging to f on S. Then, given $\epsilon > 0$, there exists $n_0(\epsilon) \in \mathfrak{N}$ such that for all $n \geqslant n_0$ and $x \in S$

$$|f_n(x) - f(x)| \leqslant \epsilon/2$$

Hence if $\min(m, n) \geqslant n_0$, we have for any $x \in S$,

$$|f_m(x) - f_n(x)| \leqslant |f_m(x) - f(x)| + |f(x) - f_n(x)| \leqslant \epsilon$$

Conversely suppose that the Cauchy condition (11.2) holds. Then for each fixed $x \in S$, the sequence $\{f_n(x)\}$ is a Cauchy sequence of numbers and so converges. We define f by $f(x) = \lim_{n \to \infty} f_n(x)$. Then given $\epsilon > 0$ we choose n_0 such that (11.2) holds whenever $\min(m, n) \geqslant n_0$. Fixing n and letting $m \longrightarrow \infty$ in (11.2), we obtain

$$|f_n(x) - f(x)| \leqslant \epsilon \qquad (n \geqslant n_0, \ x \in S).$$

This shows that the sequence $\{f_n\}$ converges uniformly to f on S. ∎

11.3. Some Properties of Uniformly Convergent Sequences

Theorem 11.3.1. Let S be the interval $[a, b]$. If the sequence $\{f_n(x)\}_{n=1}^{\infty}$ of real-valued R-integrable functions on S converges uniformly to $f(x)$ on S, then $f(x)$ is also R-integrable on S and

$$\int_a^b f(x)\, dx = \lim_{n \to \infty} \int_a^b f_n(x)\, dx$$

Proof. Since $\{f_n(x)\}_{n=1}^{\infty}$ converges uniformly to $f(x)$, given $\epsilon > 0$ we can choose $n_0(\epsilon) \in \mathfrak{N}$ such that

$$|f_n(x) - f(x)| \leqslant \frac{\epsilon}{3(b - a)} \qquad (11.3)$$

for $n \geq n_0$ and for all $x \in S$. Hence

$$|f(x)| \leq |f_n(x)| + \frac{\epsilon}{3(b-a)}.$$

This shows that $f(x)$ is bounded. To prove that f is R-integrable on S, we note that there is a partition P of $[a,b]$ such that

$$S(f_n,P) - s(f_n,P) \leq \epsilon/3$$

From (11.3) we have

$$f_n(x) - \frac{\epsilon}{3(b-a)} \leq f(x) \leq f_n(x) + \frac{\epsilon}{3(b-a)}$$

Thus

$$s(f_n,P) - \epsilon/3 < s(f,P) \leq S(f,P) < s(f_n,P) + \epsilon/3$$

and so

$$S(f,P) - s(f,P) < S(f_n,P) - s(f_n,P) + 2\epsilon/3 < \epsilon.$$

Hence f is R-integrable on $[a,b]$. Now

$$\left| \int_a^b f_n(x)\, dx - \int_a^b f(x)\, dx \right| = \left| \int_a^b (f_n(x) - f(x))\, dx \right|$$

$$\leq \int_a^b |f_n(x) - f(x)|\, dx \leq \frac{\epsilon}{b-a}(b-a) = \epsilon$$

and the theorem is proved. ∎

Corollary. If $\phi_n(x) = \displaystyle\int_a^x f_n(t)\, dt$ for $a \leq x \leq b$, then $\{\phi_n(x)\}$ is uniformly convergent on $[a,b]$.

Proof. For $n \geq n_0$,

$$\left| \int_a^x f_n(t)\, dt - \int_a^x f(t)\, dt \right| \leq \int_a^x |f_n(t) - f(t)|\, dt$$

$$< \frac{\epsilon}{3(b-a)}(x-a) \leq \frac{\epsilon}{3}. \quad ∎$$

Theorem 11.3.2. Suppose that $\{f_n(x)\}_{n=1}^{\infty}$ converges to $f(x)$ on $[a,b]$ and that each $f_n(x)$ is real-valued and has a continuous derivative on $[a,b]$. If the sequence $\{f'_n(x)\}^{\infty}_{n=1}$ converges uniformly on $[a,b]$, then $f'(x)$ exists and $\lim\limits_{n \to \infty} f'_n(x) = f'(x)$ on $[a,b]$.

Proof. Since $\{f'_n(x)\}$ converges uniformly on $[a,b]$,

$$\lim_{n \to \infty} f'_n(x) = F(x)$$

is continuous on $[a,b]$. By Theorem 11.3.1 we have, for $a \leqslant x \leqslant b$,

$$\int_a^x F(t)\, dt = \lim_{n \to \infty} \int_a^x f'_n(t)\, dt$$

$$= \lim_{n \to \infty} \{f_n(x) - f_n(a)\}$$

$$= f(x) - f(a).$$

Hence

$$f(x) = f(a) + \int_a^x F(t)\, dt.$$

Since $F(x)$ is continuous on $[a,b]$, it follows that $f(x)$ is differentiable and $f'(x) = F(x)$, that is,

$$\lim_{n \to \infty} f'_n(x) = F(x) = f'(x). \; \blacksquare$$

11.4.   Uniform Convergence of Series

We shall now consider the uniform convergence of a series $\sum\limits_{n=1}^{\infty} a_n(x)$.

Definition 11.4.1. *A series* $\sum\limits_{n=1}^{\infty} a_n(x)$ *converges uniformly to a function*

$f(x)$ *on S if the sequence of partial sums* $\{s_n(x)\}^{\infty}_{n=1}$, $s_n(x) = \sum\limits_{i=1}^{n} a_i(x)$, *converges*

uniformly to $f(x)$ on S.

The following test gives the uniform (and also absolute) convergence of a series of real or complex terms.

Theorem 11.4.1 (Weierstrass's M-Test). The series $\sum\limits_{n=1}^{\infty} a_n(x)$ is uniformly

convergent on *S* if

(i) $|a_n(x)| \leqslant M_n$ $(n = 1, 2, \ldots; x \in S)$

(ii) $\displaystyle\sum_{n=1}^{\infty} M_n$ is convergent.

Proof.   Given $\epsilon > 0$ we can find $n_0(\epsilon) \in \mathfrak{N}$ such that for $m > n \geqslant n_0(\epsilon)$,

$$M_{n+1} + M_{n+2} + \cdots + M_m < \epsilon.$$

Consequently for $x \in S$,

$$\left| \sum_{r=n+1}^{m} a_r(x) \right| \leqslant \sum_{r=n+1}^{m} |a_r(x)| \leqslant \sum_{r=n+1}^{m} M_r < \epsilon.$$

The uniform convergence of the series $\displaystyle\sum_{n=1}^{\infty} a_n(x)$ on S follows from Cauchy's criterion. ∎

EXAMPLE 11.4.1.   The series $\displaystyle\sum_{n=1}^{\infty} r^n \cos n\theta, \sum_{n=1}^{\infty} r^n \sin n\theta \ (0 < r < 1)$ converge uniformly for all real values of θ.

For $|r^n \cos n\theta| \leqslant r^n$ and $\displaystyle\sum_{n=1}^{\infty} r^n$ is convergent.

EXAMPLE 11.4.2.   The series $\displaystyle\sum_{n=1}^{\infty} \frac{x}{n(1 + nx^2)}$ is uniformly convergent on any interval $[a, b]$. For

$$\frac{\dfrac{1}{x} + nx}{2} \geqslant \sqrt{nx \frac{1}{x}} = \sqrt{n} \ (x > 0)$$

and so

$$\left| \frac{x}{n(1 + nx^2)} \right| \leqslant \frac{1}{2n^{3/2}} = M_n$$

It is easily seen that this inequality holds when $x \leqslant 0$, and the M-test applies.

Theorem 11.4.2 (Dirichlet's Test).   Let $A_n(x) = \displaystyle\sum_{k=1}^{n} a_k(x)$.   The series

$\displaystyle\sum_{k=1}^{\infty} a_k(x) b_k(x)$ is uniformly convergent on S, if on S

(i) $\{A_n(x)\}_1^\infty$ is uniformly bounded,

(ii) $\{b_n(x)\}_1^\infty$ is monotonic decreasing for every fixed x and converges uniformly to 0.

The proof of this theorem and of the next theorem are left to the reader as exercises. (See Sec. 8.6.) Note that in this and the next test all $b_n(n)$ are real.

Theorem 11.4.3 (Abel's Test). The series $\displaystyle\sum_{n=1}^{\infty} a_n(x)b_n(x)$ converges uniformly on S if the following hold:

(i) $\displaystyle\sum_{n=1}^{\infty} a_n(x)$ converges uniformly on S.

(ii) There exists a constant K such that $|b_n(x)| \leqslant K$ for every $n \in \mathfrak{N}$ and x.

(iii) $\{b_n(x)\}_{n=1}^\infty$ is a monotonic sequence for every fixed x.

EXAMPLE 11.4.3.

$$\sum_{n=1}^{\infty} \frac{\sin nx}{n^p}, \quad \sum_{n=1}^{\infty} \frac{\cos nx}{n^p}$$

When $p > 1$, M-test shows that both series are uniformly convergent for all real values of x. When $0 < p \leqslant 1$, Dirichlet's test shows that both series converge uniformly on $[\delta, 2\pi - \delta]$, where $0 < \delta < 2\pi$. Take $b_n = \dfrac{1}{n}$ and use inequality (8.9).

Theorem 11.4.4. If $\displaystyle\sum_{n=1}^{\infty} a_n(x)$ is uniformly convergent on S and if each $a_n(x)$ is continuous on S, then the sum function $f(x) = \displaystyle\sum_{n=1}^{\infty} a_n(x)$ is continuous on S.

Proof. See Theorem 11.2.2.

In the next two theorems we shall assume that $a_n(x), n = 1, 2, \ldots$, are real-valued.

Theorem 11.4.5. If the series $\displaystyle\sum_{n=1}^{\infty} a_n(x)$ is uniformly convergent to $f(x)$ on $[a,b]$ then

$$\int_a^b f(x)\,dx = \sum_{n=1}^{\infty} \int_a^b a_n(x)\,dx$$

provided each $a_n(x)$ is R-integrable on $[a,b]$.

Proof. See Theorem 11.3.1.

Theorem 11.4.6. Let $\displaystyle\sum_{n=1}^{\infty} a_n(x)$ converge to $f(x)$ on $[a,b]$. If each $a_n(x)$

has a continuous derivative on $[a,b]$ and if $\displaystyle\sum_{n=1}^{\infty} a'_n(x)$ is uniformly convergent on

$[a,b]$, then $f'(x)$ exists and

$$f'(x) = \frac{d}{dx}\left\{\sum_{n=1}^{\infty} a_n(x)\right\} = \sum_{n=1}^{\infty} a'_n(x), \qquad a \leqslant x \leqslant b.$$

Proof. See Theorem 11.3.2.

Exercise 11.4

1. Consider the uniform convergence of the series

(i) $\displaystyle\sum_{n=1}^{\infty} \frac{1}{x^2 + n^2}$

(ii) $\displaystyle\sum_{n=0}^{\infty} \frac{x^n}{n!}$

(iii) $\displaystyle\sum_{n=1}^{\infty} \left(\frac{x}{3}\right)^n$

(iv) $\displaystyle\sum_{n=2}^{\infty} \frac{x^n}{n \log^2 n}$

2. Prove that

$$\sum_{n=1}^{\infty} \frac{x}{\{1 + (n-1)x\}\,\{1 + nx\}}$$

is uniformly convergent in any interval $[a,b]$, $b > a > 0$. $\left(\textit{Hint: Here}\right.$

$$a_n(x) = \frac{1}{1 + (n-1)x} - \frac{1}{1 + nx}$$

and

$$S_n(x) = 1 - \frac{1}{1 + nx}\left.\right)$$

3. Prove that the sequence $\{x^n(1-x)\}_{n=1}^{\infty}$ converges uniformly to zero on $[0,1]$.

$\left[\textit{Hint:} \lim_{n\to\infty} f_n(x) = 0,\ 0 \leqslant x \leqslant 1;\right.$ the maximum of $f_n(x)$ occurs at $x = \left.\dfrac{n}{n+1}.\right]$

4. Suppose that both $\{f_n(x)\}_{n=1}^{\infty}$ and $\{F_n(x)\}_{n=1}^{\infty}$ converge uniformly on S. Prove that $\{f_n(x) + F_n(x)\}_{n=1}^{\infty}$ converges uniformly on S.

5. If $\displaystyle\sum_{n=1}^{\infty} |a_n(x)|$ converges uniformly on S, prove that so does $\displaystyle\sum_{n=1}^{\infty} a_n(x)$.

6. If $\displaystyle\sum_{n=1}^{\infty} a_n$ converges or oscillates finitely, show that

$$\sum_{n=1}^{\infty} \frac{a_n}{n^x}$$

is a continuous function of x on $[a,b]$ where $0 < a < b$.

$\left[\textit{Hint:}\ \text{Use Dirichlet's test with } b_n(x) = \dfrac{1}{n^x}.\ \text{Note that } b_n(x) \leqslant \dfrac{1}{n^a} \downarrow 0.\right]$

7. Consider the uniform convergence of the series $\displaystyle\sum_{n=1}^{\infty} \frac{1}{n^x}$ and prove that differentiation and integration term-by-term are permissible on $[a,b]$ where $1 < a < b$.

$\left[\textit{Hint:}\ \text{The series}\ \displaystyle\sum_{n=1}^{\infty}\frac{1}{n^x}\ \text{and}\ \sum_{n=1}^{\infty}\frac{(-\log n)}{n^x}\ \text{are both uniformly convergent}\right.$

on $[a,b]$ by M-test. $\Big]$

8. Show that the series

$$f(x) = \sum_{n=1}^{\infty} \frac{1}{n^p + n^q x^2} \qquad (p > 1, q \geqslant 0, 2p > q + 1)$$

is uniformly convergent on any interval $[a,b]$ and that $f'(x)$ is given by term-by-term differentiation.

$$\left[\text{Hint:} \right.$$

$$0 < a_n(x) \leqslant \frac{1}{n^p} = M_n$$

If $|x| \leqslant 1$, $|a'_n(x)| \leqslant \frac{2n^q}{n^{2p}}$; if $|x| \geqslant 1$ then use the arithmetic mean $\geqslant$ the geometric mean inequality to get

$$|a'_n(x)| \leqslant \frac{2n^q |x|}{4n^{p+q} |x|^2} \leqslant \frac{1}{2n^p}$$

Hence for all x

$$\left. |a'_n(x)| \leqslant \frac{2n^q}{n^{2p}} + \frac{1}{2n^p} \equiv M_n. \right]$$

9. Prove that the series $\displaystyle\sum_{n=1}^{\infty} \frac{\sin nx}{n}$ is not uniformly convergent on any interval that includes $x = 0$.

$$\left[\text{Hint: When } x = \frac{\pi}{4n}, \right.$$

$$\frac{\sin nx}{n} + \frac{\sin (n+1)x}{n+1} + \cdots + \frac{\sin 2nx}{2n}$$

$$> \frac{1}{2n} (\sin nx + \cdots + \sin 2nx)$$

$$\left. > \frac{1}{2n} (n+1) \sin \frac{\pi}{4} > \frac{1}{2\sqrt{2}} \right]$$

10. Prove that for all x and n

$$\left| \sum_{k=1}^{n} \frac{\sin kx}{k} \right| \leqslant 2\sqrt{\pi}.$$

(*Hint*: Consider first the case $0 < x < \pi$. Let m satisfy

$$m \leqslant \frac{\sqrt{\pi}}{x} < m + 1.$$

Then

$$\left| \sum_{k=1}^{n} \frac{\sin kx}{k} \right| \leqslant \sum_{k=1}^{m} \left| \frac{\sin kx}{k} \right| + \left| \sum_{k=m+1}^{n} \frac{\sin kx}{k} \right| = \Sigma_1 + \Sigma_2, \text{ say,}$$

where Σ_1 vanishes if $m = 0$ and Σ_2 vanishes if $m > n$. Since $|\sin kx| \leqslant kx$,

$$\Sigma_1 \leqslant \sum_{k=1}^{m} \frac{kx}{k} = xm \leqslant \sqrt{\pi}$$

and since $\sin x/2 \geqslant x/\pi$,

$$\Sigma_2 < \frac{1}{(m+1)} \frac{1}{\sin\left(\dfrac{x}{2}\right)} \leqslant \frac{1}{(m+1)} \frac{\pi}{x} < \sqrt{\pi}.$$

Hence the inequality follows for $0 < x < \pi$, and so for $-\pi < x < 0$. It is trivially true for $x = 0, \pm\pi$. By periodicity the inequality holds for all x.)

11. Suppose that (i) $a_n(x)$ $(n \in \mathfrak{N})$ is real on $[a,b]$, (ii) the series $\displaystyle\sum_{n=1}^{\infty} a_n(x)$ is convergent on $[a,b]$, and (iii) there exists a constant K such that

$$\left| \sum_{k=1}^{n} a_k(x) \right| \leqslant K$$

for all $n \in \mathfrak{N}$ and all $x \in [a,b]$, and (iv) the series is uniformly convergent on $[a,b]$ except in the neighborhood of a finite number of points on $[a,b]$. Prove that the series can be integrated term by term on $[a,b]$.

Section 11.5.   Uniform Convergence of Power Series

In this section we consider the uniform convergence and differentiation term-by-term of a power series

$$\sum_{n=o}^{\infty} a_n x^n \tag{11.3}$$

We restrict ourselves to real values of x and assume that the radius of convergence $\mathfrak{R}$ of the series (11.3) is positive.

Theorem 11.5.1.   If the radius of convergence of the power series (11.3) is $\mathfrak{R}(\neq 0)$, then the power series is uniformly convergent on $[-r, r]$ where $0 < r < \mathfrak{R}$.

Proof.

$$|a_n x^n| \leqslant |a_n| r^n$$

and $\Sigma \, |a_n| r^n$ is convergent. Thus by M-test the series (11.3) is uniformly convergent on $[-r,r]$. ▌

Let the sum of the series be

$$f(x) = \sum_{n=0}^{\infty} a_n x^n \qquad (11.4)$$

Then $f(x)$ is continuous in $(-\mathcal{R}, \mathcal{R})$ (Theorem 11.4.4).

Theorem 11.5.2. The radius of convergence of the power series

$$\sum_{n=1}^{\infty} n a_n x^{n-1} \qquad (11.5)$$

is also $\mathcal{R}$.

Proof. Let $\mathcal{R}'$ denote the radius of convergence of the series (11.5). Then for $n \geqslant 1$, $|a_n x^n| \leqslant |x \| n a_n x^{n-1}|$. It follows from the Comparison test that $\mathcal{R}' \leqslant \mathcal{R}$. To prove $\mathcal{R}' \geqslant \mathcal{R}$ we may suppose that $\mathcal{R} > 0$. If $0 < \epsilon < \mathcal{R}$ then $|a_n(\mathcal{R}-\epsilon)^n| < 1$ for $n \geqslant n_0(\epsilon)$ and so $n|a_n x^{n-1}| \leqslant n|x|^{n-1}/(\mathcal{R}-\epsilon)^n$. By the Ratio test, the series $\sum_{n-1}^{\infty} n|x|^{n-1}/(\mathcal{R}-\epsilon)^n$ is convergent if $|x| < \mathcal{R}-\epsilon$ and this implies, since ϵ is arbitrary, that $\mathcal{R}' \geqslant \mathcal{R}$. ▌

Corollary. The series (11.5) is uniformly convergent on $[-r,r]$ where $0 < r < \mathcal{R}$ and

$$f'(x) = \sum_{n=1}^{\infty} n a_n x^{n-1} \qquad (-r \leqslant x \leqslant r) \qquad (11.6)$$

Proof. Since the series (11.5) is uniformly convergent, and the series (11.4) is convergent, on $[-r,r]$ Theorem 11.4.6 gives (11.6).

Applying this result successively to $f', f'', \ldots$ we conclude that the power series (11.3) can be differentiated term-by-term k times to obtain

$$f^{(k)}(x) = \sum_{n=k}^{\infty} n(n-1) \ldots (n-k+1) a_k x^{n-k} (-\mathcal{R} < x < \mathcal{R}). \qquad (11.7)$$

If we put $x = 0$, we obtain

$$a_k = \frac{f^{(k)}(0)}{k!}. \qquad (11.8)$$

Hence the Taylor expansion of $f(x) = \displaystyle\sum_{n=0}^{\infty} a_n x^n$ in powers of x is the original series.

Theorem 11.5.3 (Uniqueness Theorem for Power Series). If the two power series $\displaystyle\sum_{n=0}^{\infty} a_n x^n$ and $\displaystyle\sum_{n=0}^{\infty} b_n x^n$ converge on some interval $(-\mathfrak{R}, \mathfrak{R})$ $(\mathfrak{R} > 0)$ to the same function $f(x)$ then $a_n = b_n$ for all $n \in \mathfrak{N}$.

Proof. By (11.8) we have

$$a_n = \frac{f^{(n)}(0)}{n!}, \quad b_n = \frac{f^{(n)}(0)}{n!}. \quad \blacksquare$$

Exercise 11.5

1. If the radius of convergence of $\displaystyle\sum_{0}^{\infty} a_n x^n$ is $\mathfrak{R}$ then prove that

    $$\int_0^r \sum_{n=0}^{\infty} a_n x^n \ dx = \sum_{n=0}^{\infty} a_n \frac{r^{n+1}}{n+1} \quad r < \mathfrak{R}$$

2. Prove that the series

    $$\sum_{n=1}^{\infty} (-1)^{n+1} \frac{x^n}{n}$$

    converges for $|x| < 1$ and for $x = 1$ but that it diverges for $x = -1$.

3. Let $f(x) = e^{-1/x^2}, x \neq 0, f(0) = 0$.
    (i)   Show that $f^{(n)}(0) = 0$ for all $n \in \mathfrak{N}$

    (ii) Show that the Maclaurin series $\displaystyle\sum_{n=0}^{\infty} \frac{f^{(n)}(0)}{n!} x^n$ converges everywhere and
    does not represent $f(x)$ except at $x = 0$.

    Hint: Prove, by induction on n, that there exists a polynomial p_n such that

    $$f^{(n)}(x) = p_n\left(\frac{1}{x}\right) e^{-1/x^2}, \quad x \neq 0.$$

4. Prove that if $|f^{(n)}(x)| \leqslant K$ for all $|x| < r$ $(r > 0)$ and $n = 0, 1, 2, \ldots$ then the series

$$\sum_{n=0}^{\infty} \frac{f^{(n)}(0)}{n!} x^n$$

converges to $f(x)$ for $|x| < r$.
(*Hint:* Use Taylor's formula with a remainder.)

5. Prove that the series

$$\sum_{n=0}^{\infty} \frac{x^n}{n!}, \quad \sum_{n=1}^{\infty} (-1)^{n+1} \frac{x^{2n+1}}{(2n+1)!}, \quad \sum_{n=0}^{\infty} (-1)^{n+2} \frac{x^{2n}}{(2n)!}$$

converge for all values of x.

6. Find the interval of convergence of the series

$$\sum_{n=1}^{\infty} \frac{(x-5)^n}{n^2}$$

$\left(\textit{Hint:} \text{ The series } \sum_{n=1}^{\infty} \frac{X^n}{n^2} \text{ is convergent for } X = \pm 1. \text{ Hence the given series} \right.$
is convergent for $4 \leqslant x \leqslant 6.\Big)$

11.6. Cesàro and Abel Summability Methods

In this section we generalize the concept of the sum of a convergent series. If $\sum_{k=0}^{n} a_k = S_n$ and if the sequence $\{S_n\}_0^{\infty}$ converges to S, then we associate the number S with the series $\sum_{n=0}^{\infty} a_n$ as its sum. It may happen that $\{S_n\}_0^{\infty}$ is not convergent but the sequence $\{\sigma_n\}_0^{\infty}$, where

$$\sigma_n = \frac{S_0 + S_1 + S_2 + \cdots + S_n}{n+1}, \tag{11.9}$$

converges to L, say. Then we can consider this number L as a conventional sum of the series $\sum_{n=0}^{\infty} a_n$ or as a conventional limit of the sequence $\{S_n\}_0^{\infty}$.

Definition 11.6.1. *Let $\{S_n\}_{n=0}^{\infty}$ be a sequence of real or complex numbers and let $\{\sigma_n\}_0^{\infty}$ be the sequence of arithmetic means defined by (11.9). If*

$\{\sigma_n\}_0^\infty$ *converges to* L *then* $\{S_n\}_0^\infty$ *is said to be summable by Cesàro means of the first order, or briefly summable* $(C,1)$, *to* L.

If the series $\displaystyle\sum_{n=0}^\infty a_n$ has partial sums $\{S_n\}$ and the sequence $\{S_n\}_0^\infty$ is summable

$(C,1)$ to L, then the series $\displaystyle\sum_{n=0}^\infty a_n$ is said to be summable $(C,1)$ and L is said to be

the $(C,1)$ sum.

EXAMPLE 11.6.1.
(a)   Let $S_n = \begin{cases} 1, n = 0,2,4,\ldots \\ 0, n = 1,3,5,\ldots . \end{cases}$

Then

$$\sigma_n = \frac{\left[\dfrac{n}{2}\right] + 1}{n + 1} \quad \text{or} \quad \frac{\left[\dfrac{n + 1}{2}\right]}{n + 1}$$

according as n is even or odd. Thus $\sigma_n \longrightarrow \dfrac{1}{2}$ as $n \longrightarrow \infty$. Hence the sequence

$\{S_n\}_0^\infty$ is summable $(C,1)$ to $\dfrac{1}{2}$ but it is not convergent.

(b)   Let $a_n = (-1)^{n+2}$.   Then $\displaystyle\sum_{n=0}^\infty a_n$ is not convergent $(a_n \neq o(1))$.

Since $\{S_n\} = 1, 0, 1, 0, \ldots,$ $\sigma_n \longrightarrow \dfrac{1}{2}$ as $n \longrightarrow \infty$ and so the series $\displaystyle\sum_{n=0}^\infty a_n$ is sum-

mable $(C,1)$ to $\dfrac{1}{2}$.

EXAMPLE 11.6.2.   Let

$$S_n = \begin{cases} n, \text{ if } n \text{ is odd}, \\ -(n - 1), \text{ if } n \text{ is even}. \end{cases}$$

Then $\displaystyle\liminf_{n\to\infty} \sigma_n = 0$, $\displaystyle\limsup_{n\to\infty} \sigma_n = 1$. The sequence $\{S_n\}$ is not summable $(C,1)$.

EXAMPLE 11.6.3.   Let $\displaystyle\sum_{n=0}^\infty a_n$ be summable $(C,1)$.   Then $a_n = o(n)$, $S_n =$

$o(n)$. For

$$\sigma_n - \left(1 - \frac{1}{n+1}\right)\sigma_{n-1} = \frac{S_n}{n+1}.$$

Hence $S_n = o(n)$ and $a_n = S_n - S_{n-1} = o(n)$. (Compare with Theorem 8.2.2.)
If a sequence $\{S_n\}_0^\infty$ is convergent to S then it is summable $(C, 1)$ to S.

Theorem 11.6.1. If $\lim\limits_{n\to\infty} S_n = S$ then

$$\lim_{n\to\infty} \frac{S_0 + S_1 + \cdots + S_n}{n+1} = S$$

Proof. The proof follows from Exercise 4.1(5).

We now prove a theorem on sequences and deduce that if a series of real terms diverges to $+\infty$ (or $-\infty$) then it cannot be summable $(C,1)$.

Theorem 11.6.2. If $\{b_n\}_0^\infty$ is a sequence of nonnegative numbers that increases strictly to $+\infty$ and if $\{a_n\}$ is any sequence of real numbers then

$$\liminf_{n\to\infty} \frac{a_{n+1} - a_n}{b_{n+1} - b_n} \leq \liminf_{n\to\infty} \frac{a_n}{b_n} \leq \limsup_{n\to\infty} \frac{a_n}{b_n} \leq \limsup_{n\to\infty} \frac{a_{n+1} - a_n}{b_{n+1} - b_n} \quad (11.10)$$

Proof. Let

$$\limsup \frac{a_{n+1} - a_n}{b_{n+1} - b_n} = L.$$

We may assume that $L < \infty$. Then for $n \geq n_0(\epsilon) \in \mathfrak{N}$,

$$a_{n+1} - a_n < (L + \epsilon)(b_{n+1} - b_n).$$

Adding these inequalities for $n = n_0, n_0 + 1, \ldots, n$ we have

$$a_{n+1} - a_{n_0} < (L + \epsilon)(b_{n+1} - b_{n_0})$$

or

$$\frac{a_{n+1}}{b_{n+1}} < \frac{a_{n_0}}{b_{n+1}} + (L + \epsilon) b_{n+1} - \frac{(L + \epsilon) b_{n_0}}{b_{n+1}}.$$

Letting $n \longrightarrow \infty$, we get

$$\limsup_{n\to\infty} \frac{a_{n+1}}{b_{n+1}} \leq L + \epsilon.$$

This proves that $\limsup\limits_{n\to\infty} \dfrac{a_n}{b_n} \leq L$. The remaining part of (11.10) can be proved similarly. ∎

Corollary 1.

$$\liminf_{n\to\infty} (a_{n+1} - a_n) \leqslant \liminf_{n\to\infty} \frac{a_n}{n} \leqslant \limsup_{n\to\infty} \frac{a_n}{n} \leqslant \limsup_{n\to\infty} (a_{n+1} - a_n) \tag{11.11}$$

Take $b_n = n(n = 0,1,2,\ldots)$ in (11.10).

Corollary 2.

$$\limsup_{n\to\infty} S_n \geqslant \limsup_{n\to\infty} \frac{S_0 + S_1 + \cdots + S_n}{n + 1}$$

$$\liminf_{n\to\infty} S_n \leqslant \liminf_{n\to\infty} \frac{S_0 + S_2 + \cdots + S_n}{n + 1} \tag{11.12}$$

Take in (11.10) $a_n = S_0 + S_2 + \cdots + S_n, b_n = n + 1$.

Corollary 3.   Let the series $\displaystyle\sum_{n=0}^{\infty} a_n$ be summable $(C,1)$ to S and suppose that

$a_n \geqslant 0$ $(n \in \mathfrak{N})$. Then $\displaystyle\sum_{n=0}^{\infty} a_n$ is convergent and has the same sum S.

Let $S_n = \displaystyle\sum_{n=0}^{n} a_k$. Then $S_n \uparrow$. If $S_n = O(1)$, then $\{S_n\}$ converges to L, say,

and (11.12) implies that $L = S$. If $\{S_n\}$ is unbounded then $\lim_{n\to\infty} S_n = \infty$ and (11.12)

implies that $\lim \sigma_n = \infty$, contradicting our hypothesis. $\blacksquare$

Corollary 4.   If the series $\displaystyle\sum_{n=0}^{\infty} a_n$ (a_n real) diverges to $+\infty$ (or $-\infty$), then the

series cannot be summable $(C, 1)$.

This follows immediately from (11.12). (Note that $\displaystyle\sum_{n=0}^{\infty} a_n$ diverges to $+\infty$

(or $-\infty$) if $S_n \longrightarrow +\infty$ (or $-\infty$).)

We have seen (Theorem 11.6.2) that if a series is convergent then it is summable $(C,1)$. The converse is not true. If however we impose a suitable condition on the magnitude of $|a_n|$ then the converse is true. Theorems of this type are called Tauberian theorems, after A. Tauber, who proved the first theorem of this type. [See Theorem 11.6.6].

Theorem 11.6.3.   Let $\sum\limits_{n=0}^{\infty} a_n$ be summable $(C,1)$ to S and suppose that

$$a_1 + 2a_2 + \cdots + na_n = o(n). \tag{11.13}$$

Then $\sum\limits_{n=0}^{\infty} a_n$ is convergent and has the sum S.

Proof.   Let $T_n = a_1 + 2a_2 + \cdots + na_n$. Then

$$(n+1)(S_n - \sigma_n) = (n+1)(a_0 + \cdots + a_n) - (S_0 + \cdots + S_n)$$
$$= a_1 + 2a_2 + \cdots + na_n = T_n.$$

Hence

$$S_n - \sigma_n = \frac{T_n}{n+1} \tag{11.14}$$

Since $\sigma_n \longrightarrow S$ and $T_n = o(n)$, we have

$$\lim_{n \to \infty} S_n = \lim_{n \to \infty} \sigma_n = S$$

Note that if the series is convergent then (11.14) implies that $T_n = o(n)$. ∎

Corollary.   If $\sum\limits_{n=0}^{\infty} a_n$ is summable $(C,1)$ and $na_n = o(1)$ then $\sum\limits_{n=0}^{\infty} a_n$ is convergent.

For $na_n = o(1)$ implies by Theorem 11.6.2, that $T_n = o(n)$. ∎

EXAMPLE 11.6.4.   (i)   The series

$$\frac{1}{2} + \cos x + \cos 2x + \cdots$$

is summable $(C,1)$ to 0 for $x \neq 2p\pi$.

For $x \neq 2p\pi$ (see Example 8.6.3)

$$S_n = \frac{1}{2} + \cos x + \cdots + \cos nx = \frac{\sin\left(n + \frac{1}{2}\right)x}{2 \sin \frac{x}{2}} \tag{11.15}$$

and

$$\sigma_n = \frac{S_0 + S_1 + \cdots + S_n}{n+1}$$

$$= \frac{1}{2(n+1)\sin\dfrac{x}{2}} \left\{ \sum_{k=0}^{n} \sin\left(k+\frac{1}{2}\right)x \right\}$$

$$= \frac{1}{4(n+1)\sin^2\dfrac{x}{2}} \left\{ \sum_{k=0}^{n} 2\sin\frac{x}{2}\sin\left(k+\frac{1}{2}\right)x \right\}$$

$$= \frac{1}{4(n+1)\sin^2\dfrac{x}{2}} \left\{ \sum_{k=0}^{n} \left(\cos kx - \cos(k+1)x\right) \right\}$$

$$= \frac{1}{4(n+1)\sin^2\dfrac{x}{2}} \left\{ 1 - \cos(n+1)x \right\}$$

$$= \frac{1}{2(n+1)\sin^2\dfrac{x}{2}} \sin^2\left(\frac{n+1}{2}\right)x \qquad (11.16)$$

Hence

$$\lim_{n\to\infty} \sigma_n(x) = 0, \qquad x \neq 2p\pi$$

Note that the series is divergent to $+\infty$ when $x = 2p\pi$ and so cannot be summable $(C,1)$ (see Corollary 4 of Theorem 11.6.2).

We now define Abel summability method.

Definition 11.6.2. *Given a series* $\displaystyle\sum_{n=0}^{\infty} a_n$, *we say that the series is Abel summable, or briefly A-summable, to sum S, if*

(i) $f(x) = \displaystyle\sum_{n=0}^{\infty} a_n x^n$ is convergent for $|x| < 1$

and

(ii) $\lim_{x\to 1-0} f(x) = S.$

EXAMPLE 11.6.6. Let

$$a_n = (-1)^{n+2} \qquad \text{[Example 11.6.1]}$$

Then

$$f(x) = 1 - x + x^2 - x^3 + \cdots = \frac{1}{1+x} \qquad |x| < 1$$

and

$$\lim_{x \to 1-0} f(x) = \frac{1}{2}$$

Hence the series $1 - 1 + 1 - 1 + \cdots$ is A-summable to $\frac{1}{2}$.

We now show that if the series $\sum_{0}^{\infty} a_n = S$ is convergent then it is A-summable to S.

Theorem 11.6.4 (Abel's Continuity Theorem).   If a series $\sum_{n=0}^{\infty} a_n$ is convergent to S then the series $\sum_{n=0}^{\infty} a_n x^n$ is uniformly convergent on $0 \leqslant x \leqslant 1$ and

$$\lim_{x \to 1-0} \sum_{n=0}^{\infty} a_n x^n = S \qquad (11.17)$$

Proof.   Given $\epsilon > 0$ there exists $n_0(\epsilon) \in \mathfrak{N}$ such that for all $n \geqslant m \geqslant n_0(\epsilon)$,

$$\left| \sum_{k=m}^{n} a_k \right| < \epsilon$$

Hence (see Inequality (8.8)) for $0 \leqslant x \leqslant 1$,

$$\left| \sum_{k=m}^{n} a_k x^k \right| \leqslant H(m,n) x^m \leqslant H(m,n) < \epsilon$$

This proves the uniform convergence of $\sum_{k=0}^{\infty} a_k x^k$ on $[0,1]$. Let its sum be denoted by $f(x)$. Then $f(x)$ is a continuous function of x on $[0,1]$ and so (11.17) holds. ∎

Theorem 11.6.5.   If $\sum_{n=0}^{\infty} a_n$ is summable $(C,1)$ to S then it is A-summable to the same sum S.

Proof.   Since Σa_n is summable $(C,1)$, $a_n = o(n)$ (Example 11.6.3), and so the series $f(x) = \sum_{n=0}^{\infty} a_n x^n$ is certainly convergent for $|x| < 1$. Write $S_n = \sum_{k=0}^{n} a_k$, $\sigma_n = \left(\sum_{k=0}^{n} S_k \right) / (n + 1)$. For $0 < x < 1$,

$$\frac{f(x)}{1 - x} = \sum_{n=0}^{\infty} a_n x^n \sum_{n=0}^{\infty} x^n = \sum_{n=0}^{\infty} S_n x^n$$

$$\frac{f(x)}{(1 - x)^2} = \sum_{n=0}^{\infty} S_n x^n \sum_{n=0}^{\infty} x^n = \sum_{n=0}^{\infty} (S_0 + \cdots + S_n) x^n = \sum_{n=0}^{\infty} (n + 1) \sigma_n x^n$$

Hence

$$f(x) = (1 - x)^2 \sum_{n=0}^{\infty} (n + 1) \sigma_n x^n$$

$$S = (1 - x)^2 \sum_{n=1}^{\infty} (n + 1) S x^n$$

Subtracting, we get

$$f(x) - S = (1 - x)^2 \sum_{n=0}^{\infty} (n + 1) (\sigma_n - S) x^n$$

Given $\epsilon > 0$, choose $n_0 \in \mathfrak{N}$ such that $|\sigma_n - S| < \epsilon/2$ for $n \geqslant n_0$. Let

$$\sum_{n=0}^{n_0 - 1} (n + 1) |\sigma_n - S| = K(n_0)$$

and choose $\delta = \delta(\epsilon) > 0$ such that for $1 - \delta \leqslant x < 1$

$$(1 - x)^2 K(n_0) < \epsilon/2$$

Then for $1 - \delta \leqslant x < 1$,

$$|f(x) - S| \leqslant (1 - x)^2 \sum_{n=0}^{n_0 - 1} (n + 1) |\sigma_n - S| + (1 - x)^2 \sum_{n=n_0}^{\infty} (n + 1) |\sigma_n - S| x^n$$

$$< \frac{\epsilon}{2} + (1 - x)^2 \frac{\epsilon}{2} \sum_{n_0}^{\infty} (n + 1) x^n < \frac{\epsilon}{2} + \frac{\epsilon}{2} = \epsilon. \quad \blacksquare$$

The converse of this theorem is false. There are series which are A-summable but not summable $(C, 1)$.

EXAMPLE 11.6.7. Let $a_n = (-1)^{n+2} \ n$. Since $a_n \neq o(n)$, the series is not summable $(C, 1)$ (Example 11.6.3). Consider now $f(x) = \sum\limits_{n=0}^{\infty} a_n x^n =$

$\sum\limits_{n=0}^{\infty} (-1)^{n+2} \ n x^n$ which is convergent for $|x| < 1$. Further, $f(x) = \dfrac{-x}{(1+x)^2} \longrightarrow$

$-\dfrac{1}{4}$ as $x \longrightarrow 1 - 0$. Hence the series $\sum\limits_{n=0}^{\infty} a_n$ is A-summable.

Theorem 11.6.6 (Tauber). If the series $\sum\limits_{n=0}^{\infty} a_n$ is A-summable to S and if

$a_n = o\left(\dfrac{1}{n}\right)$ then the series is convergent and has the sum S.

Proof. Let $f(x) = \sum\limits_{n=0}^{\infty} a_n x^n$, $S_n = \sum\limits_{k=0}^{n} a_k$. For $n > 0$, $0 < x < 1$ we have

$$|S_n - S| = \left| f(x) - \sum_{k=0}^{\infty} a_k x^k - S + \sum_{k=0}^{n} a_k \right|$$

$$= \left| \sum_{k=0}^{n} a_k (1 - x^k) - \sum_{k=n+1}^{\infty} a_k x^k + f(x) - S \right|$$

$$\leqslant \sum_{k=1}^{n} |a_k| (1 - x^k) + \sum_{k=n+1}^{\infty} |a_k| x^k + |f(x) - S|$$

$$= I_1 + I_2 + I_3, \text{ say}$$

Now $1 - x^k = (1 - x)(1 + x + \cdots x^{k-1}) \leqslant k(1 - x)$ and so $I_1 \leqslant$

$(1 - x) \sum\limits_{k=1}^{n} k \ |a_k|$. Let $\epsilon_m = \sup\limits_{k > m} k \ |a_k|$. Then $\epsilon_m \downarrow 0$ as $m \longrightarrow \infty$ and

$$I_2 < \sum_{k=n+1}^{\infty} k \ |a_k| \frac{x^k}{k} < \frac{\epsilon_n}{n+1} \sum_{k=0}^{\infty} x^k$$

$$= \frac{\epsilon_n}{(n+1)(1-x)}.$$

Let $x = 1 - \dfrac{1}{n}$. Given $\epsilon > 0$, choose n_0 such that for $n \geqslant n_0$ $I_1 \leqslant \dfrac{1}{n} \displaystyle\sum_{k=1}^{n} k \, |a_k| <$

$\dfrac{\epsilon}{3}$. This is possible by our hypothesis and Theorem 11.6.1. Choose n_1 such that for $n \geqslant n_1$,

$$I_2 < \frac{n\epsilon_n}{n+1} < \epsilon_n < \frac{\epsilon}{3}$$

Finally we choose n_2 such that for $n \geqslant n_2$,

$$\left| f\left(1 - \frac{1}{n}\right) - S \right| < \frac{\epsilon}{3}.$$

Then for $n \geqslant \max(n_0, n_1, n_2)$ we have

$$|S_n - S| < \epsilon. \, \blacksquare$$

Exercise 11.6

1. Prove that the series

$$\sin x + \sin 2x + \sin 3x + \cdots$$

is summable $(C,1)$ to $\dfrac{1}{2} \cos \dfrac{x}{2}$ provided $x \neq 2p\pi$, $(p \in \mathcal{I})$. (For $x = 2p\pi$ the series is obviously convergent with the sum 0.

2. Prove that if the series $\displaystyle\sum_{n=0}^{\infty} a_n$ is summable $(C,1)$ to S then the series $\displaystyle\sum_{n=1}^{\infty} a_n$ is summable $(C,1)$ to $S - a_0$.

3. Prove that if $a_n \geqslant 0$ and if the series $\displaystyle\sum_{n=0}^{\infty} a_n$ is A-summable to S then $\displaystyle\sum_{n=0}^{\infty} a_n$ converges to S. $\left(\textit{Hint:} \text{ If } \displaystyle\sum_{n=0}^{\infty} a_n \text{ diverges then show that } \lim_{x \to 1-0} \displaystyle\sum_{n=0}^{\infty} a_n x^n = +\infty. \right)$

4. Prove that if $a_n \longrightarrow a$ and $b_n \longrightarrow b$, then

$$\frac{1}{n}(a_0 b_n + a_1 b_{n-1} + \cdots + a_n b_0) \longrightarrow ab$$

5. Let $c_n = a_0 b_n + \cdots + a_n b_0$. Prove that if $\displaystyle\sum_{n=0}^{\infty} a_n = A$, $\displaystyle\sum_{n=0}^{\infty} b_n = B$, $\displaystyle\sum_{n=0}^{\infty} c_n = C$

are all convergent then $AB = C$. $\left[\text{Hint:}\right.$ The series $f(x) = \sum_{0}^{\infty} a_n x^n$, $g(x) =$

$\sum_{0}^{\infty} b_n x^n$ converge absolutely for $|x| < 1$. Hence for $|x| < 1$,

$$f(x)\, g(x) = \sum_{0}^{\infty} a_n x^n \sum_{0}^{\infty} b_n x^n = \sum_{0}^{\infty} c_n x^n$$

The last series is also absolutely convergent for $|x| < 1$ (Theorem 8.8.2). Now use Abel's continuity theorem. $\Big]$

6. Let $S_n = \sum_{0}^{n} a_k$, $r \in \mathfrak{N}$ and

$$S_n^{(r)} = S_n + r S_{n-1} + \frac{r(r+1)}{2!}\, S_{n-2} + \cdots + \frac{r(r+1)\ldots(r+n-1)}{n!}\, S_0,$$

$$D_n^{(r)} = \frac{(r+1)(r+2)\ldots(r+n)}{n!}$$

We say that $\sum_{n=0}^{\infty} a_n$ is summable (C,r) to S if

$$\lim_{n \to \infty} \frac{S_n^{(r)}}{D_n^{(r)}} = S$$

If $r = 0$ we get the convergence and if $r = 1$ we get Definition 11.6.1. Prove that $\sum_{n=0}^{\infty} (-1)^{n+2}\, n$ is summable $(C,2)$ to $-\frac{1}{4}$.

11.7. Some Special Functions

In this section we define the logarithmic and exponential functions and give some of their properties.

Definition 11.7.1. *We define* $\log x$, *the natural logarithm of* x, *for* $x > 0$, *by*

$$\log x = \int_{1}^{x} \frac{dt}{t}, \qquad x > 0$$

It is easily seen that $\log 1 = 0$ and that $\log x > 0$ if $x > 1$.

Theorem 11.7.1. For $x > 0, y > 0$,

(a) $\log (xy) = \log x + \log y$.

(b) $\dfrac{d}{dx} \log x = \dfrac{1}{x}$.

(c) $\log x$ is strictly increasing on $(0, \infty)$.

(d) $\log x \longrightarrow \infty$ as $x \longrightarrow \infty$ and $\log x \longrightarrow -\infty$ as $x \longrightarrow 0+$.

(e) $\dfrac{x}{1 + x} < \log (1 + x) < x$.

Proof. (a) $\log (xy) = \displaystyle\int_1^{xy} \frac{dt}{t} = \int_1^x \frac{dt}{t} + \int_x^{xy} \frac{dt}{t}$.   Let $t = xu$ in the last integral. We get

$$\log (xy) = \log x + \int_1^y \frac{du}{u} = \log x + \log y.$$

In particular we have, on taking $y = \dfrac{1}{x}$,

$$\log \frac{1}{x} = -\log x \tag{11.19}$$

(b) Let $0 < |h| < x$. Then

$$\frac{\log (x + h) - \log x}{h} = \frac{1}{h} \int_x^{x+h} \frac{dt}{t}$$

and the integral lies between $\dfrac{1}{x}$ and $\dfrac{1}{x + h}$. Now let $h \longrightarrow 0$.

(c) For $0 < x_1 < x_2$,

$$\log x_2 - \log x_1 = \int_{x_1}^{x_2} \frac{dt}{t} > 0$$

(d) Let $n = [x]$ (integer part of x). Then

$$\log n = \int_1^n \frac{dt}{t} > \sum_{j=2}^n \frac{1}{j}$$

and $\displaystyle\lim_{n \to \infty} \sum_{j=2}^n \frac{1}{j} = \infty$. (Example 8.2.1). The second part follows from (11.19).

(e) The integrand in $\displaystyle\int_1^{1+x} \frac{dt}{t}$ lies between 1 and $\displaystyle\frac{1}{1+x}$.

Hence (e) follows.∎

Exponential Function. Since the function $f(x) = \log x$ is continuous and strictly increasing for $x > 0$ and takes all real values, it has a unique continuous inverse function which is strictly increasing for all values of its variable. We denote this function by $\exp x$. The domain of $\exp x$ is $\{-\infty < x < \infty\}$ and the range is $(0, \infty)$. Note that $y = \exp x$ is equivalent to $x = \log y (y > 0)$ and $\exp(\log y) = y, y > 0$; $\log(\exp x) = x$. We write $\exp 1 = e$. The properties in the following theorem can be easily deduced.

Theorem 11.7.2.
(a) $\exp 0 = 1$.

(b) $\exp x$ has derivatives of all orders for all x and $\displaystyle\frac{d}{dx} \exp x = \exp x$.

(c) $\exp(x + y) = \exp x \, \exp y$.
(d) $\exp(-x) = 1/\exp x$.
(e) $\exp n = e^n$, n any positive integer.
(f) $\exp r = e^r$, r any rational number.
The proof is left as an exercise.

The function $\exp x$ coincides with e^x whenever x is rational. Since $\exp x$ is continuous on $(-\infty, \infty)$ we define e^x for nonrational x to be $\exp x$.

Definition 11.7.2. *For nonrational* x, e^x *is defined to be* $\exp x$. *Thus, for all* x,

$$\exp x = e^x$$

By Taylor's Theorem with a remainder (Theorem 9.2.8) we get

$$e^x = 1 + x + \frac{x^2}{2!} + \cdots + \frac{x^{n-1}}{(n-1)!} + R_n$$

where

$$|R_n| = \left| \frac{x^n}{(n-1)!} \int_0^1 (1-t)^{n-1} e^{tx} \, dt \right|$$

$$< \frac{|x|^n e^{|x|}}{n!} \longrightarrow 0 \text{ as } n \longrightarrow \infty,$$

whatever be the value of x. Hence, making $n \longrightarrow \infty$, we have

$$e^x = 1 + x + \frac{x^2}{2!} + \cdots + \frac{x^n}{n!} + \cdots \tag{11.20}$$

We proved earlier that the power series on the right converges for all x (Example 8.7.2). In particular we have

$$e = 1 + 1 + \frac{1}{2!} + \cdots + \frac{1}{n!} + \cdots . \qquad (11.21)$$

Exercise 11.7

1. Prove that $2 < e < 3$ and that e is not a rational number.
2. Prove that if $c > 0$,

$$\lim_{n \to \infty} \frac{e^x}{x^c} = \infty, \quad \lim_{x \to \infty} \frac{\log x}{x^c} = 0$$

By (11.20),

$$\frac{x^n}{n!} < e^x \text{ for all } x > 0$$

Choose $n > c$.)

4. Prove that for $-1 < x \leqslant 1$

$$\log (1 + x) = \sum_{n=1}^{\infty} \frac{(-1)^{n+1} x^n}{n} .$$

(Consider the remainder in Taylor's theorem when (a) $0 \leqslant x \leqslant 1$ and (b) $-1 < x < 0$.)

5. Prove that for $-1 \leqslant x < 1$,

$$\log \frac{1}{1 - x} = \sum_{n=1}^{\infty} \frac{x^n}{n} .$$

6. (a) Prove that

$$\sum_{n=2}^{\infty} \frac{1}{n^p (\log n)^q}$$

is convergent when $p > 1$ or $p = 1$ and $q > 1$.

(b) Consider the convergence or divergence of the improper integral

$$\int_2^{\infty} \frac{dx}{x^p (\log x)^q} .$$

11.8. Hyperbolic and Circular Functions

We define for all $x \in R$

$$\sinh x = \frac{1}{2}(e^x - e^{-x}) = \frac{x}{1!} + \frac{x^3}{3!} + \cdots + \frac{x^{2n+1}}{(2n+1)!} + \cdots$$

$$\cosh x = \frac{1}{2}(e^x + e^{-x}) = 1 + \frac{x^2}{2!} + \cdots + \frac{x^{2n}}{(2n)!} + \cdots$$

$$\sin x = \frac{1}{2i}(e^{ix} - e^{-ix}) = x - \frac{x^3}{3!} + \frac{x^5}{5!} - \frac{x^7}{7!} + \cdots$$

$$\cos x = \frac{1}{2}(e^{ix} + e^{-ix}) = 1 - \frac{x^2}{2!} + \frac{x^4}{4!} - \cdots, (i^2 = -1).$$

The series on the right are power series convergent for all x. Note that $\sinh x$ and $\sin x$ are odd functions (that is, $f(-x) = -f(x)$) and $\cosh x$ and $\cos x$ are even functions (that is, $f(-x) = f(x)$). Further, $\cosh^2 x - \sinh^2 x = 1$, that is the point whose coordinates are $x = \cosh t$, $y = \sinh t$ lies, for all $t \in R$, on the hyperbola $x^2 - y^2 = 1$. Similarly $\cos^2 x + \sin^2 x = 1$ and the point $(\cos t, \sin t)$ lies for all $t \in R$ on the circle $x^2 + y^2 = 1$. From these two relations we have, for all $x \in R$,

$$-1 \leqslant \sin x \leqslant 1$$

$$-1 \leqslant \cos x \leqslant 1$$

$$\cosh x \geqslant 1$$

Exercise 11.8

1. Prove that for all $x, y \in R$
   (a) $\cosh(x + y) = \cosh x \cosh y + \sinh x \sinh y$
   (b) $\sinh(x + y) = \sinh x \cosh y + \cosh x \sinh y$
   (These formula hold also when x and y are complex. A similar remark applies to some results stated below.)
2. Prove that for all $x \in R$

$$\frac{d \sin x}{dx} = \cos x, \qquad \frac{d \cos x}{dx} = -\sin x$$

3. Prove that for all $x, y \in R$
   $\cos ix = \cosh x$, $\sin ix = i \sinh x$, where $i = \sqrt{-1} = \langle 0,1 \rangle$ in R^2 (Cf. Sec. 2.5)
   $\cos(x + y) = \cos x \cos y - \sin x \sin y$

$\sin (x + y) = \sin x \cos y + \cos x \sin y$

$\cos 2x = 1 - 2 \sin^2 x, \sin 2x = 2 \sin x \cos x$

4. Prove the following:

   (i) If $0 < x \leqslant 2, \sin x > 0$.

   $$\left(\sin x = x\left(1 - \frac{x^2}{6}\right) + \frac{x^5}{5!}\left(1 - \frac{x^2}{42}\right) + \cdots > x\left(1 - \frac{x^2}{6}\right) \geqslant \frac{x}{3} > 0\right)$$

   (ii) $\sin 1 > \dfrac{5}{6}$.

   (iii) $\sin 4 = 2 \sin 2(1 - 2 \sin^2 1) < 0$.

5. Prove that there exists a unique real number π such that $2 < \pi < 4$ and $\sin \pi = 0$. (*Uniqueness.* Suppose if possible $2 < \mu < 4$ and $\sin \mu = 0$. Then $\sin (\mu - \pi) = 0$ (Use Exercise 3 above), and $|\mu - \pi| < 2$. If $\mu > \pi$, we use Exercise 4 (i) to arrive at a contradiction. Similarly if $\mu < \pi$.)

6. Prove that for $x \in R$

   (i) $\sin (x + 2\pi) = \sin x$.

   (ii) $\cos (x + 2\pi) = \cos x$ (Use Exercise 3).

   (iii) $\sin \dfrac{\pi}{2} = 1, \cos \dfrac{\pi}{2} = 0, \cos \pi = -1$.

7. Prove that

   (i) $\sin x = o(x), x \longrightarrow \infty$.

   (ii) $\sin x \sim x, x \longrightarrow 0$.

   $\sin x < x, x > 0$.

   $$\left(\text{Note that } x - \sin x = \int_0^x (1 - \cos t)\, dt > 0\right)$$

   (iii) $x \leqslant \dfrac{\sin x}{\cos x}, 0 \leqslant x < \dfrac{\pi}{2}$. $\left(\text{Consider } \displaystyle\int_0^x \left(1 - \dfrac{1}{\cos^2 t}\right) dt = x - \dfrac{\sin x}{\cos x}.\right)$

   (iv) $\dfrac{\sin x}{x} > \dfrac{2}{\pi}, 0 < x \leqslant \dfrac{\pi}{2}$

   $\left(\text{Consider } \dfrac{dy}{dx} \text{ where } y(x) = \dfrac{\sin x}{x} - \dfrac{2}{\pi}. \text{ See also Exercise 10.2(7)}\right)$

12

Elementary Measure Theory

12.1. Lengths of Bounded Open Sets and Bounded Closed Sets

The simplest notion of "measure" is that of "length of an interval." If a set S of reals is the union of a class of pairwise disjoint intervals then the sum of lengths of these intervals may be called the measure of S. The measure of some unbounded sets may be $+\infty$. However, if we wish to extend this idea of "measure" to an arbitrary set of reals we run into some difficulty.

If $m(A)$ denotes the measure of an arbitrary set A, we would desire to have the following properties:

1. $m(A)$ is defined for every set of real numbers; it may be $+\infty$.
2. $m(A) \geqslant 0$.
3. If A is an interval then $m(A)$ is the length of the interval.
4. If $\{A_n\}$ is a countable class of pairwise disjoint sets, then $m\left(\bigcup_n A_n\right) = \sum_n m(A_n)$. (This property is known as "countable additivity.")
5. If $S = \{x + x_0 : x \in A\}$, x_0 being a fixed point; then $m(S) = m(A)$. (This property is known as "translation invariance.")

Unfortunately, it is impossible to find a "measure" which satisfies all the five conditions. As a matter of fact, if the continuum hypothesis is assumed, then it would be impossible to find a measure which would satisfy the properties 1, 3, and 4. It is, therefore, necessary that one of these conditions be weakened. Following Henri Lebesgue (1875–1941), who made many contributions in measure theory and integration, we drop the first condition.

For the purpose of developing the ideas of external and internal measure, we start with the length of an interval.

Definition 12.1.1. *The length of an interval I (open, closed, or semiopen),* denoted by $l(I)$, is defined as the difference between its end-points. Thus if $I = (a,b)$ then $l(I) = b - a$.

Now we prove the following theorem.

Theorem 12.1.1. Let $\{I_k\}$ be a class of pairwise disjoint intervals contained in an interval J. Then $\sum_k l(I_k) \leqslant l(J)$.

Proof. We first prove this result for a finite case. Let $\{I_k : k = 1,2,\ldots,n\}$ be a finite class of pairwise disjoint intervals, and a_1, b_1 be the end-points of the first interval and a_2, b_2 be the end-points of the second, and so on. We can assume without any loss of generality that these intervals are listed in the order of increasing left end-points, which implies $a_1 < a_2 < \cdots < a_n$; and $b_k \leqslant a_{k+1}$. Now if α and β are the end-points of J, then $\alpha \leqslant a_1$ and $b_n \leqslant \beta$. Thus the expression $(\beta - b_n) + (a_n - b_{n-1}) + \cdots + (a_2 - b_1) + (a_1 - \alpha) \geqslant 0$ or $\beta - \alpha \geqslant$

$$(b_1 - a_1) + (b_2 - a_2) + \cdots + (b_n - a_n) \text{ which means } l(J) \geqslant \sum_{k=1}^{n} l(I_k).$$

Next, if $\{I_k : k = 1,2,\ldots\}$ is a denumerable class of pairwise disjoint intervals contained in J, then every partial sum $\sum_{k=1}^{n} l(I_k) \leqslant l(J)$, and as such

$$\sum_{k=1}^{\infty} l(I_k) = \lim_{n \to \infty} \sum_{k=1}^{n} l(I_k) \leqslant l(J). \;\blacksquare$$

We are now ready to define the length of a bounded open set.

Definition 12.1.2. *Let G be a bounded open set with component intervals $\{(a_k, b_k) : k = 1,\ldots\}$. The length of G, written as $l(G)$, is defined as follows:* $l(G) = \sum_k (b_k - a_k)$; *that is, $l(G)$ is the sum of the lengths of component intervals of G.*

This definition takes into account the fact that the series $\sum_k (b_k - a_k)$ converges because of the last theorem.

We now give an obvious result whose proof is left as an exercise.

Theorem 12.1.2. If G_1 and G_2 are two bounded open sets such that $G_1 \subset G_2$, then $l(G_1) \leqslant l(G_2)$.

The following is an important corollary to this theorem.

Corollary. If $\{G_\alpha\}$ is the class of all bounded open sets containing G then $l(G) = \text{glb} \{l(G_\alpha)\}$.

Theorem 12.1.3. If an interval I is the union of a class of intervals $\{J_k\}$ then $l(I) \leqslant \sum_k l(J_k)$.

The proof of this theorem is left as an exercise.

Lemma. Let G be a bounded open set and let $\{J_m\}$ be a countable class of open intervals such that $G = \bigcup_m J_m$. Then $l(G) \leqslant \sum_m l(J_m)$.

Proof. For every component interval (a,b) of G ∃ a subclass $\{J_k\}$ of $\{J_m\}$ such that $(a,b) = \bigcup_k J_k$, implying $(b - a) \leqslant \sum_k l(J_k)$ by the last theorem. Furthermore, a and b cannot belong to any J_m. Doing this for all components, we get the result $l(G) \leqslant \sum_m l(J_m)$. ∎

Theorem 12.1.4. Let G be a bounded open set and $\{G_k\}$ be a countable class of open sets such that $G = \bigcup_k G_k$. Then $l(G) \leqslant \sum_k l(G_k)$.

Proof. Let $G_k = \bigcup_m I_{km}$ where I_{km}'s are the intervals of G_k. Then $l(G_k) = \sum_m l(I_{km})$, and

$$G = \bigcup_k \bigcup_m I_{km}$$

The result now follows from the lemma just proved. ∎

We now define the length of a bounded closed set.

Definition 12.1.3. *Let F be a bounded closed set. If F is empty, or if it consists of one element only, then we define $l(F)$ to be zero. If F consists of more than one point then let $\Delta = [a,b]$ be the smallest closed interval containing F [which we denote by* sci (F)]. *We know that $[a,b] - F$ is open, and as such it would have some length (may be even zero). We define the length of F as follows:*

$$l(F) = (b - a) - l([a,b] - F)$$

It can be shown that $l(F) \geqslant 0$. (Why?)

EXAMPLE 12.1.1. Let

$$F = [a,b]$$

Then

$$l(F) = b - a$$

EXAMPLE 12.1.2. Let

$$F = \left\{0, 1, \frac{1}{2}, \dots, \frac{1}{n}, \dots\right\}$$

Then

$$\Delta = [0,1]$$

and

$$\Delta - F = \bigcup_{n=1}^{\infty} \left(\frac{1}{n+1}, \frac{1}{n}\right)$$

and

$$l(\Delta - F) = \sum_{n=1}^{\infty} \left(\frac{1}{n} - \frac{1}{n+1}\right) = 1$$

Thus

$$l(F) = 0.$$

EXAMPLE 12.1.3. Let F be the Cantor ternary set. Then $\Delta = [0,1]$, and

$$\Delta - F = \left(\frac{1}{3}, \frac{2}{3}\right) \cup \left\{\left(\frac{1}{9}, \frac{2}{9}\right) \cup \left(\frac{7}{9}, \frac{8}{9}\right)\right\} \cup \cdots$$

Thus $l(\Delta - F) = \dfrac{1}{3} + \dfrac{2}{9} + \dfrac{4}{27} + \dfrac{8}{81} + \cdots = 1$, and it follows that $l(F) = 1 - 1 = 0$.

The last example is very significant, since the Cantor ternary set is a non-denumerable set yet has length equal to zero.

EXAMPLE 12.1.4. Let

$$F = \bigcup_{k=1}^{n} [a_k, b_k]$$

such that

$$\{[a_k, b_k] : k = 1, 2, \dots, n\}$$

be a class of pairwise disjoint closed intervals.

Then it can be easily shown that $l(F) = \displaystyle\sum_{k=1}^{n} (b_k - a_k)$.

We now prove the following result.

Theorem 12.1.5. Let F be a bounded closed set contained in an open interval I. Then

$$l(I - F) = l(I) - l(F)$$

Proof. Let $I = (a,b)$ and let $\Delta = [\lambda,\mu]$ be sci (F). It is obvious from the hypothesis that $a < \lambda < \mu < b$.

$I - F$ is an open set and as such $l(I - F)$ has a meaning [and so has $l(\Delta - F)$].

Furthermore, $I - F = (\Delta - F) \cup (a,\lambda) \cup (\mu,b)$, and since $(\Delta - F)$, (a,λ) and (μ,b) are pairwise disjoint $l(I - F) = l(\Delta - F) + (\lambda - a) + (b - \mu)$.

From Def. 13.1.4 it follows that

$$l(\Delta - F) = l(\Delta) - l(F)$$

Hence,

$$l(I - F) = l(\Delta) - l(F) + (\lambda - a) + (b - \mu)$$
$$= (\mu - \lambda) - l(F) + (\lambda - a) + (b - \mu)$$
$$= (b - a) - l(F)$$

Thus the result is proved. ∎

Theorem 12.1.6. Let F_1 and F_2 be two bounded closed sets such that $F_1 \subset F_2$. Then $l(F_1) \leq l(F_2)$.

Proof. Let I be an open interval containing F_2 and hence also containing F_1. Then

$$I - F_2 \subset I - F_1$$

By Theorem 12.1.2,

$$l(I - F_2) \leq l(I - F_1)$$

Using Theorem 12.1.5, we get

$$l(I) - l(F_2) \leq l(I) - l(F_1)$$

Thus,

$$l(F_1) \leq l(F_2)$$

which completes the proof. ∎

Corollary. For a bounded closed set F, $l(F) = $ lub $\{l(F_\alpha): F_\alpha \subset F, F_\alpha$ is closed$\}$.

Compare this corollary with that of Theorem 12.1.2.

Theorem 12.1.7. If a bounded open set G is the union of a countable number of pairwise disjoint open sets $\{G_k\}$, then

$$l(G) = \sum_k l(G_k)$$

The proof is left as an exercise.

Theorem 12.1.8. Let F and G be bounded closed and bounded open sets, respectively; such that $F \subset G$. Then

$$l(F) \leqslant l(G)$$

Proof. Let I be an open interval containing G and hence, also containing F. It is then easy to verify that $I = G \cup (I - F)$.

Using Theorem 12.1.4, we have $l(I) \leqslant l(G) + l(I - F)$, and from Theorem 12.1.5 we have $l(I) \leqslant l(G) + l(I) - l(F)$. Thus $l(F) \leqslant l(G)$, which completes the proof.

We now present two significant results in Theorems 12.1.9 and 12.1.10.

Theorem 12.1.9. Let G be a bounded open set. If $\{F_\alpha\}$ is the class of all closed sets contained in G, then $l(G) = $ lub $\{l(F_\alpha)\}$.

Proof. That $l(G)$ is an upper bound of $\{l(F_\alpha)\}$ follows from the last theorem. To prove that it is the lub, we proceed as follows:

Let $G = \bigcup_k (a_k, b_k)$, where (a_k, b_k), $k = 1, \ldots$ are component intervals of G, so that

$$l(G) = \sum_k (b_k - a_k)$$

The series on the right, if infinite, converges. Therefore, for $\epsilon > 0$ there is a natural number n, such that

$$\sum_{j=1}^{n} (b_j - a_j) > l(G) - \frac{\epsilon}{2} \tag{12.1}$$

For every (a_j, b_j) it is possible to construct a closed interval $[c_j, d_j]$ such that

$$(d_j - c_j) > (b_j - a_j) - \frac{\epsilon}{2n} \tag{12.2}$$

This construction may be easily done by choosing

$$c_j < \min \left\{ \left(a_j + \frac{\epsilon}{4n}\right), \frac{a_j + b_j}{2} \right\},$$

and

$$d_j > \max \left\{ \left(b_j - \frac{\epsilon}{4n}\right), \frac{a_j + b_j}{2} \right\}$$

with $c_j > a_j$ and $d_j < b_j$.

Now let $F_\epsilon = \bigcup_{j=1}^{n} [c_j, d_j]$. It is obvious that $F_\epsilon \subset G$ and that $l(F_\epsilon) = \sum_{j=1}^{n} (d_j - c_j)$. Since $[c_j, d_j]$ are pairwise disjoint, we obtain with the aid of inequalities (12.1) and (12.2) the following:

$$l(F_\epsilon) = \sum_{j=1}^{n} (d_j - c_j) > \sum_{j=1}^{n} (b_j - a_j) - \frac{\epsilon}{2} > l(G) - \epsilon.$$

Thus for $\epsilon > 0$, we have $F_\epsilon \subset G$, such that $l(F_\epsilon) > l(G) - \epsilon$, which means $l(G) = \text{lub } \{l(F_\alpha)\}$ and that proves the result.

Theorem 12.1.10. Let $\{G_\alpha\}$ be a class of bounded open sets containing a closed set F. Then $l(F) = \text{glb } \{l(G_\alpha)\}$.

Proof. From Theorem 12.1.8 it follows that $l(F)$ is a lower bound of $\{l(G_\alpha)\}$. To show that it is the greatest lower bound, we construct an open interval $I \supset F$. Obviously, $I - F$ is open and since $l(I - F)$ is the lub of lengths of all closed sets contained in $I - F$ (Theorem 12.1.9), it follows that for $\epsilon > 0$ there is a closed set $F_\epsilon \subset I - F$ such that $l(F_\epsilon) > l(I - F) - \epsilon$.

Now let $G_\epsilon = I - F_\epsilon$, and since $F_\epsilon \subset I$, from Theorem 12.1.5, it follows that $l(G_\epsilon) = l(I) - l(F_\epsilon)$. Thus

$$l(I) - l(G_\epsilon) = l(F_\epsilon) > l(I - F) - \epsilon$$
$$= l(I) - l(F) - \epsilon$$

Therefore,

$$l(G_\epsilon) < l(F) + \epsilon$$

Furthermore,

$$G_\epsilon \supset F$$

Hence $l(F)$ is the glb $\{l(G_\alpha)\}$.

Exercise 12.1

1. Prove Theorem 12.1.2.
2. Prove Theorem 12.1.3.
3. Prove Theorem 12.1.7.
4. Construct a nonperfect closed set whose length is 1.
5. Is there any open set whose length is 0?

12.2. External and Internal Measures of a Bounded Set; Measurable Sets

We now introduce the concepts of *external measure* and *internal measure* of a set by using length of open sets and closed sets. In this section we shall limit our discussion to bounded sets only.

Definition 12.2.1. *Let S be a bounded set, and let $\{G_\alpha\}$ be the class of all bounded open sets covering S. The external measure of S, written as $m_e(S)$ is defined to be the* glb $\{l(G_\alpha): G_\alpha \supset S\}$.

Definition 12.2.2. *The internal measure of a bounded set S, written as $m_i(S)$, is defined as the* lub $\{l(F_\alpha): F_\alpha \subset S\}$, F_α *being a closed set.*

It must be noticed that both the external measure and the internal measure of a set are nonnegative set functions.

Given an arbitrary bounded set S, we could find a bounded open $G \supset S$ and a bounded closed set $F \subset S$ such that $l(F) \leqslant l(G)$. This would mean the set $\{l(G_\alpha): G_\alpha \supset S\}$ would be bounded below and the set $\{l(F_\alpha): F_\alpha \subset S\}$ would be bounded above. By the least upper bound property and the greatest lower bound property, we infer that $m_e(S)$ and $m_i(S)$ would always exist as long as S is bounded.

We now define measurability of a bounded set.

Definition 12.2.3. *A bounded set S is measurable if $m_e(S) = m_i(S)$. In this case we refer to the measure of S, written $m(S)$, as this common value.*

If for any bounded set $S, m_e(S) \neq m_i(S)$, then S is called nonmeasurable.

Most of the sets we know are measurable. Nonmeasurable sets exist, however. The proof of the existence of such a set will be given later on, and is based on the axiom of choice. Those mathematicians (especially intuitionists) who reject this famous axiom are not convinced of the existence of non-measurable sets. (Some probability theorists wish that every set would be measurable in the sense of Lebesgue.)

We now discuss some theorems dealing with these concepts.

Theorem 12.2.1. For any bounded set S, $m_i(S) \leqslant m_e(S)$.

Proof. Let G be an arbitrary open set containing S, and let $F \subset S$ be an arbitrary closed set contained in S. Then $F \subset S \subset G$.

We know that $l(F) \leqslant l(G)$ for all $F \subset S$. Thus lub $\{l(F) : F \subset S, F$ being closed$\} \leqslant l(G)$. Or $m_i(S) \leqslant l(G)$ which is true for all bounded open sets containing S. Therefore, $m_i(S) \leqslant$ glb $\{l(G) : G$ being a bounded open set containing $S\}$. Or, $m_i(S) \leqslant m_e(S)$, which completes the proof. $\mathbf{I}$

We now establish the measurability of a bounded open set and that of a bounded closed set.

Theorem 12.2.2. Every bounded open set G is measurable, and $m(G) = l(G)$.

Proof. From Theorem 12.1.9 and the corollary of Theorem 12.1.2, it can be easily shown that $m_e(G) = l(G)$ and $m_i(G) = l(G)$. Hence, the result. $\mathbf{I}$

Theorem 12.2.3. Every bounded closed set F is measurable, and $m(F) = l(F)$.

Again, the proof follows directly from the corollary of Theorems 12.1.6, and 12.1.10.

From now on, $l(F)$ and $l(G)$ will be replaced by $m(F)$ and $m(G)$, respectively.

It is obvious from Definitions 12.2.1, 12.2.2 and 12.2.3 that every semiopen interval is measurable and its measure is the length.

Theorem 12.2.4. If A and B are two bounded sets such that $A \subset B$, then $m_i(A) \leqslant m_i(B)$ and $m_e(A) \leqslant m_e(B)$.

The proof of this theorem is an easy exercise.

We now discuss some sets of measure zero, but first the following theorem, which is useful.

Theorem 12.2.5. If a bounded set S is such that its external measure is zero, then S is measurable and $m(S) = 0$.

Proof. For any set S, we know $m_i(S) \geqslant 0$. Now, if $m_e(S) = 0$, that would mean $m_i(S) \geqslant m_e(S)$. But $m_e(S) \geqslant m_i(S)$ (Theorem 12.2.1). Thus $m_i(S) = m_e(S) = 0$. Hence, S is measurable, and, of course, $m(S) = 0$. The proof is complete.

From this result one can easily prove that a finite set is measurable. We have a stronger result, however, in the following theorem.

Theorem 12.2.6. Every denumerable bounded set is measurable and its measure is zero.

Proof. Let $S = \{x_1, x_2, \ldots x_n, \ldots\}$ be a bounded and denumerable set. Let $\epsilon > 0$. For every $x_n \in S$, construct an open interval $\left(x_n - \dfrac{\epsilon}{2^{n+1}}, x_n + \dfrac{\epsilon}{2^{n+1}}\right)$ the length of which is, of course, $\dfrac{\epsilon}{2^n}$.

Now let $G = \bigcup_{n=1} \left(x_n - \dfrac{\epsilon}{2^{n+1}}, x_n + \dfrac{\epsilon}{2^{n+1}}\right)$. Obviously, G contains S and

$l(G) \leqslant \sum_n \dfrac{\epsilon}{2^n} = \epsilon$. Since ϵ can be arbitrarily small, $m_e(S) = 0$, and from Theorem 12.2.5 the result is established.

Not every set whose measure is zero is countable, however. For example, the Cantor ternary set is a closed set whose measure (length) is zero, but is nondenumerable.

We now prove a property of external measure.

Theorem 12.2.7. Let a bounded set S be the union of a countable class $\{S_k\}$ of sets; then $m_e(S) \leqslant \sum_k m_e(S_k)$.

Proof. The result is trivial if the series $\sum_k m_e(S_k)$ diverges. We shall therefore assume here that the series $\sum_k m_e(S_k)$ converges. By the definition of external measure, given $\epsilon > 0$ there exists a bounded open set $G_k \subset S_k$ such that

$$m(G_k) < m_e(S_k) + \dfrac{\epsilon}{2^k}, \quad k = 1, 2, \ldots$$

Thus,

$$\sum_k m(G_k) < \sum_k m_e(S_k) + \epsilon$$

since $\sum_k \dfrac{\epsilon}{2^k} \leqslant \epsilon$.

Now the series $\sum_k m_e(S_k)$ converges; therefore, the series $\sum_k m(G_k)$ must converge also.

Let $G = \bigcup_k G_k$. From Theorem 12.1.7, $m(G) \leqslant \sum_k m(G_k)$ (It can be shown that G is bounded.)

This implies

$$m_e(S) \leqslant m(G) < \sum_k m_e(S_k) + \epsilon \implies m_e(S) < \sum_k m_e(S_k) + \epsilon$$

Since ϵ is arbitrary, $m_e(S) \leqslant \sum_k m_e(S_k)$, and the proof is complete. |

Theorem 12.2.8. Let $\{F_k\}$ be a finite class of nonempty pairwise disjoint closed sets. Then

$$m\left(\bigcup_{k=1}^{n} F_k\right) = \sum_{k=1}^{n} m(F_k)$$

Proof. It suffices to prove this result for the case of two disjoint closed sets. The general result then follows from the finite induction.

From Theorem 12.2.2, it follows that F_1 and F_2 are measurable and so is their union.

Thus $m_e(F_1) = m(F_1)$, $m_e(F_2) = m(F_2)$ and $m_e(F_1 \cup F_2) = m(F_1 \cup F_2)$.

From the preceding theorem it follows that $m(F_1 \cup F_2) = m_e(F_1 \cup F_2) \leqslant m_e(F_1) + m_e(F_2) = m(F_1) + m(F_2)$.

Hence

$$m(F_1 \cup F_2) \leqslant m(F_1) + m(F_2) \tag{12.3}$$

Now to prove the reverse inequality we use the separation theorem (Theorem 5.5.3), which implies that there are two disjoint open sets O_1 and O_2 containing F_1 and F_2, respectively.

For $\epsilon > 0$ ∃ a bounded open set $G \supset F_1 \cup F_2$ such that $m(F_1 \cup F_2) > m(G) - \epsilon$.

Let $G_1 = G \cap O_1$ and $G_2 = G \cap O_2$.

It is obvious that $G_1 \supset F_1$ and $G_2 \supset F_2$; moreover, G_1 and G_2 are disjoint open sets and $G_1 \cup G_2 \subset G$.

Therefore, from Theorem 12.1.2 and Theorem 12.1.7 it follows $m(G) \geqslant m(G_1) + m(G_2)$, and, finally, that $m(F_1 \cup F_2) > m(G) - \epsilon \geqslant m(G_1) + m(G_2) - \epsilon \geqslant m(F_1) + m(F_2) - \epsilon$. Since ϵ is arbitrary, $m(F_1 \cup F_2) \geqslant m(F_1) + m(F_2)$. Combining this with (12.3), we get the required result.

Theorem 12.2.9. If $\{S_k\}$ is a countable class of pairwise disjoint sets such that $\bigcup_k S_k$ is bounded, then $m_i\left(\bigcup_k S_k\right) \geqslant \sum_k m_i(S_k)$.

Proof. First we shall prove the result for the finite case. Let $A_n = \bigcup\limits_{k=1}^{n} S_k$, n being a natural number. For $\epsilon > 0$ and for each S_k, there exists a closed set $F_k \subset S_k$ such that $m(F_k) > m_i(S_k) - \dfrac{\epsilon}{n}$.

Let $F = \bigcup\limits_{k=1}^{n} F_k$; then $F \subset A_n$. Therefore, $m_i(A_n) \geqslant m(F) = \sum\limits_{k=1}^{n} m(F_k)$. (Cf. Theorem 12.2.8.)

We conclude that

$$m_i(A_n) \geqslant m(F) = \sum_{k=1}^{n} m(F_k) > \sum_{k=1}^{n} m_i(S_k) - \epsilon$$

and since ϵ is arbitrary, that

$$m_i(A_n) \geqslant \sum_{k=1}^{n} m_i(S_k)$$

To prove the result for a denumerable case, it suffices to note that $\bigcup\limits_{k=1}^{\infty} S_k \supset A_n$, for this allows us to conclude that

$$m_i\left(\bigcup_{k=1}^{\infty} S_k \right) \geqslant m_i(A_n) \geqslant \sum_{k=1}^{n} m_i(S_k)$$

Thus,

$$\sum_{k=1}^{n} m_i(S_k) \leqslant m_i(S), \qquad \text{for every } n$$

where

$$S = \bigcup_{k=1}^{\infty} S_k$$

Taking the limit of left-hand side as $n \longrightarrow \infty$, we infer

$$\sum_{k=1}^{\infty} m_i(S_k) \leqslant m_i\left(\bigcup_{k=1}^{\infty} S_k \right),$$

which proves the result. ∎

This result is certainly not true if the sets are not pairwise disjoint as can easily be seen from the following trivial example:

Let $S_1 = (0,1)$, $S_2 = \left(\dfrac{1}{2},2\right)$, $(S_1 \cup S_2) = (0,2)$. In this case, $m_i(S_1 \cup S_2) = 2$ and $m_i(S_1) + m_i(S_2) = 1 + \dfrac{3}{2} = \dfrac{5}{2}$.

Theorem 12.2.10. Let a bounded set S be the union of a countable number of *pairwise disjoint* sets $\{S_k\}$. Then S is measurable and $m(S) = \sum_k m(S_k)$.

Proof. Since each S_k is measurable, $m_i(S_k) = m_e(S_k) = m(S_k)$.

Using Theorem 12.2.7, we get $m_e(S) \leqslant \sum_k m_e(S_k) = \sum_k m(S_k)$. Using

Theorem 12.2.9, we have $\sum_k m_i(S_k) \leqslant m_i(S)$. Thus

$$m_e(S) \leqslant \sum_k m_e(S_k) = \sum_k m(S_k) = \sum_k m_i(S_k) \leqslant m_i(S) \leqslant m_e(S).$$

From these inequalities, it follows that

$$m_e(S) = m_i(S) = \sum_k m(S_k) \quad \blacksquare$$

Corollary 1. If a closed set F is contained in an open set G, then $m(G) = m(F) + m(G - F)$.

Corollary 2. If an open set $G \subset F$, then

$$m(F) = m(G) + m(F - G)$$

Corollary 3. If F is a closed set contained in $[a,b]$ then $[a,b] - F$ is measurable and $m([a,b] - F) = (b - a) - m(F)$.

We now give a theorem which gives a necessary and sufficient condition for the measurability of a bounded set.

Theorem 12.2.11. A bounded set S is measurable $\Longleftrightarrow$ for $\epsilon > 0$ there is a bounded open set $G \supset S$ and a bounded closed set F contained in S such that $m(G) - m(F) < \epsilon$.

The proof is left as an exercise.

Theorem 12.2.12. Let $\{S_k : k = 1, 2, \ldots, n\}$ be a finite class of measurable sets, then $S = \bigcup\limits_{k=1}^{n} S_k$ is measurable.

Proof. Since each S_k is measurable, using the last theorem, we have for $\epsilon > 0$ and for every k, a bounded open set $G_k \supset S_k$ and a closed set $F_k \subset S_k$ such that $m(G_k) - m(F_k) < \dfrac{\epsilon}{n}, k = 1, 2, \ldots, n$. Using Corollary 1 of Theorem 12.2.10, we get $m(G_k - F_k) = m(G_k) - m(F_k) < \dfrac{\epsilon}{n}$.

Now let $G = \bigcup\limits_{k=1}^{n} G_k$ and $F = \bigcup\limits_{k=1}^{n} F_k$. Obviously, $G \supset S$ and $F \subset S$. G being the finite union of bounded sets, is bounded; and, of course, F is a bounded closed set. Using elementary set theory, it may be readily seen that

$$(G - F) \subset \bigcup_{k=1}^{n} (G_k - F_k)$$

Thus

$$m(G - F) \leqslant \sum_{k=1}^{n} m(G_k - F_k) = \sum_{k=1}^{n} [m(G_k) - m(F_k)] < n\,\frac{\epsilon}{n} = \epsilon;$$

and by the last theorem, S is measurable, which completes the proof. ∎

Theorem 12.2.13. If a closed set $F \subset [a,b]$ then $[a,b] - F$ is measurable and $m([a,b] - F) = (b - a) - m(F)$.

The proof is left as an exercise.

We now discuss a theorem which is very useful for bounded sets.

Theorem 12.2.14. If a bounded set S is contained in a closed interval $[a,b]$, then

$$m_e(S) + m_i([a,b] - S) = b - a$$

Proof. For $\epsilon > 0$ there exists a bounded open set $G \supset S$ such that

$$m(G) < m_e(S) + \epsilon \qquad (12.4)$$

In that case, $[a,b] - G$ is a closed set contained in $[a,b] - S$, and by the definition of internal measure

$$m([a,b] - G) \leqslant m_i([a,b] - S) \qquad (12.5)$$

From (12.4) and (12.5) we have

$$m(G) + m([a,b] - G) < m_e(S) + m_i([a,b] - S) + \epsilon \qquad (12.6)$$

But $[a,b] \subset G \cup ([a,b] - G)$, which implies

$$b - a \leqslant m(G) + m([a,b] - G)$$

Using (12.6), we get

$$(b - a) < m_e(S) + m_i([a,b] - S) + \epsilon$$

Since ϵ is arbitrary,

$$b - a \leqslant m_e(S) + m_i([a,b] - S) \qquad (12.7)$$

To establish the reverse inequality, we note that there exists a closed set $F \subset [a,b] - S$ such that

$$m_i([a,b] - S) - \epsilon < m(F) \qquad (12.8)$$

It is obvious that the set $[a,b] - F \supset S$, which implies

$$m_e(S) \leqslant m([a,b] - F)$$
$$= (b - a) - m(F)$$

Using it with (12.8), we obtain

$$m_e(S) + m_i([a,b] - S) \leqslant b - a \qquad (12.9)$$

Combining (12.6) and (12.9), we get

$$m_e(S) + m_i([a,b] - S) = b - a$$

which proves the theorem.

Corollary 1. If $S \subset [a,b]$ then S is measurable if and only if $m_e(S) + m_e([a,b] - S) = b - a$.

Corollary 2. A bounded set $S \subset [a,b]$ is measurable if and only if $[a,b] - S$ is measurable.

The proofs of these corollaries are left as exercises.

In Theorem 12.2.13, we could replace the closed interval $[a,b]$ by an open interval, but the proof would be different.

Theorem 12.2.15. Let $S \subset (a,b)$; then

$$m_e(S) + m_i\{(a,b) - S\} = b - a$$

The proof is left as an exercise.

We can now prove that the intersection of a finite number of measurable sets is measurable.

Theorem 12.2.16. Let $\{S_k : k = 1, 2, \ldots, n\}$ be a finite class of bounded and measurable sets. Then $\bigcap\limits_{k=1}^{n} S_k$ is measurable.

Proof. Let $S = \bigcap_{k=1}^{n} S_k$, and let $[a,b]$ be a closed interval containing every S_k.* Then from the Corollary 2 of Theorem 12.2.14, $[a,b] - S_k$ is measurable for every k. Using Theorem 12.2.12, $\bigcup_{k=1}^{n} \{[a,b] - S_k\}$ is measurable. By De-Morgan's formula,

$$\bigcup_{k=1}^{n} \{[a,b] - S_k\} = [a,b] - \bigcap_{k=1}^{n} S_k = [a,b] - S$$

Hence $[a,b] - S$ is measurable, and thus S is measurable, again by the use of Corollary 2 of Theorem 12.2.14. This completes the proof.

The following is a very useful corollary of the last theorem.

Corollary. If S_1 and S_2 are two bounded and measurable sets, then $S_1 - S_2$ is measurable. Furthermore, if $S_1 \subset S_2$, then $m(S_1 - S_2) = m(S_1) - m(S_2)$.

The proof of this corollary is not difficult and could be easily handled by the reader.

Using the last corollary and Theorem 12.2.12, we would obtain the following result.

Theorem 12.2.17. Let a bounded set S be the union of a countable number of measurable sets $\{S_k\}$. Then S is measurable.

Proof. If S is the union of a finite number of measurable sets, then S is measurable by Theorem 12.2.12. Therefore, we assume that S is the union of a denumerable number of measurable sets. We write $S = \bigcup_{k=1}^{\infty} S_k$.

Now let

$$T_1 = S_1$$
$$T_2 = S_2 - S_1$$
$$T_3 = S_3 - (S_1 \cup S_2)$$
.
.
.
$$T_n = S_n - (S_1 \cup S_2 \cup \cdots \cup S_{n-1})$$

Using Theorem 12.2.12 and the last corollary, we find that each T_k is measurable. Furthermore, $\{T_k\}$ is a class of pairwise disjoint sets, and, therefore,

---

*It is possible to construct such an interval since the union of a finite number of bounded sets is bounded.

by Theorem 12.2.10, $\bigcup\limits_{k=1}^{\infty} T_k$ is measurable. But $\bigcup\limits_{k=1}^{\infty} T_k = \bigcup\limits_{k=1}^{\infty} S_k$. Hence $\bigcup\limits_{k=1}^{\infty} S_k$ is measurable, and the proof is complete. $\blacksquare$

Theorem 12.2.18. If $S = \bigcap\limits_{k=1}^{\infty} S_k$, where each S_k is bounded and measurable then S is measurable.

Proof. It is obvious that S is bounded. Let $[a,b]$ be a closed interval containing S, and let $A_k = [a,b] - S_k$. Then from Theorem 12.2.16, A_k is measurable for every k, and of course $A_k \subset [a,b]$. Then

$$S = S \cap [a,b] = \left(\bigcap\limits_{k=1}^{\infty} S_k\right) \cap [a,b] = \bigcap\limits_{k=1}^{\infty} (S_k \cap [a,b]) = \bigcap\limits_{k=1}^{\infty} A_k$$

Now the proof can be completed by using the argument* of Theorem 12.2.16. $\blacksquare$

A very important corollary of Theorems 12.2.17 and 12.2.18 is the following.

Corollary. All Borel sets (bounded), F_σ, $F_\sigma^\delta, \ldots$; G^δ, $G_\sigma^\delta, \ldots$, are measurable.

Later on, we shall remove the restriction of "boundedness" in Borel sets. We now prove a result for an arbitrary bounded set.

Theorem 12.2.19. For any bounded set S (measurable or not), there is a set A of type G^δ containing S, and a set B of type F_σ contained in S, such that $m_e(S) = m(A)$ and $m_i(S) = B$.

Proof. For $\epsilon = \dfrac{1}{n}$ (n being a natural number), there is an open set $G_n \supset S$ and a closed set $F_n \subset S$ such that

$$m_e(S) > m(G_n) - \frac{1}{n} \text{ and } m_i(S) < m(F_n) + \frac{1}{n}$$

---

*It is important to note that we could not use that argument from the very beginning, since there might not exist a closed interval $[a,b]$ containing each S_k. Indeed, $\bigcup\limits_{k=1}^{\infty} S_k$ may not be even bounded, though each S_k is bounded. The construction of auxiliary sets A_k's is significant.

Let

$$A = \bigcap_{n=1}^{\infty} G_n \quad \text{and} \quad B = \bigcup_{n=1}^{\infty} F_n$$

Obviously, A is of type G^δ and B is of type F_σ; moreover, since each $G_n \supset S$, it follows that $A \supset S$. Similarly, each $F_n \subset S$ and, therefore, $B \subset S$. Thus

$$m_e(S) > m(G_n) - \frac{1}{n} \geqslant m(A) - \frac{1}{n} \quad \text{for every } n,$$

with

$$G_n \supset A,$$

and

$$m_i(S) < m(F_n) + \frac{1}{n} \leqslant m(B) + \frac{1}{n} \quad \text{for every } n \, (F_n \subset B)$$

Therefore,

$$m_e(S) > m(A) - \frac{1}{n}$$

and

$$m_i(S) < m(B) + \frac{1}{n}$$

Since $\frac{1}{n}$ can be arbitrarily small, we next conclude that $m_e(S) \geqslant m(A)$ and $m_i(S) \leqslant m(B)$. But $m_e(S) \leqslant m_e(A) = m(A)$ and $m_i(S) \geqslant m_i(B) = m(B)$. Hence

$$m_e(S) = m(A) \quad \text{and} \quad m_i(S) = m(B)$$

which completes the proof. ∎

We shall now give a few results which give necessary and sufficient conditions for the measurability of a bounded set, and are sometimes given as alternative definitions of "measurability."

Theorem 12.2.20. (de La Vallée-Poussin's criterion): A bounded set S is measurable $\iff$ For every $\epsilon > 0$ there is a closed set $F \subset S$ such that $m_e(S - F) < \epsilon$.

Proof. If S is measurable, then $m_e(S) = m_i(S) = m(S)$, and for $\epsilon > 0$ there is a closed set $F \subset S$ such that $m(S) = m_i(S) < m(F) + \epsilon$ which means $m(S) - m(F) < \epsilon$.

Using the corollary to Theorem 12.2.16, we get $m(S - F) < \epsilon$, and since $S - F$ is measurable, we can write $m_e(S - F) < \epsilon$.

To prove the converse, we assume that for $\epsilon > 0$ there is a closed set $F \subset S$ such that $m_e(S - F) < \epsilon$. We can then write $S = F \cup (S - F)$, which would imply

$$m_e(S) \leqslant m_e(F) + m_e(S - F) < m(F) + \epsilon$$

But $m(F) \leqslant m_i(S)$, and, therefore, $m_e(S) < m_i(S) + \epsilon$.

Since ϵ is arbitrary, we get $m_e(S) \leqslant m_i(S)$, implying $m_i(S) = m_e(S)$, and hence S is measurable. The proof is now complete. $\blacksquare$

The next theorem is a criterion of measurability according to Caratheodory.

Theorem 12.2.21. A bounded set S is measurable if and only if for an arbitrary bounded set A,

$$m_e(A) = m_e(A \cap S) + m_e(A \cap cS)$$

Proof. Let S be measurable and let A be an arbitrary bounded set. For $\epsilon > 0$, there is a bounded open set $G \supset A$ such that $m_e(A) > m(G) - \epsilon$.

Now $G = (G \cap S) \cup (G \cap cS)$, and we know that $G \cap S$ and $G \cap cS$ are measurable (S being measurable) and disjoint.

Therefore, $m(G) = m(G \cap S) + m(G \cap cS)$. But $G \cap S \supset A \cap S$ and $G \cap cS \supset A \cap cS$, so that

$$m_e(A) > m(G) - \epsilon = m(G \cap S) + m(G \cap cS) - \epsilon$$
$$\geqslant m_e(A \cap S) + m_e(A \cap cS) - \epsilon$$

Since ϵ is arbitrary, $m_e(A) \geqslant m_e(A \cap S) + m_e(A \cap cS)$. But $A = (A \cap S) \cup (A \cap cS)$, and therefore,

$$m_e(A) \leqslant m_e(A \cap S) + m_e(A \cap cS)$$

Combining the two inequalities, we get

$$m_e(A) = m_e(A \cap S) + m_e(A \cap cS)$$

Conversely, if this equation is satisfied for any bounded set A, then S is measurable. To see this, let A be a closed interval $[a,b] \supset S$. Then $m_e(A) = b - a$, and $A \cap S = S$, $A \cap cS = [a,b] - S$.

Therefore, $(b - a) = m_e(S) + m_e([a,b] - S)$, which implies that S is measurable.

The proof is complete. $\blacksquare$

Exercise 12.2

1. Show that the external measure of a set is a translation invariant, that is, if A is any set and $S = \{x + x_0 : x \in A\}$, x_0 being a fixed real number then $m_e(S) = m_e(A)$.

2. Prove that $m_e(B) = 0 \Longrightarrow m_e(A \cup B) = m_e(A)$.

3. Prove Theorem 12.2.4.

4. Prove Theorem 12.2.11.

5. Prove Theorem 12.2.14.

6. Prove Theorem 12.2.15.

7. Prove the corollary of Theorem 12.2.16.

8. Show by an example that the converse of the result in Exercise 2 is not necessarily true.

9. Let $\{S_n\}$ be a sequence of bounded measurable sets of real numbers. Prove that:

   (i) If $S_n \subset S_{n+1}$ for every n, and $\displaystyle\bigcup_{n=1}^{\infty} S_n$ is bounded, then $\displaystyle\lim_{n\to\infty} m(S_n) = m\left(\bigcup_{n=1}^{\infty} S_n\right)$.

   (ii) If $S_n \supset S_{n+1}$ for every n then $\displaystyle\lim_{n\to\infty} m(S_n) = m\left(\bigcap_{n=1}^{\infty} S_n\right)$.

12.3. Translation Invariance of Measure; A Nonmeasurable Set

In the beginning of this chapter, we mentioned the "translation invariance" of a measure as one of the desirable properties. This is quite true with the Lebesgue measure as we shall prove in this section. First, we give the definition of "translation."

Definition 12.3.1. *A mapping on the real line given by the function $f(x) = x + \lambda$, where λ is real number is called a translation. Such a translation is said to be characterized by λ.*

It is obvious that f is a bijective mapping, and that the inverse of f is also a translation ($f^{-1}(x) = x - \lambda$).

The following results are easily established.

Theorem 12.3.1. The image of an open interval (closed interval) under a translation is an open interval (closed interval); disjoint intervals are mapped onto disjoint intervals. Finally, a translation is length preserving, and as such it would preserve the structures and the lengths of bounded open sets and bounded closed sets.

The proof is left as an exercise.

If f is a translation defined by $f(x) = x + \lambda$, then the image of a set S under this translation is denoted by $S + \lambda$, that is $S + \lambda = f(S)$.

Now we establish the translation invariance of external and internal measure of a bounded set in the following theorem.

Theorem 12.3.2. The external and internal measures of a bounded set remain invariant under a translation.

Proof. Let f be a translation and S be a bounded set. For $\epsilon > 0$ there is an open set $G \supset S$ such that $m(G) < m_e(S) + \epsilon$, and a closed set $F \subset S$ such that $m(F) > m_i(S) - \epsilon$. It follows that $f(G) \supset f(S)$ and $f(F) \subset f(S)$.

Also, from Theorem 12.3.1, $m(f(G)) = m(G)$, and $m(f(F)) = m(F)$.

Therefore

$$m_e(f(S)) \leqslant m(f(G)) = m(G) < m_e(S) + \epsilon$$

and

$$m_i(f(S)) \geqslant m(f(F)) = m(F) > m_i(S) + \epsilon$$

Restating the above inequalities,

$$m_e(f(S)) < m_e(S) + \epsilon, \text{ and } m_i(f(S)) > m_i(S) - \epsilon.$$

Since ϵ is arbitrary,

$$m_e(f(S)) \leqslant m_e(S) \qquad (i)$$

and

$$m_i(f(S)) \geqslant m_i(S)$$

These inequalities are true for any arbitrary translation f. From the Definition 12.3.1, it follows that f^{-1} is also a translation, and we let $T = f(S)$. Since f is bijective, $f^{-1}(T) = S$.

From (i),

$$m_e(f^{-1}(T)) \leqslant m_e(T)$$

implying

$$m_e(S) \leqslant m_e(f(S)) \qquad (ii)$$

and similarly,

$$m_i(S) \geqslant m_i(f(S))$$

From (i) and (ii),

$$m_e(S) = m_e(f(S)) \quad \text{and} \quad m_i(S) = m_i(f(S))$$

which completes the proof. ∎

An immediate corollary of this theorem is as follows.

Corollary. Measurability and measure of a set are translation invariant.

It has already been shown that every Borel set is measurable, and that there are c Borel sets (Sec. 6.4). The following theorem establishes the existence of sets which are Lebesgue measurable but are not Borel sets.

Theorem 12.3.3. In any interval, there are 2^c measurable sets.

Proof. Without any loss of generality, we can consider the interval $[0,1]$. This interval contains the Cantor ternary set which is measurable and has the zero measure. Yet it is nondenumerable and as such contains 2^c subsets. Each one of these subsets would have external measure zero and, therefore, would be measurable. Thus there are at least 2^c measurable sets in $[0,1]$. Since the cardinal number of subsets of $[0,1]$ cannot exceed 2^c, we claim that there are 2^c measurable sets in $[0,1]$. Moreover, every interval would have a subset like Cantor set, and the proof is complete.

Unfortunately, this result fails to prove the existence of nonmeasurable set. To prove that, we introduce the following concepts.

Definition 12.3.2. *Let $[0,a)$ be a semiopen interval, $a > 0$. Let $\lambda \in [0,a)$. A mapping ϕ defined on a subset S of $[0,a)$ by*

$$\phi(x) = x + \lambda, \quad \text{if } x < a - \lambda \quad \text{and} \quad x \in S$$

$$\phi(x) = x + \lambda - a, \quad \text{if } x \geq a - \lambda \quad \text{and} \quad x \in S$$

is called "translation modulo a" of the set S.

For the sake of convenience, we write

$$\phi(x) = x \oplus \lambda \pmod{a}$$

and

$$\phi(S) = S \oplus \lambda \pmod{a}$$

It must be noticed here that for every x in $[0,a)$ and some fixed λ of $[0,a)$, $\phi(x) \in [0,a)$. The purpose of this mapping is to translate all points of S by λ and retreat those points back into $[0,1)$ which move out of it simply by translating them by $-a$. This way, the range of the "translation modulo a" remains in $[0,a)$.

The following theorem proves an interesting fact; namely, that a translation modulo a ($a > 0$) preserves both "measurability" and the "measure" of a set.

Theorem 12.3.4. Let S be a measurable subset of $[0,a)$ ($a > 0$). Then for every $\lambda \in [0,a)$, $S \oplus \lambda \pmod{a}$ is measurable and $m \{S \oplus \lambda \pmod{a}\} = m(S)$.

Proof. If $\lambda = 0$, then $S + \lambda \pmod{a} = S$, and then the result is obvious.

We may assume $0 < \lambda < a$; let $S_1 = S \cap [0, a - \lambda)$ and $S_2 = S \cap [a - \lambda, a)$. It is obvious that S_1 and S_2 are disjoint and measurable (since S is measurable), and $S_1 \cup S_2 = S$. Therefore,

$$m(S_1) + m(S_2) = m(S)$$

Now, according to Definition 12.3.2,

$$S_1 \oplus \lambda \pmod{a} = S_1 + \lambda \text{ (since } x \in S_1 \Longrightarrow x < a - \lambda)$$

and $S_2 \oplus \lambda \pmod{a} = S_2 + \lambda - a$ (since $x \in S_2 \Longrightarrow x \geqslant a - \lambda$). But $S_1 + \lambda$ is the image of S_1 under the translation $f(x) = x + \lambda$; hence, $S_1 + \lambda$ is measurable and so is $S_2 \oplus \lambda \pmod{a}$. Moreover, $m(S_1 + \lambda \pmod{a}) = m(S_1 + \lambda) = m(S_1)$. Similarly, $S_2 \oplus \lambda \pmod{a}$ is measurable and $m(S_2 \oplus \lambda \pmod{a}) = m(S_2)$.

It is easy to show that

$$S \oplus \lambda \pmod{a} = (S_1 \oplus \lambda \pmod{a}) \cup (S_2 \oplus \lambda \pmod{a}),$$

and that $S_1 \oplus \lambda \pmod{a}$ and $S_2 \oplus \lambda \pmod{a}$ are disjoint. From this, we conclude that $S \oplus \lambda \pmod{a}$ is measurable and that

$$m(S \oplus \lambda \pmod{a}) = m(S_1 + \lambda \pmod{a})$$
$$+ m(S_2 + \lambda \pmod{a}) = m(S_1) + m(S_2) = m(S)$$

The proof is now complete. **|**

We are going to make use of this result in our next very important theorem which establishes the existence of a non-measurable set. There, we shall let $a = 1$ for the sake of convenience.

Theorem 12.3.5. There exists a nonmeasurable set.

Proof. Let us partition $[0,1)$ into a set of equivalence classes $\{C_\alpha\}$ such that two real numbers x and y of $[0,1)$ belong to the same class if and only if $x - y$ is rational. If this is the case, we write $x \sim y$ (Clearly, $\sim$ is an equivalence relation).*

Now, using the axiom of choice, we find a set S which contains *exactly one* element of each class. We shall prove that S is nonmeasurable.

Let $0 = r_1, r_2, \ldots r_n, \ldots$ be an enumeration of the set of all rational numbers in $[0,1)$. Let $S_k = S \oplus r_k \pmod{1}$, so that $S_1 = S$.

First we show that S_k's are pairwise disjoint. If $x_0 \in S_j \cap S_k (j \neq k)$, then $x_0 \in S_j$ and $x_0 \in S_k$. Now, $x_0 \in S_j \Longrightarrow x_0 = a + r_j$ or $x_0 = a + r_j - 1$, where $a \in S$; and $x_0 \in S_k \Longrightarrow x_0 = b + r_k$ or $x_0 = b + r_k - 1$, where $b \in S$. Since $r_j \neq$

---

*It follows that each class is denumerable since the set of all rationals is denumerable, and that any two distinct classes are disjoint. From this, it is obvious that there are c such disjoint classes.

r_k, and $r_j \neq r_k - 1$ ($r_k - 1$ is negative) and $r_k \neq r_j - 1$, thus $a \neq b$, yet $a - b$ is rational, but that means S contains two distinct elements from the same class.

Next, we show that $\bigcup_{k=1}^{\infty} S_k = [0,1)$. Obviously, $\bigcup_{k=1}^{\infty} S_k \subset [0,1)$ since each S_k is a subset of $[0,1)$. Now, if $x \in [0,1)$, then $x \in C_\alpha$ for some α. But there exists a representative y_α of C_α such that $y_\alpha \in S$. In that case $|x - y_\alpha|$ is a rational number in $[0,1)$, say $|x - y_\alpha| = r_n$.

Then, either $x = y_\alpha + r_n \Longrightarrow x \in S_n \Longrightarrow x \in \bigcup_{k=1}^{\infty} S_k$, or $y_\alpha = x + r_n \Longrightarrow x =$

$y_\alpha + (1 - r_n) - 1 \Longrightarrow x = y_\alpha + r_m - 1 \Longrightarrow x \in S_m \Longrightarrow x \in \bigcup_{k=1}^{\infty} S_k$, where $r_m = 1 - r_n$.

Thus we establish that $\bigcup_{k=1}^{\infty} S_k = [0,1)$. Using Theorems 12.2.7 and 12.2.9, we get

$$1 = m_e\{[0,1)\} \leqslant \sum_{k=1}^{\infty} m_e(S_k)$$

and

$$1 = m_i\{[0,1)\} \geqslant \sum_{k=1}^{\infty} m_i(S_k)$$

But $m_e(S_k) = m_e(S)$ and $m_i(S_k) = m_i(S)$ for all k. Hence

$$1 \leqslant m_e(S) + m_e(S) + \cdots \text{ ad inf.} \Longrightarrow m_e(S) > 0$$
$$1 \geqslant m_i(S) + m_i(S) + \cdots \text{ ad inf.} \Longrightarrow m_i(S) = 0$$

Therefore S is nonmeasurable and the proof is complete. ∎

Remark. In the proof of the last theorem, we used the properties 2, 3, 4, and 5 listed in Sec. 12.1, and, of course, axiom of choice. It is interesting to note that if we allow the measure of an interval to be 0 or $+\infty$ (instead of 3), then nonmeasurable sets would not exist.

It may be added that no example of nonmeasurable sets is given without the aid of the axiom of choice (or any of its equivalent forms).

Exercise 12.3

1. Give an example of a countable class $\{S_k\}$ of pairwise disjoint measurable sets such that

(a) $m_e(\cup_k S_k) < \sum_k m(S_k)$

(b) $m_i(\cup_k S_k) > \sum_k m_i(S_k)$

(*Hint*: Examine the proof of Theorem 12.3.5.)

2. Let S be the nonmeasurable set as constructed in the proof of Theorem 12.3.5. Show that every measurable subset of S must have the zero measure.

3. Let S be a set such that $m_e(S) > 0$. Prove that there exists a nonmeasurable subset of S.

12.4. Measurability of Unbounded Sets

The concept of measurability can easily be extended to "unbounded sets" if we include the possibility of a measure being $+\infty$.

Definition 12.4.1. *If S is an unbounded set, then S is said to be measurable if $S \cap [-n,n]$ is measurable for every natural number n. In that case, we define*

$$m(S) = \lim_{n \to \infty} m(S \cap [-n,n])$$

This limit may be finite or $+\infty$ (if it does not exist).

We give some examples of unbounded measurable sets.

EXAMPLE 12.4.1. The set $\mathfrak{N}$ of all natural numbers is measurable and its measure is zero.

EXAMPLE 12.4.2. The sets $(-\infty,a)$, $(-\infty,a]$, (a,∞) and $[a,\infty)$ are all measurable, and the measure in each case is $+\infty$.

EXAMPLE 12.4.3. Let

$$S = \bigcup_{n=1}^{\infty} \left[n - \frac{1}{3n}, n \right]$$

then S is measurable and

$$m(S) = \lim_{n \to \infty} S \cap [-n,n]$$

$$= \lim_{n \to \infty} \sum_{i=1}^{n} \frac{1}{3^i} = \frac{1}{2}$$

Most of the results of Sec. 12.2 are valid for unbounded sets. We list some of them.

Theorem 12.4.1. Every unbounded open set and every unbounded closed set are measurable, and the measure of such sets can be finite or infinite.

The proof is left as an exercise.

Theorem 12.4.2. The union of a countable number of measurable sets S_k is measurable, and if S_k's are pairwise disjoint, then $m\left(\bigcup_k S_k\right) = \sum_k m(S_k)$.

The proof is left as an exercise.

Also, we can prove that the intersection of a countable number of measurable sets is measurable.

Theorem 12.4.3. Let S be a measurable; then the complement of S is also measurable.

Proof. The proof follows from the fact that

$$cS \cap [-n,n] = \{c(S \cap [-n,n])\} \cap [-n,n]$$

and then, by use of the theorem for bounded sets. ∎

We can extend de La Vallée-Poussin's criterion and that of Caratheodory to any arbitrary set. Furthermore, we can show that all Borel sets are measurable.

In conclusion of this chapter, we briefly discuss Vitali covering.

Definition 12.4.2. *Let S be a set of reals. A class of closed intervals $\{I_\alpha\}$ is said to cover S in the Vitali sense if for every x in S and for $\epsilon > 0$ there is a member I_β of $\{I_\alpha\}$ such that $l(I_\beta) < \epsilon$ and $x \in I_\beta$.*

We now state Vitali covering theorem *without proof.*

Theorem 12.4.4. (Vitali Covering Theorem). Let S be a set of finite external measure and let $\{I_\alpha\}$ be a class of closed intervals which is also a Vitali covering of S. For $\epsilon > 0$ there is a finite subclass $\{I_1, I_2, \ldots, I_n\}$ of $\{I_\alpha\}$ such that the intervals in this subclass are pairwise disjoint and this subclass covers a subset S_1 of S, such that $m_e(S - S_1) < \epsilon$.

The most elegant proof of this theorem is given by Banach. We omit this proof.

The advantage of the "Vitali covering theorem" over "Heine-Borel theorem" and "Lindeloff covering theorem" is that a finite number of intervals selected are pairwise disjoint. The disadvantage is that the same intervals do not necessarily cover the entire set.

Exercise 12.4

1. Prove Theorem 12.4.1.
2. Prove Theorem 12.4.2.

3. Prove Caratheodary's criterion for an unbounded measurable set.
4. Construct an example of an unbounded measurable set whose measure is a fixed positive number k.

Miscellaneous Exercises on Chapter 12

1. Prove that a set S is measurable if and only if for $\epsilon > 0$, there is an open set $G \supset S$ such that $m_e(G - S) < \epsilon$.
2. Show that a set S is measurable if and only if there is a set A of type G^δ containing S such that $m_e(A - S) = 0$.
3. Prove that a set S is measurable if and only if for $\epsilon > 0$ there is a set $B \subset S$ of type F_σ such that $m_e(S - B) = 0$.
4. Give an example of a "measure" such that it satisfies all the desirable properties except that the measure of an interval is 0 or ∞.
5. Prove that a bounded function f is Riemann-integrable on $[a,b]$ if and only if the set of points of discontinuities of f has measure zero.

13

Measurable Functions and Lebesgue Integration

13.1. Measurable Functions

In this section we discuss measurable functions which are defined in terms of measurable sets.

Definition 13.1.1. *Let* $f: D \longrightarrow R$ *be a function such that its domain D is measurable. Then f is said to be measurable if the set* $\{x: f(x) \leqslant d\}$ *is measurable, for every real number d.*

From a property of continuous functions it follows that every continuous function is measurable, if its domain is a measurable set. We shall see, however, that there are many functions which are measurable but not continuous.

The following theorem gives various alternatives to Def. 13.1.1.

Theorem 13.1.1. A function $f: D \longrightarrow R$ (D being a measurable set) is measurable $\Longleftrightarrow$ (a) $\{x: f(x) < d\}$ is measurable for every real number $d \Longleftrightarrow$ (b) $\{x: f(x) > d\}$ is measurable for every real number $d \Longleftrightarrow \{x: f(x) \geqslant d\}$ is measurable for every real number d.

Proof. Since $\{x: f(x) > d\}$ is the complement of $\{x: f(x) \leqslant d\}$ with respect to D, (b) is logically equivalent to Def. 13.1.1. (Corollary 2, Theorem 12.2.14). Also, the set $\{x: f(x) \geqslant d\}$ is the complement of $\{x: f(x) < d\}$ with respect to D. Therefore, it will be enough to prove that the measurability of f according to Def. 13.1.1 is equivalent to the measurability of $\{x: f(x) < d\}$. To this end, we show that

$$\{x: f(x) < d\} = \bigcup_{n=1}^{\infty} \left\{x: f(x) \leqslant d - \frac{1}{n}\right\}$$

Since $f(x) \leqslant d - \dfrac{1}{n} \Longrightarrow f(x) < d$ for every n; and $f(x) < d$ implies that there is a natural number n such that $f(x) \leqslant d - \dfrac{1}{n}$.

Now if f is measurable according to Def. 13.1.1 each $\left\{ x : f(x) \leqslant d - \dfrac{1}{n} \right\}$ is measurable, and then $\{x : f(x) < d\}$ would be measurable.

Conversely, if $\{x : f(x) < d\}$ is measurable for every real number d, then by using the argument given above we can show that

$$\{x : f(x) \leqslant d\} = \bigcap_{n=1}^{\infty} \left\{ x : f(x) < d + \dfrac{1}{n} \right\}$$

which implies that f is measurable (according to Def. 13.1.1). The proof is now complete. ∎

From this theorem it is possible to replace $\{x : f(x) \leqslant d\}$ by any of the three sets $\{x : f(x) < d\}$, $\{x : f(x) > d\}$ and $\{x : f(x) \geqslant d\}$ in Def. 13.1.1.

Theorem 13.1.2. If a function $f : D \longrightarrow R$ is measurable then $\{x : f(x) = d\}$ is measurable where d is a real number.

Proof. If d is a real number, then we can write

$$\{x : f(x) = d\} = \{x : f(x) \leqslant d\} \cap \{x : f(x) \geqslant d\}$$

and since f is measurable, the two sets on the right hand side are measurable. Therefore, $\{x : f(x) = d\}$ is measurable. The proof is complete. ∎

The converse of this theorem is not necessarily true as follows from the following example:

EXAMPLE 13.1.1. Let S be a nonmeasurable set of $[0,1]$, then $c_{\Delta}S = [0,1] - S$ is also nonmeasurable ($\Delta = [0,1]$).

Furthermore, each one of these sets must be nondenumerable, and as such there is 1-1 correspondence between S and $\left[0, \dfrac{1}{2} \right)$ and a 1-1 correspondence between $c_{\Delta}S$ and $\left[\dfrac{1}{2}, 1 \right]$. Let f be a function which maps S bijectively onto $\left[0, \dfrac{1}{2} \right)$ and maps $c_{\Delta}S$ bijectively onto $\left[\dfrac{1}{2}, 1 \right]$.

It must be noticed that f is a bijective mapping of $[0,1]$ onto $[0,1]$ such that $f(x) \in \left[0, \dfrac{1}{2} \right)$ if $x \in S$, and $f(x) \in \left[\dfrac{1}{2}, 1 \right]$ if $x \in c_{\Delta}S$.

Now

$$\{x : f(x) = d\} = \phi \text{ if } d \notin [0,1]$$

and $\{x : f(x) = d\}$ consists of exactly one point if $d \in [0,1]$. In either case $\{x : f(x) = d\}$ is measurable.

However, $\left\{ x : f(x) < \dfrac{1}{2} \right\} = S$, which is nonmeasurable.

This, incidentally, is also an example of a nonmeasurable function defined on the measurable set $[0,1]$.

Theorem 13.1.3. A function $f : D \longrightarrow R$ is measurable
$\iff$ the set $\{x : a < f(x) < b\}$ is measurable for all a and b such that $a < b$.
$\iff$ the set $\{x : a \leqslant f(x) \leqslant b\}$ is measurable for all a and b, such that $a < b$.
The proof of this theorem is left as an exercise.

In the next two theorems we discuss some properties of measurable functions.

Theorem 13.1.4. Let $f : D \longrightarrow R$ be a measurable function (D being measurable). Then
    (a) $f(x) + \lambda$ is measurable for every real number λ.
    (b) λf is measurable for every real number λ.
    (c) f^2 is measurable.
    (d) $|f|$ is measurable.
    (e) $\dfrac{1}{f}$ is measurable, if $f \neq 0$ on D.

Proof. Let d be an arbitrary real number. Then $\{x : f(x) + \lambda \leqslant d\} = \{x : f(x) \leqslant d - \lambda\}$, and this proves (a).

To prove (b), consider the case when $\lambda > 0$ and then $\{x : \lambda f(x) \leqslant d\} = \left\{ x : f(x) \leqslant \dfrac{d}{\lambda} \right\}$ which is measurable.

If $\lambda < 0$ then $\{x : \lambda f(x) \leqslant d\} = \{x : f(x) \geqslant d/\lambda\}$, which is again measurable.
If $\lambda = 0$, then, of course, λf is measurable.
Therefore, (b) is established.
To prove (c), we write

$$\{x : [f(x)]^2 \leqslant d\} = \{x : -\sqrt{d} \leqslant f(x) \leqslant \sqrt{d}\} \text{ if } d \geqslant 0$$

The set on the right-hand side is measurable by virtue of Theorem 13.1.3, and thus f^2 is measurable.

For $d < 0$ the set $\{x : [f(x)]^2 \leqslant d\}$ is empty, and obviously measurable. (c) is proved.

The proof of (d) is quite similar.
Now if $f(x) \neq 0$ on D, then

$$\left\{x:\frac{1}{f(x)}<d\right\}=\begin{cases}\{x:f(x)<0\} & \text{if } d=0\\[2mm]\{x:f(x)<0\}\cup\left(\{x:f(x)>0\}\cap\left\{x:f(x)>\frac{1}{d}\right\}\right), & \text{if } d>0.\\[3mm]\{x:f(x)<0\}\cap\left\{x:f(x)>\frac{1}{d}\right\}\text{ if } d<0\end{cases}$$

In either case f is measurable.

The proof of the theorem is now complete. ∎

Theorem 13.1.5. Let f and g be two functions defined on a measurable set D, then

(i) the set $\{x:f(x)<g(x)\}$ is measurable

(ii) $f+g$ and $f-g$ are measurable

(iii) $f(x)\cdot g(x)$ is measurable

(iv) $\dfrac{f(x)}{g(x)}$ is measurable if $g(x)\neq 0$ on D

Proof. To prove (i) we enumerate the set of all rational numbers as follows: $\{r_1,r_2,\ldots,r_n,\ldots\}$; and write

$$A_k = \{x:f(x)<r_k\}\cap\{x:g(x)>r_k\}$$

Then

$$\{x:f(x)<g(x)\}=\bigcup_{k=1}^{\infty} A_k$$

To see this we note that if $x_0\in\{x:f(x)<g(x)\}$ then $f(x_0)<g(x_0)\Longrightarrow$ there is a rational number r_n such that $f(x_0)<r_n<g(x_0)$ implying $x_0\in A_n$, and then $x_0\in\bigcup_{k=1}^{\infty} A_k$.

Conversely, if $x_0\in\bigcup_{k=1}^{\infty} A_k$, then $x_0\in A_n$ for some n implying $f(x_0)<r_n<g(x_0)\Longrightarrow x_0\in\{x:f(x)<g(x)\}$. Now since each A_k is measurable, the set $\{x:f(x)<g(x)\}$ is measurable.

To prove (ii) we only have to notice that

$$\{x:f(x)+g(x)<d\}=\{x:f(x)<d-g(x)\}$$

and since $d-g(x)$ is a measurable function by virtue of Theorem 13.1.4, the result follows by the application of (i).

(iii) is established by writing

$$f(x)\cdot g(x)=\frac{1}{2}(\{f(x)\}^2+\{g(x)\}^2-[f(x)-g(x)]^2)$$

and applying Theorem 13.1.4 and part (ii) of this theorem.

(iv) is a direct consequence of (iii) and Theorem 13.1.4. The proof is now complete. ∎

We next establish a result for a sequence of measurable functions.

Theorem 13.1.6. Let $\{f_n\}$ be a sequence of measurable functions defined on the measurable set D. Then $\liminf\limits_{n\to\infty} f_n$, $\limsup\limits_{n\to\infty} f_n$, $\lim\limits_{n\to\infty} f_n$ are measurable.

Proof. Let

$$g_m = \text{glb } \{f_1, f_2, \ldots, f_m\}$$

Then

$$\{x : g_m(x) \geqslant d\} = \bigcap_{k=1}^{m} \{x : f_k(x) \geqslant d\}$$

This implies that $g_m : D \longrightarrow R$ is measurable. Similarly glb $\{f_n\}$ is measurable, and also lub $\{f_n\}$. Now we can write $\liminf\limits_{n\to\infty} f_n = \text{lub } \{g_m\} =$ lub [glb $\{f_k : k \geqslant m\}$] which establishes the measurability of $\liminf f_n$.

A similar argument holds for $\limsup\limits_{n\to\infty} f_n$. The measurability of $\lim\limits_{n\to\infty} f_n$ is now obvious, and the proof is complete. ∎

We now give a definition.

Definition 13.1.2. *Let $f : D \longrightarrow R$ be a function defined in such a way that D can be decomposed into a finite number of pairwise disjoint measurable sets $D_1, D_2, \ldots, D_n$, with $f(x) = c_k$ for $x \in D_k$ (c_k being a constant). Then f is called a simple function.*

We now prove that every simple function is measurable.

Theorem 13.1.7. If $f : D \longrightarrow R$ is a simple function, then f is measurable.

Proof. Let $\{c_1, c_2, \ldots, c_n\}$ be the range of this simple function. Let $\{d_1, d_2, \ldots, d_n\} = \{c_1, c_2, \ldots, c_n\}$ such that $d_1 \leqslant d_2 \cdots \leqslant d_n$. (We only require that $d_i = c_j$, i and j may not necessarily be equal.)

Let $\{E_1, E_2, \ldots, E_n\}$ be the decomposition of D such that $f(x) = d_k$ for $x \in E_k$.

Now for an arbitrary real number d

$$\{x : f(x) \leqslant d\} = \begin{cases} \phi \text{ if } d < d_1 \\ D \text{ if } d \geqslant d_n \\ \bigcup\limits_{k=1}^{m} E_k \text{ if } d_m \leqslant d < d_{m+1} \\ \qquad\qquad m = 1, 2, \ldots, (n-1) \end{cases}$$

Thus the set $\{x : f(x) \leqslant d\}$ is measurable for every real number d, and the result follows.

A special case of a simple function is the characteristic function of a measurable set. A characteristic function is defined as follows:

Definition 13.1.3. *A function* $f : D \longrightarrow R$ *is called the characteristic function of a set* $S (S \subset D)$. *if*

$$f(x) = 1 \quad \text{for} \quad x \in S$$

and

$$f(x) = 0 \quad \text{for} \quad x \in D - S$$

It is sometimes convenient to denote the characteristic function of a set S by $f_s(x)$ or $\phi_s(x)$.

We now prove the following result for a characteristic function.

Theorem 13.1.8. The characteristic function of a set S defined on a measurable domain D $(D \supset S)$ is measurable if and only if S is measurable.

Proof. Let $f_s : D \longrightarrow R$ be the characteristic function of S. If f_s is measurable then by the Theorem 13.1.2 the set $\{x : f_s(x) = 1\}$ is measurable. But that set is S, and as such it is measurable.

Conversely, if S is measurable then $D - S$ is measurable and by Theorem 13.1.7, f_s is measurable (being a simple function). That completes the proof. ∎

Now a terminology.

Definition 13.1.4. *A property* $\mathcal{P}$ *is said to hold almost everywhere on a set* S *if the set of points of S where $\mathcal{P}$ fails to hold has measure zero.*

For example, if a function f is discontinuous on a subset of S whose measure is zero and is continuous everywhere else, then we can say that f is continuous almost everywhere on S.

Interestingly enough, if S has measure zero, then a property $\mathcal{P}$ may not hold anywhere on S but we can still claim that $\mathcal{P}$ holds almost everywhere on S.

We now prove a theorem which is due to Egoroff (1869-1931).

Theorem 13.1.9. Let $\{f_n\}$ be a sequence of measurable functions defined and convergent on a measurable set D of finite measure. Then for $\epsilon > 0$ there is a measurable set $S \subset D$ such that $m(D - S) < \epsilon$ and $\{f_n\}$ is uniformly convergent on S.

Proof. Let $f = \lim_{n \to \infty} f_n$. By Theorem 13.1.6, f is measurable.

Write

$$F_n(x) = |f_n(x) - f(x)|$$

Clearly, F_n is measurable for every n. We now define

$$S_{nm} = \left\{ x : F_k(x) < \frac{1}{m}, \quad k \geqslant n \right\}$$

Obviously, every S_{nm} is measurable.

Since for $x_0 \in D$ and $m \in \mathfrak{N}$ there is $n \in \mathfrak{N}$ such that

$$F_k(x_0) = |f_k(x_0) - f(x_0)| < \frac{1}{m} \quad \text{for } k \geqslant n$$

Therefore

$$D = \bigcup_{n=1}^{\infty} S_{nm}$$

Also, it is an easy matter to show that

$$S_{1m} \subset S_{2m} \subset \cdots$$

so that $D = \lim_{n \to \infty} (S_{nm})$.

We may now conclude that

$$m(D) = \lim_{n \to \infty} (m(S_{nm}))$$

This means that for $\epsilon > 0$ and for every m there is a natural number n_0 such that $m(D) - m(S_{nm}) < \epsilon/2\,m$ for $n \geqslant n_0$

$$m(D - S_{nm}) < \frac{\epsilon}{2^m} \quad \text{for } n \geqslant n_0$$

Let

$$S = \bigcap_{m=1}^{\infty} S_{n_0 m}$$

For every $x \in S$,

$$F_k(x) < \frac{1}{m}, \quad \begin{array}{l} m = 1, 2, \ldots \\ k \geqslant n_0 \end{array}$$

Since n_0 does not depend upon x, $\{F_k\}$ converges uniformly to zero on S; i.e., $\{f_n\}$ converges uniformly to f on S.

Now, using De Morgan's formula we have

$$D - S = \bigcup_{m=1}^{\infty} (D - S_{nm})$$

Therefore

$$m(D - S) \leqslant \sum_{m=1}^{\infty} m(D - S_{n_0 m}) < \sum_{m=1}^{\infty} \frac{\epsilon}{2^m} = \epsilon$$

and the theorem is proved.

Exercise 13.1

1. Prove Theorem 13.1.3.
2. Prove that if a function f is equal to g almost everywhere on D and if one of these two functions is measurable, then the other is measurable.
3. Prove that if $\{f_n\}$ is a sequence of measurable functions converging to f almost everywhere on D then f is measurable.
4. Define *convergence in measure* as follows: Let $\{f_n\}$ be a sequence of measurable functions defined on D. Let $f: D \longrightarrow R$ be a function such that for every $\lambda > 0$,

$$\lim_{n \to \infty} [m \{x : |f_n(x) - f(x)| \geqslant \lambda\}] = 0$$

then $\{f_n\}$ is said to converge in measure to f.

   Prove that the "convergence" implies "convergence almost everywhere" which further implies "convergence in measure."
5. Prove the following generalized version of Egoroff's theorem. Let $\{f_n\}$ be a sequence of measurable functions defined and *convergent in measure* to f on D. For $\epsilon > 0$ there is a measurable set $S \subset D$ such that $m(D - S) < \epsilon$ and $\{f_n\}$ is uniformly convergent to f on S.
6. In Example 13.1.1, the continuum hypothesis is used. Explain where.
7. Prove that every continuous function of a measurable function is measurable. (*Hint*: Use Theorems 7.2.7, 6.1.2, and 13.1.3.)
8. Show by an example that a measurable function of a measurable function (or even a continuous function) may not be measurable. (Challenging exercise.)

13.2. Lebesgue Integration

The theory of Riemann integration as developed in Chapter 9, though very useful, is not free from defects. First of all, the Riemann integral of a function is defined on a closed interval and cannot be defined on an arbitrary measurable set. Second and more important is the fact that the Riemann integrability depends upon the continuity of the function. Of course there are functions which are discontinuous and yet Riemann-integrable, but these functions are not too discontinuous, *since a necessary and sufficient condition that a bounded function f is Riemann-integrable over* $[a,b]$ *is that it must be continuous almost everywhere on* $[a,b]$ (cf. Miscellaneous Exercise 5, Chapter 12). Henri Lebesgue introduced the concept of another integral which depends upon the measurability of the function rather than its continuity and therefore includes a much larger class of functions than the one of continuous functions.

We now start with some definitions.

Definition 13.2.1. *Let* D *be a measurable set. A finite class* $P = \{D_1, D_2, \ldots, D_n\}$ *of pairwise disjoint measurable subsets of* D *is called a decomposition or a generalized partition of* D *if* $D = \bigcup_{k=1}^{n} D_k$. *The* D_k*'s are called components of* D.

It may be recalled that the term "partition" mentioned in Chapter 9 is a special case of "decomposition."

Definition 13.2.2. *A decomposition* $Q = \{E_j\}$ *of* D *is called a refinement of* $P = \{D_k\}$ *(also a decomposition of* D*) if for every* E_j *of* Q *there is a* D_k *of* P, *such that* $E_j \subset D_k$ *(and, of course* $E_j \cap D_l = \phi, k \neq l$*).*

We now define upper and lower Lebesgue sums.

Definition 13.2.3. *Let f be a bounded function defined on a measurable set* D *which has a finite measure. Let* $P = \{D_1, D_2, \ldots, D_n\}$ *be a generalized partition of* D. *Let*

$$m_k = \text{glb} \ \{f(x) : x \in D_k\},$$

and $M_k = \text{lub} \ \{f(x) : x \in D_k\}, k = 1, 2, \ldots, n$

Let

$$s(f,P) = \sum_{k=1}^{n} m_k \cdot m(D_k),$$

and

$$S(f,P) = \sum_{k=1}^{n} M_k \cdot m(D_k).$$

Then $s(f,P)$ is called the *lower Lebesgue sum* of f for the decomposition P, and $S(f,P)$ is called the *upper Lebesgue sum* of f for the decomposition P.

In the next two theorems we shall assume that f is a *bounded function* defined on a measurable set of *finite measure*.

Theorem 13.2.1. Let $Q = \{E_j\}$ be a refinement of a generalized partition $P = \{D_k\}$ of D. If $s'(f,Q)$ and $S'(f,Q)$ are the lower and upper Lebesgue sums, respectively, for Q and $s(f,P)$ and $S(f,P)$ represent similar sums for the decomposition P, then $s(f,P) \leqslant s'(f,Q)$ and $S(f,P) \geqslant S'(f,Q)$. (In other words, by refining a generalized partition the lower sum does not decrease and the upper sum does not increase.)

Proof. If $Q = P$, the result is obvious. So let us assume there exists at least one set D_k of P which has been decomposed into nonempty disjoint measurable sets $E_{k_1}, E_{k_2}, \ldots, E_{km}$ all of which belong to the partition Q. Here we can assume $m \geq 2$.

Let

$$m_{k_p} = \text{glb } \{f(x) : x \in E_{k_p}\}, \quad p = 1, 2, \ldots, m$$

Obviously

$$m_k \leq m_{k_p} \text{ for } p = 1, 2, \ldots, m$$

and

$$\sum_{k=1}^{m} m(E_{k_p}) = m(D_k)$$

Now if we consider the term $m_k \cdot m(D_k)$ in $s(f,D)$ we notice that

$$m_k \cdot m(D_k) \leq \sum_{p=1}^{m} m_{k_p} \cdot m(E_{k_p})$$

Summing on k, we get

$$s(f,P) \leq s'(f,Q)$$

Similarly, we can prove that $S(f,P) \geq S'(f,Q)$, and the theorem is now proved.

The next theorem establishes that every lower Lebesgue sum is less than or equal to every upper Lebesgue sum.

Theorem 13.2.2. If P_1 and P_2 are two generalized partitions of D, then $s_1(f,P_1) \leq S_2(f,P_2)$ and $s_2(f,P_2) \leq S_1(f,P_1)$.

The proof of this theorem is exactly similar to that of Theorem 9.2.3 and is left as an exercise.

What we have seen here is that the class of all lower sums of a bounded measurable function is bounded above, since every upper Lebesgue sum is an upper bound. A similar statement can be made about the upper Lebesgue sums.

We can now define lower Lebesgue integral and upper Lebesgue integral.

Definition 13.2.4. *The least upper bound of $\{s(f,P)\}$ for all different generalized partitions of D is called the "lower Lebesgue integral" of f over D and is written as $L \underline{\int_D} f$.*

In other words, $L \displaystyle\int_{\underline{D}} f = \text{lub } \{s(f,P): \quad P \in \mathfrak{D}\}$

where $\mathfrak{D}$ is the class of all decompositions of D.

Definition 13.2.5. *The greatest lower bound of $\{s(f,P)\}$ for all different generalized partitions of D is called the "upper Lebesgue integral" of f over D and is written as $L \displaystyle\int_D^{\overline{}} f$. That is,*

$$L \int_D^{\overline{}} f = \text{glb } \{S(f,P): \quad P \in \mathcal{D}\}$$

The existence of these integrals for a bounded function over a set of finite measure is guaranteed in view of what we remarked following Theorem 13.1.2.

Definition 13.2.6. *A bounded function f is said to be Lebesgue integrable over D (sometimes written as L-integrable over D) if $L \displaystyle\int_D^{\overline{}} f = L \displaystyle\int_{\underline{D}} f$.*

Their common value is called the Lebesgue integral of f over D and is simply written as $L \displaystyle\int_D f$.

From these definitions and Theorem 13.2.2 it follows that

$$s(f,P) \leqslant L \int_{\underline{D}} f \leqslant L \int_D^{\overline{}} f \leqslant S(f,P)$$

where P is any decomposition of D.

We shall show very soon that the only condition we need for the Lebesgue integrability of a bounded function is that it must be measurable. To prove it we construct a special type of decomposition of D as follows:

Let f be a bounded measurable function defined over a set D of finite measure.

Let $A \leqslant f(x) < B$ for $x \in D$. We write

$$A = y_0 < y_1 < \cdots < y_n = B$$

Now let

$$d_k = \{x : y_k \leqslant f(x) < y_{k+1}\}, \quad k = 0, 1, \ldots, (n-1)$$

Since f is measurable, each d_k is measurable. Clearly, $\{d_k : k = 0, 1, \ldots, (n-1)\}$ is a decomposition of D.

Definition 13.2.7. *We shall call such a partition as constructed in the preceding paragraph an* L-*partition of D (in deference to Lebesgue) for the function* f.

It may be realized that the set d_k could be any type of sets in D, not necessarily a semi open interval.

Now, we prove the most fundamental theorem for Lebesgue integration.

Theorem 13.2.3. If f is a bounded and measurable function defined over a set D of finite measure, then f is Lebesgue integrable over D.

Proof. Let $A \leqslant f(x) < B$ over D, and let $\rho = \{d_k : k = 0, 1, \ldots, (n-1)\}$ be an L-partition of D for f as constructed in Def. 13.2.7. For this L-partition we define the following sums:

$$t(f,\rho) = \sum_{k=1}^{n-1} y_k \cdot m(d_k)$$

$$T(f,\rho) = \sum_{k=1}^{n-1} y_{k+1} \cdot m(d_k)$$

It must be noticed here that t and T are not exactly the lower and upper sums for the partition ρ, because y_k, y_{k+1} are not necessarily the glb and lub of f in d_k. In fact,

$$y_k \leqslant \text{glb } \{f(x) : x \in d_k\} \quad \text{if } d_k \neq \emptyset,$$

and

$$y_{k+1} \geqslant \text{lub } \{f(x) : x \in d_k\} \quad \text{if } d_k \neq \emptyset$$

$$\text{If } d_k = \emptyset \text{ then } y_k m(d_k) = 0 = y_{k+1} m(d_k).$$

In any case $t(f,\rho) \leqslant s(f,\rho)$; and $T(f,\rho) \geqslant S(f,\rho)$ (s and S are, as usual, the lower and upper sums, respectively).

Now

$$T(f,\rho) - t(f,\rho) = \sum_{k=0}^{n-1} (y_{k+1} - y_k) m(d_k) \leqslant \mu_n \sum_{k=0}^{n-1} m(d_k) = \mu_n \cdot m(D)$$

where

$$\mu_n = \max \{(y_{k+1} - y_k) : k = 0, 1, \ldots, (n-1)\}$$

By increasing the number of points appropriately, μ_n can be made arbitrarily small. That is to say, for $\epsilon > 0$ there is an L-partition ρ such that

$$T(f,\rho) - t(f,\rho) < \epsilon \qquad \left(\text{Let } n > \frac{(B-A) \cdot m(D)}{\epsilon} \text{ and } \mu_n < \frac{\epsilon}{m(D)}\right)^*$$

But

$$t(f,\rho) \leqslant s(f,\rho) \leqslant L \underline{\int_D} f \leqslant L \overline{\int_D} f \leqslant S(f,\rho) \leqslant T(f,\rho)$$

Therefore,

$$L \overline{\int_D} f - L \underline{\int_D} f \leqslant T(f,\rho) - t(f,\rho) < \epsilon$$

Since ϵ is arbitrary, $L \overline{\int_D} f - L \underline{\int_D} f = 0$, and the result now follows. ∎

Conversely, if a *bounded* function is Lebesgue integrable then it must be measurable (cf. Exercise 13.2(3)).

Another way of defining the Lebesgue integral of a function f is to construct an L-partition of D and consider the sum $\sum_{k=0}^{n-1} f(\xi_k) \cdot m(d_k)$, where ξ_k is an arbitrary point in d_k. Then, simply let

$$L \int_D f = \lim_{\substack{n \to \infty \\ \mu_n \to 0}} \sum_{k=0}^{n-1} f(\xi_k) \cdot m(d_k)$$

The last limit exists, in view of the fact that

$$t(f,\rho) \leqslant \sum_{k=0}^{n-1} f(\xi_k) \cdot m(d_k) \leqslant T(f,\rho)$$

In the next theorem we prove that the Riemann-integrability of a function f is a sufficient condition for the existence of Lebesgue integral of f.

Theorem 13.2.4. If a function f is Riemann integrable over $[a,b]$, then it is Lebesgue integrable on $[a,b]$ and its Lebesgue integral is the same as the Riemann integral.

Proof. Let f be a function R-integrable over $[a,b]$. Then according to Def. 9.2.3,

---

*If $m(D) = 0$, then $T(f,\rho)$, and $t(f,\rho) = 0 < \epsilon$.

$$R \int_a^{\overline{b}} f = \text{glb } \{S(f,P): \ P \in \mathcal{P}\}$$

$$R \int_{\underline{a}}^b f = \text{lub } \{s(f,P): \ P \in \mathcal{P}\}$$

Since the class $\mathcal{P}$ of all partitions of $[a,b]$ is a subclass of $\mathcal{D}$ (all decompositions of $[a,b]$), we see that:

$$R \int_{\underline{a}}^b f \leqslant L \int_{\underline{a}}^b f \leqslant L \int_a^{\overline{b}} f \leqslant R \int_a^{\overline{b}} \qquad (13.1)$$

But f is R-integrable, thus

$$R \int_{\underline{a}}^b f = R \int_a^{\overline{b}} f$$

which implies that all the four expressions in (13.1) are equal, and that concludes the result. ∎

From now on the sign F will stand for Lebesgue integral unless otherwise specified.

We now discuss some properties of Lebesgue integration.

Theorem 13.2.5. If f is a bounded measurable function defined on a measurable set D and if $l \leqslant f(x) \leqslant u$ on D, then

$$l \cdot m(D) \leqslant \int_D f \leqslant u \cdot m(D)$$

Proof. Let P be any decomposition of D, described by the sets $\{D_k : k = 0,1,\ldots,(n-1)\}$. Then

$$l \leqslant m_k \leqslant M_k \leqslant u$$

and thus

$$l \cdot m(D) \leqslant s(f,P) \leqslant S(f,P) \leqslant u \cdot m(D)$$

Since this is true for all partitions

$$l \cdot m(D) \leqslant \int_{\underline{D}} f = \int_D^{\overline{}} f \leqslant u \cdot m(D)$$

which proves the result. ∎

Corollary 1. If $f(x) \geq 0$ on D, then $\displaystyle\int_D f \geq 0$, and if $f(x) \leq 0$ on D, then $\displaystyle\int_D f \leq 0$ on D.

Corollary 2. If $m(D) = 0$ then $\displaystyle\int_D f = 0$.

Theorem 13.2.6. Let $D = \bigcup_j D_j$, where $\{D_j\}$ is a countable class of pairwise disjoint measurable sets, and D has a finite measure. If f is a bounded measurable function defined on D, then

$$\int_D f = \sum_j \int_{D_j} f$$

Proof. First we shall prove this result for a finite case. Let $D = D_1 \cup D_2$ with D_1 and D_2 measurable and disjoint. Let $\rho = \{d_k : k = 0, 1, \ldots, n\}$ be an L-partition of D constructed as in the Definition 13.2.7. Write $d_k^1 = D_1 \cap d_k$ and $d_k^2 = D_2 \cap d_k$. It follows that d_k^1 and d_k^2 are measurable, and $d_k^1 \cup d_k^2 = d_k$, and $d_k^1 \cap d_k^2$ is empty. This implies $m(d_k^1) + m(d_k^2) = m(d_k)$. It must also be noticed that $\{d_k^1 : k = 0, 1, \ldots, (n-1)\}$ and $\{d_k^2 : k = 0, 1, \ldots, (n-1)\}$ form L-partitions of D_1 and D_2, respectively, which means

$$\sum_{k=0}^{n-1} y_k \cdot m(d_k) = \sum_{k=0}^{n-1} y_k \cdot m(d_k^1) + \sum_{k=0}^{n-1} y_k \cdot m(d_k^2)$$

If now we let $n \longrightarrow \infty$ and $\mu_n \longrightarrow 0$, we get

$$\int_D f = \int_{D_1} f + \int_{D_2} f$$

By induction the result can be proved for a finite case.

We shall now prove this result for a denumerable case; that is to say when $D = \bigcup_{j=1}^{\infty} D_j$, where $\{D_j\}$ is a class of pairwise disjoint measurable sets.

We can write $D = \left(\bigcup_{m=1}^{n} D_m\right) \cup S_n$, where $S_n = \bigcup_{k=n+1}^{\infty} D_k$. Since we have already established the result for a finite case;

$$\int_D f = \bigcup_{m=1}^{n} \left(\int_{D_m}\right) f + \int_{S_n} f = \sum_{m=1}^{n} \int_{D_m} f + \int_{S_n} f$$

Now, as $n \longrightarrow \infty, m(S_n) \longrightarrow 0$. If $f(x) \leqslant u$ on D, then

$$\int_{S_n} f \leqslant u \cdot m(S_n) \Longrightarrow \lim_{n \to \infty} \int_{S_n} f = 0$$

Therefore,

$$\int_D f = \sum_{m=1}^{\infty} \int_{D_m} f$$

which completes the proof. ∎

If $D = [a,b]$ or (a,b) or $[a,b)$ or $(a,b]$, it is sometimes convenient to write $\int_D f = \int_a^b f$. The next theorem gives a useful result for a simple function.

Theorem 13.2.7. Every bounded simple function f is integrable over a set D of finite measure, and

$$\int_D f = \sum_{k=1}^{n} c_k \cdot m(D_k), \qquad \text{where } f(x) = c_k \text{ for } x \in D_k$$

The proof of this theorem follows easily by the application of Theorems 13.2.5 and 13.2.6, and is left as an exercise.

Theorem 13.2.7 gives an easy way to compute the integral of a simple function.

Theorem 13.2.8. If f and g are two bounded measurable functions defined on a set D (of finite measure) and if f is equal to g almost everywhere on D then

$$\int_D f = \int_D g.$$

Proof. Let

$$D_1 = \{x : f(x) = g(x)\}$$
$$D_2 = \{x : f(x) \neq g(x)\}$$

It follows that $D_1 \cap D_2$ is empty and $D_1 \cup D_2 = D$. Furthermore, $m(D_2) = 0$. Therefore,

$$\int_D f = \int_{D_1} f + \int_{D_2} f = \int_{D_1} f$$

and

$$\int_D g = \int_{D_1} g + \int_{D_2} g = \int_{D_1} g = \int_{D_1} f = \int_D f$$

which completes the proof. ∎

Corollary. If $f(x) = c$ almost everywhere on D then $\int_D f = c \cdot m(D)$. In particular, if f is zero almost everywhere on D then $\int_D f = 0$.

The result of Theorem 13.2.8 is very significant in as much as it allows us to change the values of a function f on a subdomain of measure zero, provided f remains bounded on D.

The converse of Theorem 13.2.8 is false as shown by the following example.

EXAMPLE 13.2.1. Let

$$f(x) = 2 \quad \text{if} \quad x \in [-1,0]$$
$$f(x) = 0 \quad \text{if} \quad x \in (0,1]$$

Then

$$\int_{-1}^{1} f = \int_{-1}^{0} f + \int_{0}^{1} f = 2 \cdot 1 + 0 \cdot 1 = 2$$

Now consider $g(x) = 1$ for $x \in [-1,1]$. Obviously,

$$\int_{-1}^{1} g = 2$$

Therefore,

$$\int_{-1}^{1} f = \int_{-1}^{1} g$$

Yet $f \neq g$ almost everywhere on $[-1,1]$. One may construct numerous similar examples.

We now prove the following Lemma.

Lemma. If f is a bounded and measurable function defined on D such that $f(x) \geqslant 0$ on D, and if $\int_D f = 0$ then the measure of all sets of the type $\left\{ x : f(x) > \dfrac{1}{n} \right\}$, n being a natural number, is zero.

Proof. Suppose by way of contradiction that there is a natural number n_0 such that $m\left(\left\{x:f(x)>\dfrac{1}{n_0}\right\}\right)$ is positive, say λ.

Let $D_1 = \left\{x:f(x)>\dfrac{1}{n_0}\right\}$ and $D_2 = \left\{x:f(x)\leqslant\dfrac{1}{n_0}\right\}$. Then D_1 and D_2 are disjoint measurable sets such that $D = D_1 \cup D_2$.

Therefore,

$$\int_D f = \int_{D_1} f + \int_{D_2} f$$

But

$$\int_{D_1} f \geqslant \lambda \cdot \frac{1}{n_0} + 0 > 0$$

which contradicts the hypothesis of the lemma; hence the result. ∎

Theorem 13.2.9. If $\displaystyle\int_D f = 0$ and $f(x) \geqslant 0$ on D, then $f(x) = 0$ almost everywhere on D.

Proof. We show here that $\{x:f(x)>0\} = \displaystyle\bigcup_{n=1}^{\infty} \left\{x:f(x)>\dfrac{1}{n}\right\}$. Indeed, $x_0 \in \{x:f(x)>0\} \Longleftrightarrow f(x_0)>0$.

$$\Longleftrightarrow \text{There is a natural number } n_0 \text{ such that } f(x_0) > \frac{1}{n_0}$$

$$\Longleftrightarrow x_0 \in \bigcup_{n=1}^{\infty} \left\{x:f(x)>\frac{1}{n}\right\}$$

The conclusion now follows by the direct applications of the last lemma and the countable subadditivity of the Lebesgue measure. ∎

Theorem 13.2.10. Let f and g be two bounded measurable functions defined over a set D of finite measure; then

$$\int_D (f+g) = \int_D f + \int_D g$$

Proof. Let $A \leqslant f(x) < B$ and $U \leqslant g(x) < V$ for every $x \in D$.

We consider an L-partition $\rho_1 : A = y_0 < y_1 < \cdots < y_m = B$ with $d_k = \{x:y_k \leqslant f(x) < y_{k+1}\}, k = 0,1,\ldots,(m-1)$; and an L-partition $\rho_2 : U = z_0 < z_1 < \cdots < z_n = V$ with $e_l = \{x:z_l \leqslant g(x) < z_{l+1}\}, l = 0,1,\ldots,(n-1)$. Now we write $S_{kl} = d_k \cap e_l$. It can be easily shown that

$$D = \bigcup_{k=0}^{m-1} \bigcup_{l=0}^{n-1} S_{kl}$$

Also, since d_k's are pairwise disjoint and so are e_l's.

$$\int_D (f+g) = \sum_{k=0}^{m-1} \sum_{l=0}^{n-1} \int_{S_{kl}} (f+g)$$

On the set S_{kl} we have, by using Theorem 13.2.5,

$$(y_k + z_l) \cdot m(S_{kl}) \leq \int_{S_{kl}} (f+g) \leq (y_{k+1} + z_{l+1}) \cdot m(S_{kl})$$

Writing all these inequalities for $k = 0, 1, 2, \ldots (m-1)$; $l = 0, 1, \ldots, (n-1)$ and adding, we get

$$\sum_{k=0}^{m-1} \sum_{l=0}^{n-1} (y_k + z_l) \cdot m(S_{kl}) \leq \int_D (f+g) \leq \sum_{k=0}^{m-1} \sum_{l=0}^{n=1} (y_{k+1} + z_{l+1}) m(S_{kl})$$

Now

$$\sum_{k=0}^{m-1} \sum_{l=0}^{n-1} y_k \cdot m(S_{kl}) = \sum_{k=0}^{m-1} y_k \sum_{l=0}^{n-1} m(S_{kl}) = \sum_{k=0}^{m-1} y_k \cdot m\left(\bigcup_{l=0}^{n-1} S_{kl} \right)$$

$$= \sum_{k=0}^{m-1} y_k \cdot m\left[d_k \cap \left(\bigcup_{l=0}^{n-1} e_l \right) \right] = \sum_{k=0}^{m-1} y_k \cdot m(d_k \cap D)$$

$$= \sum_{k=0}^{m-1} y_k \cdot m(d_k) = t_1(f, \rho_1)$$

where t_1 stands for sum discussed in the Theorem 13.2.3 for f and ρ_1.

Using similar arguments and notations we get

$$\sum_{k=0}^{m-1} \sum_{l=0}^{n-1} z_l \cdot m(S_{kl}) = t_2(g, \rho_2)$$

$$\sum_{k=0}^{m-1} \sum_{l=0}^{n-1} y_{k+1} \cdot m(S_{kl}) = T_1(f, \rho_1)$$

$$\sum_{k=0}^{m-1} \sum_{l=0}^{n-1} z_{l+1} \cdot m(S_{kl}) = T_2(g, \rho_2).$$

Thus

$$t_1(f,\rho_1) + t_2(g,\rho_2) \leqslant \int_D (f+g) \leqslant T_1(f,\rho_1) + T_2(g,\rho_2)$$

Increasing m and n indefinitely so that $\mu_n \longrightarrow 0$ and $\mu_m \longrightarrow 0$, we get

$$\int_D f + \int_D g \leqslant \int_D (f+g) \leqslant \int_D f + \int_D g$$

which gives

$$\int_D (f+g) = \int_D f + \int_D g$$

That completes the proof. ▮

Theorem 13.2.11. If $f: D \longrightarrow R$ is a bounded measurable function, and D has a finite measure, then

$$\int_D cf = c \int_D f, \quad (c \text{ being a constant})$$

cf denotes the function whose value at the point x is $cf(x)$.

Proof. If $c = 0$, the result follows trivially. If $c > 0$ and if $l \leqslant f(x) \leqslant u$ on D, then consider an L-partition $\{d_k\}$ of D, as usual.

Evidently, $cy_k \cdot m(d_k) \leqslant \int_{d_k} cf \leqslant cy_{k+1} m(d_k)$. With the argument used in the proof of the last theorem, we get

$$\int_D cf = c \int_D f$$

Finally, if $c < 0$ then $(-c) > 0$ and we therefore have

$$\int_D (-c)f = (-c) \int_D f$$

Now

$$\int_D [cf + (-c)f] = 0.$$

Using the last theorem,

$$\int_D cf + \int_D (-c)f = 0$$

or

$$\int_D cf + (-c)\int_D f = 0,$$

which, obviously, means

$$\int_D cf = c\int_D f$$

and that completes the proof. ▌

Corollary.

$$\int_D (f - g) = \int_D f - \int_D g$$

Theorem 13.2.13. If f is integrable over D then $\left|\int_D f\right| \leqslant \int_D |f|.$

Proof. Let $D_1 = \{x : f(x) > 0\}$ and $D_2 = \{x : f(x) < 0\}$. On $D_1, f = |f|$ and on $D_2, f = -|f|$.

Now

$$\int_D |f| = \int_{D_1} |f| + \int_{D_2} |f|$$

and

$$\int_D f = \int_{D_1} f + \int_{D_2} f = \int_{D_1} |f| + \int_{D_2} -|f| = \int_{D_1} |f| - \int_{D_2} |f| \quad (13.2)$$

Since $\int_{D_1} |f|$ and $\int_{D_2} |f|$ are both nonnegative, it is an easy matter to see that

$$\left|\int_{D_1} |f| - \int_{D_2} |f|\right| \leqslant \int_{D_1} |f| + \int_{D_2} |f|$$

and the result follows from the substitutions in (13.2). The proof is complete. ▌

Suppose now that there is a sequence $\{f_n\}$ of Lebesgue integrable functions which is convergent to f on its domain D. We are interested in Lebesgue integrability of f. The question then arises: "Under what conditions does $\displaystyle\int_D f_n$ converge to $\displaystyle\int_D f$? Is the convergence uniform?" We discuss first the following examples:

EXAMPLE 13.2.2. Let $\{f_n\}$ be defined as follows on $[0,1]$:

$$f_n(x) = \lambda n \text{ for } x \in \left(0, \frac{1}{n}\right), \lambda \text{ being a nonzero constant.}$$

$$f_n(x) = 0 \text{ for } x \in \left[\frac{1}{n}, 1\right] \text{ or when } x = 0$$

Now,

$$\int_0^1 f_n = \lambda n \cdot \frac{1}{n} + 0 = \lambda \neq 0$$

Therefore, $\displaystyle\lim_{n\to\infty} \int_0^1 f_n = \lambda$.

However, $\displaystyle\lim_{n\to\infty} f_n(x) = 0$ and, therefore

$$\int_0^1 \lim_{n\to\infty} f_n(x) = 0 \neq \lim_{n\to\infty} \int_0^1 f_n$$

It is interesting to note that if in the above example we define $f_n(x) = \lambda \cdot n^2$, for $x \in \left(0, \frac{1}{n}\right)$ and $f_n(x) = 0$ if $x \in \left[\frac{1}{n}, 1\right]$ or $x = 0$ then $\displaystyle\lim_{n\to\infty} \int_D f_n$ does not even exist, whereas $\displaystyle\int_D \lim_{n\to\infty} f_n$ would still be zero $(D = [0,1])$.

EXAMPLE 13.2.3. Let a sequence $\{f_n\}$ be defined as follows:

$$f_n(x) = \frac{1}{n} \text{ for } x \in \left(0, \frac{1}{n}\right)$$

$$f_n(x) = 0 \text{ for } x \in \left[\frac{1}{n}, 1\right] \text{ or when } x = 0$$

Here,

$$f(x) = \lim_{n \to \infty} f_n(x) = 0 \quad \text{and} \quad \int_0^1 f = 0$$

Also,

$$\int_0^1 f_n = \frac{1}{n^2} \implies \lim_{n \to \infty} \int_0^1 f_n = 0$$

which means

$$\lim_{n \to \infty} \int_0^1 f_n = \int_0^1 \lim_{n \to \infty} f_n$$

Next we prove a theorem which is due to Lebesgue and is known as "Bounded Convergence theorem."

Theorem 13.2.12 (Bounded Convergence Theorem). Let $\{f_n\}$ be a sequence of measurable functions defined and convergent on D. If there is a positive number M such that $|f_n(x)| < M$ for all n and for every $x \in D$, then $\lim_{n \to \infty} \int_D f_n = \int_D \lim_{n \to \infty} f_n$.

Proof. Let $f = \lim_{n \to \infty} f_n$. Then $|f(x)| \leqslant M$ for every $x \in D$.

Using Egoroff's theorem, we have for $\epsilon > 0$ a measurable subset S of D such that $m(D - S) < \dfrac{\epsilon}{4M}$, and $\{f_n\}$ is uniformly convergent on S.

In case $m(S) = 0$ then

$$\left| \int_D f_n - \int_D f \right| = \left| \int_D (f_n - f) \right| \leqslant \int_D |f_n - f|$$

$$= \int_S |f_n - f| + \int_{D-S} |f_n - f| < 0.2M + 2M \cdot \frac{\epsilon}{4M} = \epsilon/2 < \epsilon,$$

Thus,

$$\left| \int_D f_n - \int_D f \right| < \epsilon \quad \text{for every } n$$

which means

$$\lim_{n\to\infty} \int_D f_n = \int_D f$$

If $m(S) \neq 0$ we proceed as follows: Since $\{f_n\}$ is uniformly convergent on S, and $\dfrac{\epsilon}{2\,m(S)} > 0$ $(m(S) > 0)$ there is a natural number n_0 such that $|f_n(x) - f(x)| < \dfrac{\epsilon}{2\,m(S)}$ for $n \geq n_0$ and for every $x \in S$.

Now

$$\left| \int_D f_n - \int_D f \right| \leq \int_D |f_n - f| = \int_S |f_n - f| + \int_{D-S} |f_n - f|$$

$$\leq m(S)\frac{\epsilon}{2m(S)} + 2M \cdot \frac{\epsilon}{4M} = \epsilon \quad \text{for } n \geq n_0$$

which again shows that

$$\lim_{n\to\infty} \int_D f_n = \int_D f$$

The proof is now complete. █

The above theorem does not necessarily hold for Riemann integration, as shown in the next example.

EXAMPLE 13.2.4. Let $r_1, r_2, \ldots, r_n, \ldots$ be an enumeration of the set of all rational numbers in $[a, b]$.
We write $S_n = \{r_1, r_2, \ldots, r_n\}$.
Now consider $\{f_n\}$ defined as follows: Let

$$f_n(x) = 0 \text{ for } x \in S_n$$
$$f_n(x) = 1 \text{ if } x \notin S_n \text{ and } x \in [a, b]$$

Each $\{f_n\}$ is discontinuous only at n points, and therefore is Riemann-integrable on $[a, b]$.

Thus $R \displaystyle\int_a^b f_n = L \int_a^b f_n = 0 \cdot m(S_n) + 1 \cdot m([a, b] - S_n) = (b - a)$.

(Note that it is easier to evaluate Lebesgue integral of f_n even though the Riemann integral exists.)

Now if $f(x) = \lim\limits_{n \to \infty} f_n(x)$, then

$$f(x) = 0 \text{ if } x \text{ is a rational number in } [a,b]$$

and

$$f(x) = 1 \text{ if } x \text{ is an irrational in } [a,b]$$

Therefore, f is the characteristic function of the set of all irrationals in $[a,b]$. It is an easy matter to show that f is discontinuous everywhere on $[a,b]$ and as such is *not* Riemann-integrable on $[a,b]$.

Indeed, the Lebesgue integral of f exists on $[a,b]$ and

$$L \int_a^b f = b - a = \lim_{n \to \infty} \int_a^b f_n$$

It may be observed that $\{f_n\}$ satisfies the hypothesis of the last theorem.

Of course, if $\{f_n\}$ happens to be a uniformly convergent sequence of continuous functions, then the Riemann integral of $\lim\limits_{n \to \infty} f_n$ exists. This is because Riemann integrability depends upon the idea of "continuity."

Another defect of Riemann integration is that a function f may be differentiable on $[a,b]$ and, therefore, continuous on $[a,b]$, yet its derivative may be discontinuous on a set of positive measure making f' nonintegrable in the Riemann sense.

Exercise 13.2

1. Prove Theorem 13.2.2.
2. Prove Theorem 13.2.7.
3. Prove that if a bounded function $f : D \longrightarrow R$ is Lebesgue integrable, then it must be measurable.
4. Let $f : D \longrightarrow R$ be a bounded measurable function (the set D having a finite measure) and $S = \bigcup\limits_k S_k$ be the union of a countable class of measurable sets such that $S \subset D$. If $f_s : D \longrightarrow R$ is the characteristic function of S, show that

   $$\int_D f_s \leqslant \sum_k m(S_k). \text{ Give an example where the strict inequality holds.}$$

13.3. Integrals of Unbounded Functions

In this section we discuss the integrability of unbounded functions. First of all we consider nonnegative unbounded functions. To define the Lebesgue integral of these functions we construct a special type of sequences of functions as follows.

Definition 13.3.1. *Let D be a set of finite measure and* $f : D \longrightarrow R$ *be an unbounded nonnegative function.*

Write

$$f_n^t(x) = f(x) \qquad \text{for } x \in \{x : f(x) < n\}$$

and

$$f_n^t(x) = n \qquad \text{for } x \in \{x : f(x) \geqslant n\}$$

The sequence $\{f_n^t\}$ is called the *truncating sequence* of f.

Each f_n^t is bounded and measurable if f is measurable, and as such it is Lebesgue integrable.

Definition 13.3.2. *Let* $f : D \longrightarrow R$ *be a nonnegative, unbounded measurable function and* $\{f_n^t\}$ *be its truncating sequence. If* $\displaystyle\lim_{n \to \infty} \int_D f_n^t$ *exists (and is finite) then we say that* f *is integrable and the value of the limit is* $\displaystyle\int_D f$.

We consider a couple of examples.

EXAMPLE 13.3.1. Let

$$f(x) = \frac{1}{\sqrt{x}} \qquad \text{for } x \in (0,1]$$

and

$$f(0) = 0$$

Then

$$f_n^t(x) = \frac{1}{\sqrt{x}} \qquad \text{for } x \in \left(\frac{1}{n^2}, 1\right]$$

$$f_n^t(x) = n \qquad \text{for } x \in \left(0, \frac{1}{n^2}\right]$$

$$f_n^t(0) = 0$$

and,

$$\int_0^1 f = \lim_{n \to \infty} \int_0^1 f_n^t = \lim_{n \to \infty} \left[n \cdot \frac{1}{n^2} + \int_{1/n^2}^1 \frac{1}{\sqrt{x}} \right] = 2.*$$

Next, we discuss an example where the integral does not exist.

_____

*It must be observed that $\dfrac{1}{\sqrt{x}}$ is not Riemann-integrable over $[0,1]$.

$R \displaystyle\int_0^1 \frac{1}{\sqrt{x}} dx$ is only an improper Riemann integral; whereas, the same function is Lebesgue integrable over $[0,1]$. This is yet another advantage of Lebesgue integration over Riemann integration.

EXAMPLE 13.3.2. Let

$$f(x) = \frac{1}{x} \qquad \text{for } x \in (0,1]$$

$$f(0) = 0$$

Here,

$$f_n^t(x) = \frac{1}{x} \qquad \text{for } x \in \left(\frac{1}{n}, 1\right]$$

$$f_n^t(x) = n \qquad \text{for } x \in \left(0, \frac{1}{n}\right]$$

$$f_n^t(0) = 0$$

In this case $\lim\limits_{n \to \infty} \int_0^1 f_n^t = \lim\limits_{n \to \infty} \left[n \cdot \frac{1}{n} - \log \frac{1}{n} \right]$, which does not exist.

It may be noticed that for a bounded nonnegative function f,

$$f_n^t(x) = f(x) \qquad \text{for } n \geq \sup \ \{f(x) : x \in D\}$$

and hence we can describe every nonnegative function in terms of limit of its truncating sequence.

Theorem 13.3.1. If $f : D \longrightarrow R$ and $g : D \longrightarrow R$ are measurable and $0 \leq g(x) \leq f(x)$ for $x \in D$, and if f is integrable then g is integrable.

The proof is left as an exercise.

To discuss the integrability of *any* arbitrary function we introduce the following terminology.

Definition 13.3.3. *Let $f : D \longrightarrow R$ be a function. Consider $f^+ : D \longrightarrow R$ and $f^- : D \longrightarrow R$ such that $f^+ = \max \ \{f, 0\}$ and $f^- = \max \ \{-f, 0\}$. They are called positive part function and negative part function of f, respectively.*

It must be understood that both positive part function and negative part function are nonnegative and have the same domain as the original function. Furthermore, f is measurable if and only if f^+ and f^- are measurable. It follows that

$$f(x) = f^+ (x) - f^- (x)$$

and

$$|f(x)| = f^+ (x) + f^- (x)$$

Definition 13.3.4. *A function $f : D \longrightarrow R$ (bounded or unbounded) defined on a set of finite measure is integrable if its positive part $f^+ : D \longrightarrow R$ and the* *negative part* $f^-: D \longrightarrow R$ *are integrable, and*

$$\int_D f = \int_D f^+ - \int_D f^-$$

Now we have defined the Lebesgue integrability of all types of functions whose domains have finite measures.

One theorem which follows easily from the definition is the next one. It must be observed that this theorem is not true for Riemann integration.

Theorem 13.3.2. $f: D \longrightarrow R$ is integrable over D if and only if $|f|$ is integrable.

The proof is trivial.

The next theorem uses Theorem 13.3.1.

Theorem 13.3.3. If $f: D \longrightarrow R$ is integrable, and $g: D \longrightarrow R$ is measurable, and if $|g(x)| \leqslant f(x)$ for $x \in D$, then g is also integrable.

Proof. $0 \leqslant g^+(x) \leqslant f(x)$ and $0 \leqslant g^-(x) \leqslant f(x)$ for $x \in D$. Applying Theorem 13.3.1 and Def. 13.3.4 enables us to complete the proof. ▌

Some of the fundamental properties of integration which we discussed in the Sec. 13.2 remain valid for unbounded functions also. For example:

$$\int_D (f + g) = \int_D f + \int_D g.$$

We consider here some results which deal with the equality (or inequality) of $\lim_{n \to \infty} \int_D f_n$ and $\int_D \lim_{n \to \infty} f_n$.

Unless otherwise specified the set D will be assumed to be of *finite measure.*

Theorem 13.3.4 (Lebesgue Monotone Convergence Theorem). Let $\{f_n\}$ be a nondecreasing sequence of nonnegative functions defined over D. Then

$$\lim_{n \to \infty} \int_D f_n = \int_D \lim_{n \to \infty} f_n$$

The proof of this theorem is omitted.

The next theorem, known as Fatou's Lemma, has a remarkable quality of having a very weak hypothesis.

Theorem 13.3.5 (Fatou's Lemma). Let $\{f_n\}$ be a sequence of nonnegative measurable functions integrable on D, and let $f = \lim_{n \to \infty} f_n$ almost everywhere on D.

Then

$$\int_D f \leqslant \liminf_{n \to \infty} \int_D f_n$$

Proof. Let $g_n(x) = \inf \{f_k(x): k \geqslant n\}$. Since each f_k is bounded below (by zero), g_n is defined for every n on D. Furthermore, $\{g_n\}$ is nondecreasing, and since $0 \leqslant g_n(x) \leqslant f_n(x)$ for every x, g_n is integrable.

Using the monotone convergence theorem

$$\int_D \lim_{n \to \infty} g_n = \lim_{n \to \infty} \int_D g_n$$

Now it can be easily seen that

$$\lim_{n \to \infty} g_n = \liminf_{n \to \infty} f_n = \lim_{n \to \infty} f_n = f \text{ a.e. on } D$$

Thus,

$$\int_D f = \int_D \lim_{n \to \infty} g_n = \lim_{n \to \infty} \int_D g_n$$

and since

$$\int_D g_n \leqslant \int_D f_n \qquad \text{for every } n$$

$$\int_D f = \lim_{n \to \infty} \int_D g_n \leqslant \liminf_{n \to \infty} \int_D f_n$$

and this completes the proof.

Next theorem is the Lebesgue dominated convergence theorem.

Theorem 13.3.6. Let $\{f_n\}$ be a sequence of measurable functions defined on D such that $|f_n| \leqslant g$ almost everywhere on D. If g is integrable on D, then

$$\int_D \lim_{n \to \infty} f_n = \lim_{n \to \infty} \int_D f_n$$

The proof of this theorem makes use of Fatou's lemma and is left as an exercise.

Exercise 13.3

1. Prove Theorem 13.3.1.
2. Prove Theorem 13.3.4. (*Hint*: Use Egoroff's Theorem.)

3. Prove Theorem 13.3.6.

4. Prove that $\displaystyle\int_D (f+g) = \int_D f + \int_D g.$

5. Prove that $\displaystyle\int_D cf = c \int_D f.$

6. Prove that if $\{f_n\}$ is a sequence of nonpositive measurable functions defined

   on D, and $f = \lim\limits_{n\to\infty} f_n$; then $\displaystyle\lim_{n\to\infty} \sup \int_D f_n \leqslant \int_D f.$

7. Give an example of a sequence of nonnegative measurable functions which satisfies the strict inequality of the Fatou's Lemma, that is,

$$\int_D f < \lim_{n\to\infty} \inf \int_D f_n$$

13.4. Square Integrable Functions

In this section we introduce a special class of functions which are called square integrable functions. This topic is very useful in higher analysis, though in this book we shall use it only in Chapter 14.

Definition 13.4.1. *Let f be a measurable function defined on a set D of finite measure. We say f is square integrable (in the Lebesgue sense) on D if f^2 is Lebesgue integrable on D. Sometimes we write $f \in L^2$ (D).*

Throughout the remainder of this section we shall assume that the set D (domain of the functions concerned) is measurable and has a finite measure.

Next, we define the norm of a square integrable function.

Definition 13.4.2. *Let $f: D \longrightarrow R$ be a square integrable function then the norm of f, written as $\|f\|$.*

$$\|f\| = \left[\int_D f^2 \right]^{1/2}$$

One may suspect that $\|f\|$ possesses some properties of length of a vector. In fact, it does. Obvious from the definition are the following two properties:

$\|f\| \geqslant 0$ and $\|f\| = 0$ if and only if $f = 0$ almost everywhere on D

Before we come around some other properties, we need a couple of results which are analogues of Schwarz and Minkowski inequalities.

Theorem 13.4.1. Let $f: D \longrightarrow R$ and $g: D \longrightarrow R$ be two square integrable functions, then the product fg is Lebesgue integrable, and

$$\int_D |fg| \leqslant \left[\int_D f^2 \right]^{1/2} \left[\int_D g^2 \right]^{1/2}$$

Proof. Since $(|f(x)| - |g(x)|)^2 \geqslant 0$ for every x, therefore, $|f(x) g(x)| \leqslant \frac{1}{2}(f^2 + g^2)$.

As f^2 and g^2 are integrable, therefore, $|f(x) g(x)|$ is integrable and so is fg.

Now $\int_D (|f| + c |g|)^2 \geqslant 0$ for every real c, which means

$$\int_D f^2 + c^2 \int_D g^2 + 2c \int |fg| \geqslant 0 \tag{13.3}$$

If $\int_D g^2 = 0$ then $g = 0$ a.e. on D and then

$$\int_D |fg| = \| f \| \| g \|$$

If

$$\int_D g^2 \neq 0$$

then let

$$c = - \frac{\displaystyle\int_D |fg|}{\displaystyle\int_D g^2}$$

Substituting in the inequality (13.3) we get

$$\int_D f^2 + \frac{\left[\displaystyle\int_D |fg| \right]^2}{\displaystyle\int_D g^2} - 2 \frac{\left[\displaystyle\int_D |fg| \right]^2}{\displaystyle\int_D g^2} \geqslant 0$$

Therefore,

$$\int_D |fg| \leqslant \left[\int_D f^2 \right]^{1/2} \left[\int_D g^2 \right]^{1/2}$$

and the theorem is proved.

The converse of the last theorem (the first part) is not necessarily true as illustrated by the following example:

EXAMPLE 13.4.1. Let

$$f(x) = x^{-1/2}$$

and

$$g(x) = x^{-1/4}$$

Then fg is integrable on $[0,1]$, but f^2 is not integrable on the same interval. Next we prove the analogue of Minkowski's inequality.

Theorem 13.4.2. Let $f: D \longrightarrow R$ and $g: D \longrightarrow R$ be two square integrable functions. Then

$$\| f + g \| \leqslant \| f \| + \| g \|$$

Proof. From the last theorem we have

$$\left| \int_D fg \right| \leqslant \int_D |fg| \leqslant \| f \| + \| g \|$$

Therefore,

$$(\| f + g \|)^2 = \int_D (f+g)^2 = \int_D f^2 + \int_D g^2 + 2 \int_D fg$$

$$\leqslant (\| f \|)^2 + (\| g \|)^2 + 2 \| f \| \| g \| = (\| f \| + \| g \|)^2$$

Thus

$$\| f + g \| \leqslant \| f \| + \| g \|,$$

and the proof is complete.

Compare the last theorem with Theorem 2.4.1(d).

Now if we define on the set of all square integrable functions a metric

$$\rho(f,g) = \| f - g \| = \sqrt{\int_D (f-g)^2},$$ then we run into one difficulty, and that is

$\| f - g \| = 0$ does not necessarily imply $f = g$ on D; rather, it infers that $f = g$

almost everywhere on D. As such, the condition $M.2$ (cf. Sec. 5.3) of the metric space is not satisfied. To overcome this we consider the equivalence classes of square integrable functions as elements of L^2 space (instead of the individual functions). Any two of them belong to the same class if and only if they are equal to each other almost everywhere on D.

It is then easy to see that the set of all equivalent classes of square integrable functions form a metric space. One of the most significant things about this metric space is that it is *complete*. This is due to a famous result known as Riesz-Fischer theorem which may be stated as follows:

Theorem 13.4.3.   The metric space of L^2 on a set of finite measure is complete.

The proof of this theorem is omitted.

The completeness property is certainly not true for the class of those functions which are square integrable in the Riemann sense. This is illustrated by the following example:

EXAMPLE 13.4.2.   Let $\{r_1, r_2, \ldots, r_n, \ldots\}$ be an enumeration of the set of all rational numbers in $[a, b]$ with the exceptions of a and b. Enclose every r_n by an open interval I_n whose length is less than $\dfrac{b - a}{2^n}$ such that $I_n \subset [a, b]$.

Write

$$S_n = \bigcup_{k=1}^{n} I_k$$

Now define f_n to be the characteristic function of S_n. Then each f_n is Riemann integrable since the only points of discontinuities of f_n on S_n are its boundary points, and they are finite in number (prove it). Furthermore, $f_n^2 = f_n$, and thus f_n is square integrable in the *Riemann sense* over $[a, b]$.

Writing $f = \lim\limits_{n \to \infty} f_n$, we notice that f is the characteristic function of S, where

$S = \bigcup\limits_{k=1}^{\infty} I_k$. Also, $f^2 = f$. But f is not Riemann-integrable, since $[a, b] - S$ con-

stitutes the set of points of discontinuities of S (prove it); and $m(S) \leqslant \sum\limits_{k=1}^{\infty} l(I_k) <$

$b - a$, and, as such, the $m([a, b] - S)$ is positive.

It is an easy matter to show that the $\lim\limits_{n \to \infty} \left(R \displaystyle\int_a^b (f_n - f)^2 \right) = 0$, and f does

not belong to the metric space of *Riemann square integrable functions*.

Exercise 13.4

1. In Example 13.4.2, show that the only points of discontinuities of S_n are the boundary points of S_n and that they cannot exceed $2n$.
2. Show that f in Example 13.4.2 is discontinuous at every point of $[a, b] - S$.
3. Give an example of a nonmeasurable function whose square is Lebesgue integrable on $[0,1]$. (*Hint*: Consider P-a nonmeasurable subset of $[0,1]$, and its complement.)
4. Show that if a measurable function is square integrable on D (whose measure is finite) then it must be Lebesgue integrable on D.
5. Show by means of an example that the result of Exercise 4 is not true if D is not of finite measure. $\left(\textit{Hint}: \text{Consider } \dfrac{1}{2 + |x - 3|} \text{ on the set of all reals.}\right)$

14

Fourier Series

14.1. Trigonometric Series and Fourier Series

A series of the form

$$\frac{1}{2}a_0 + \sum_{n=1}^{\infty} (a_n \cos nx + b_n \sin nx) \tag{14.1}$$

where the coefficients a_n, b_n are independent of x, is called a *trigonometric series*. In this chapter we consider the problem of expanding a given function $f(x)$ in a trigonometric (trig) series. If the series (14.1) is convergent to a function $f(x)$ for all $x \in R$, then since $\cos nx$ and $\sin nx$ are periodic functions of period 2π, we have

$$f(x + 2\pi) = f(x).$$

Hence in the following we will suppose that f is defined on $[0, 2\pi)$ (or equivalently on $[-\pi, \pi)$) and then extended to R by periodicity, period 2π.

The trig series (14.1) were first studied in problems of theoretical physics, chiefly in acoustics, electro dynamics and the theory of heat. Fourier started the first thorough study of these series in 1822, and the series (14.1) when a_n, b_n are defined in the manner stated below, is called a Fourier series for f.

We now define the Fourier series of f.

Definition 14.1.1. *If $f \in L[-\pi, \pi]$ (integrable in the Lebesgue sense over $[-\pi, \pi]$) then the Fourier series for $f(x)$ is the series*

$$\frac{1}{2}a_0 + \sum_{n=1}^{\infty} (a_n \cos nx + b_n \sin nx) \tag{14.2}$$

where*

$$a_n = \frac{1}{\pi} \int_{-\pi}^{\pi} f(x) \cos nx \, dx \qquad (n = 0,1, \ldots),$$

$$(14.3)$$

$$b_n = \frac{1}{\pi} \int_{-\pi}^{\pi} f(x) \sin nx \, dx \qquad (n = 1,2, \ldots).$$

The numbers a_n and b_n are called the Fourier coefficients of f. We shall denote the Fourier series (14.2) by $S(f)$ and write

$$f(x) \simeq \frac{a_0}{2} + \sum_{n=1}^{\infty} (a_n \cos nx + b_n \sin nx). \qquad (14.4)$$

Note that the symbol $\simeq$ implies nothing about the convergence of $S(f)$ and much less about the convergence of the series to $f(x)$.

(Some writers use $\sim$ to denote the Fourier series for f. We shall reserve $\sim$ to denote asymptotic equivalence. See Sec. 10.3.)

If $f(x)$ is even, that is, $f(-x) = f(x)$ for $x \in [-\pi, \pi)$, then $b_k = 0$ and the corresponding Fourier series is

$$f(x) \simeq \frac{a_0}{2} + \sum_{n=1}^{\infty} a_n \cos nx$$

where

$$a_n = \frac{2}{\pi} \int_0^{\pi} f(x) \cos nx \, dx.$$

If $f(x)$ is odd, that is, $f(-x) = -f(x)$ for $x \in [-\pi, \pi)$, then $a_k = 0$ and the corresponding series is

$$f(x) \simeq \sum_{n=1}^{\infty} b_n \sin nx$$

where

---

*In this chapter we shall write $\int_a^b f(x) \, dx$ instead of $\int_a^b f$. This will prevent confusion when we come across integrals where the integrand involves a parameter. See, for instance, (14.19).

$$b_n = \frac{2}{\pi} \int_0^\pi f(x) \sin nx \, dx.$$

The following theorem shows that every uniformly convergent trigonometric series is the Fourier series of its sum.

Theorem 14.1.1. If the trigonometric series (14.1) is uniformly convergent on $[-\pi, \pi]$ and if $f(x)$ is its sum, then it is the Fourier series of $f(x)$.

Proof. By uniform convergence of the series (14.1) it follows that the sum $f(x) \in L[-\pi, \pi]$. We integrate both sides of the equation

$$f(x) = \frac{1}{2} a_0 + \sum_{n=1}^\infty (a_n \cos nx + b_n \sin nx) \tag{14.5}$$

over $[-\pi, \pi]$. Since the series is uniformly convergent we can integrate term-by-term, and obtain

$$\int_{-\pi}^\pi f(x) \, dx = \frac{1}{2} a_0 \int_{-\pi}^\pi dx + \sum_{n=1}^\infty \left\{ a_n \int_{-\pi}^\pi \cos nx \, dx + b_n \int_{-\pi}^\pi \sin nx \, dx \right\}$$

$$= \pi a_0.$$

Now multiply both sides of the equation by $\cos mx$, where $m \geqslant 1$ is a fixed integer, and observe that by our hypothesis the series

$$\frac{1}{2} a_0 \cos mx + \sum_{n=1}^\infty (a_n \cos nx + b_n \sin nx) \cos mx$$

is uniformly convergent on $[-\pi, \pi]$ to $f(x) \cos mx$. Hence we can integrate this series term-by-term over $[-\pi, \pi]$. Since

$$\int_{-\pi}^\pi \cos mx \cos nx \, dx = \begin{cases} 0 & m \neq n \\ \pi & m = n \neq 0 \end{cases}$$

$$\int_{-\pi}^\pi \sin nx \cos mx \, dx = 0$$

and

$$\int_{-\pi}^\pi \sin mx \, \sin nx \, dx = \begin{cases} 0 & m \neq n \\ \pi & m = n \neq 0 \end{cases}$$

we find

$$\int_{-\pi}^{\pi} f(x) \cos mx \, dx = a_m \int_{-\pi}^{\pi} \cos^2 mx \, dx = \pi a_m \qquad (m = 1, 2, \ldots)$$

Similarly,

$$\int_{-\pi}^{\pi} f(x) \sin mx \, dx = \pi b_m.$$

Hence a_m and b_m are the Fourier coefficients of f and the series (14.5) is the Fourier series of $f(x)$. ∎

EXAMPLE 14.1.1. We compute the Fourier coefficients of the function

$$f(x) = x(-\pi \leqslant x < \pi).$$

Here

$$a_n = \frac{1}{\pi} \int_{-\pi}^{\pi} x \cos nx \, dx = 0,$$

$$b_n = \frac{1}{\pi} \int_{-\pi}^{\pi} x \sin nx \, dx = -\frac{2}{n} \cos n\pi.$$

The corresponding series is

$$f(x) \simeq 2 \left(\sin x - \frac{\sin 2x}{2} + \frac{\sin 3x}{3} - \cdots \right).$$

Note that the extended function is discontinuous at the points $x = (2k + 1)\pi$ $(k = 0, \pm 1, \pm 2, \ldots)$. The series is convergent for all x.

EXAMPLE 14.1.2. Let $f(x)$ be defined by

$$f(x) = \begin{cases} 0 & \text{if } x = 0 \\ \dfrac{1}{2}(\pi - x) & \text{if } 0 < x \leqslant \pi \end{cases}$$

and let $f(x)$ be odd. We compute the Fourier coefficients of f. We have

$$b_n = \frac{2}{\pi} \int_0^{\pi} \frac{1}{2} (\pi - x) \sin nx \, dx = \frac{1}{n}$$

and so

$$f(x) \simeq \sin x + \frac{\sin 2x}{2} + \frac{\sin 3x}{3} + \cdots.$$

The series is convergent for all (real) x (see Example 8.6.3). It is not absolutely convergent except when $x = k\pi$. It is uniformly convergent in any closed interval not including a multiple of 2π (see Example 11.4.3 and Exercise 11.4(9)).

If $f(x)$ is defined on $[0,\pi]$ as above and is even, then we have

$$a_0 = \frac{2}{\pi} \int_0^\pi f(x)\, dx = \frac{\pi}{2}$$

$$a_n = \frac{2}{\pi} \int_0^\pi f(x) \cos nx\, dx = \frac{(-1)^{n+1}}{\pi n^2}$$

and

$$f(x) \simeq \frac{\pi}{4} - \sum_{n=1}^{\infty} \frac{(-1)^n \cos nx}{\pi n^2}.$$

The series is absolutely and uniformly convergent over any interval $[A,B]$.

Exercise 14.1

1. Expand the following functions in Fourier series:
   (i) $f(x) = x^2$ $(-\pi \leqslant x < \pi)$,
   (ii) $f(x) = \cos \alpha x$ $(-\pi \leqslant x < \pi, \alpha \text{ not an integer})$,
   (iii) $f(x) = |x|$ $(-\pi \leqslant x < \pi)$. ($|x|$ denotes, as usual, absolute value of x.)
2. Write the Fourier series for the function
   $$f(x) = x, \quad 0 \leqslant x \leqslant \pi, f(x) \text{ even}$$

14.2. Orthogonal Systems

Definition 14.2.1. *A system of functions* $\phi_n(x) \in L^2[a,b]$, $n = 1, 2, \ldots$ *is orthogonal on* $[a,b]$ *if* $(\phi_m, \phi_n) = \int_a^b \phi_m(x) \overline{\phi_n(x)}\, dx = 0$, $m \neq n; m = 1, 2, \ldots;$ $n = 1, 2, \ldots.$ *If in addition, the norm* $\|\phi_n\|$ *defined by*

$$\|\phi_n\|^2 = \int_a^b |\phi_n(x)|^2\, dx$$

is equal to one for $n = 1, 2, \ldots,$ *then the system* $\{\phi_n\}_1^\infty$ *is orthonormal on* $[a,b]$.

EXAMPLE 14.2.1. (i) The system of functions $\frac{1}{2}$, $\cos x$, $\sin x, \ldots,$

$\cos nx$, $\sin nx$, ... is orthogonal on $[-\pi, \pi]$. The system

$$\frac{1}{\sqrt{2\pi}}, \frac{\cos x}{\sqrt{\pi}}, \frac{\sin x}{\sqrt{\pi}}, \ldots, \frac{\cos nx}{\sqrt{\pi}}, \frac{\sin nx}{\sqrt{\pi}}, \ldots \qquad (14.6)$$

is orthonormal on $[-\pi, \pi]$. Here $a = -\pi$, $b = \pi$ and

$$\phi_1(x) = \frac{1}{\sqrt{2\pi}}, \phi_{2n}(x) = \frac{\cos nx}{\sqrt{\pi}}, \phi_{2n+1}(x) = \frac{\sin nx}{\sqrt{\pi}} \quad n = 1, 2, \ldots.$$

(ii) The system of functions

$$\left\{\frac{e^{inx}}{\sqrt{2\pi}}\right\}, \quad n = 0, \pm1, \pm2, \ldots \qquad (14.7)$$

is orthonormal on $[-\pi, \pi]$. Here

$$\phi_1(x) = \frac{1}{\sqrt{2\pi}}, \phi_{2n+1}(x) = \frac{e^{inx}}{\sqrt{2\pi}}, \phi_{2n}(x) = \frac{e^{-inx}}{\sqrt{2\pi}} \quad n = 1, 2, \ldots.$$

Definition 14.2.2. *Let $\{\phi_n(x)\}_1^\infty$ be an orthonormal system on $[a,b]$ and let $f(x) \in L^2[a,b]$. If*

$$c_n = \int_a^b f(x)\,\overline{\phi_n(x)}\,dx, \quad n = 1, 2, \ldots \qquad (14.8)$$

then c_n is the nth Fourier coefficient of f with respect to $\{\phi_n\}_1^\infty$. The Fourier series of f with respect to $\{\phi_n\}_1^\infty$ is written as

$$f(x) \simeq \sum_{n=1}^\infty c_n \phi_n(x) \qquad (14.9)$$

EXAMPLE 14.2.2. Let $P_n(x)$ denote the Legendre polynomial of degree n and $\phi_n(x) = \left(\frac{2n+1}{2}\right)^{1/2} P_n(x), n = 0, 1, 2, \ldots.$ Since

$$\int_{-1}^1 P_n(x) P_m(x)\, dx = 0, \quad m \neq n$$

and

$$\int_{-1}^1 P_n^2(x)\, dx = \frac{2}{2n+1}$$

the functions $\{\phi_n(x)\}_0^1$ form an orthonormal system on $[-1, 1]$. [See E. T. Copson, *Theory of Functions of a Complex Variable*, Chapter 11.) Furthermore,

$$f(x) \simeq \sum_{n=0}^{\infty} c_n P_n(x)$$

where

$$c_n = \frac{2n+1}{2} \int_{-1}^{1} f(x) P_n(x) \, dx$$

Minimal property of the partial sums of a Fourier series. Let

$$S_n(x) = \frac{a_0}{2} + \sum_{k=1}^{n} (a_k \cos kx + b_k \sin kx)$$

be the nth partial sum of the trig series (14.1), and suppose that we want to choose the constants a_j and b_j such that the trig polynomial $S_n(x)$ makes the best approximation to $f(x)$ in the sense of "least squares," that is the integral

$$I = I(a_0, a_k, b_k) = \int_{-\pi}^{\pi} |f(x) - S_n(x)|^2 \, dx$$

is a minimum. We show below that I is minimized when a_k, b_k are Fourier coefficients.

Note that when $f, g, h \in L^2[a,b]$ and α is a complex number, then $(f,g) = \overline{(g,f)}$, $(f+h,g) = (f,g) + (h,g)$, $(\alpha f, g) = \alpha(f,g)$ and $(f,f) > 0$ if $f \neq 0$ almost everywhere (a.e.).

Theorem 14.2.1. Let $f \in L^2[a,b]$ and let $\{\phi_n\}_1^{\infty}$ be an orthonormal system on $[a,b]$. Let $\{\alpha_n\}$ be any sequence of numbers. If $c_n = (f, \phi_n) = \int_a^b f\overline{\phi}_n$ then,

$$\left\| f - \sum_{k=1}^{n} c_k \phi_k \right\| \leqslant \left\| f - \sum_{k=1}^{n} \alpha_k \phi_k \right\| \tag{14.10}$$

with equality if and only if $\alpha_k = c_k$, $k = 1, 2, \ldots n$. Also,

$$\sum_{n=1}^{\infty} |c_n|^2 \leqslant \|f\|^2. \quad \text{(Bessel's inequality)} \tag{14.11}$$

Proof. Consider

$$\left\| f - \sum_{k=1}^{n} \alpha_k \phi_k \right\|^2 = \left(f - \sum_{k=1}^{n} \alpha_k \phi_k, f - \sum_{k=1}^{n} \alpha_k \phi_k \right)$$

$$= \left(f, f - \sum_{k=1}^{n} \alpha_k \phi_k \right) - \sum_{k=1}^{n} \alpha_k \left(\phi_k, f - \sum_{p=1}^{n} \alpha_p \phi_p \right)$$

$$= \overline{\left(f - \sum_{k=1}^{n} \alpha_k \phi_k, f \right)} - \sum_{k=1}^{n} \alpha_k \overline{\left(f - \sum_{p=1}^{n} \alpha_p \phi_p, \phi_k \right)}$$

$$= \overline{(f, f) - \sum_{k=1}^{n} \alpha_k (\phi_k, f)} - \sum_{k=1}^{n} \alpha_k \overline{\left\{ (f, \phi_k) - \sum_{p=1}^{n} \alpha_p (\phi_p, \phi_k) \right\}}$$

$$= \| f \|^2 - \sum_{k=1}^{n} \overline{\alpha}_k (f, \phi_k) - \sum_{k=1}^{n} \alpha_k (\overline{f, \phi_k}) + \sum_{k=1}^{n} |\alpha_k|^2$$

$$= \| f \|^2 + \sum_{k=1}^{n} |c_k|^2 - \sum_{k=1}^{n} \overline{\alpha}_k c_k - \sum_{k=1}^{n} \alpha_k \overline{c}_k + \sum_{k=1}^{n} |\alpha_k|^2 - \sum_{k=1}^{n} |c_k|^2.$$

Since

$$|c_k|^2 - \overline{\alpha}_k c_k - \alpha_k \overline{c}_k + |\alpha_k|^2 = |c_k - \alpha_k|^2$$

we obtain

$$\left\| f - \sum_{k=1}^{n} \alpha_k \phi_k \right\|^2 = \| f \|^2 + \sum_{k=1}^{n} |c_k - \alpha_k|^2 - \sum_{k=1}^{n} |c_k|^2. \qquad (14.12)$$

In particular, when $\alpha_k = c_k$ we have

$$\left\| f - \sum_{k=1}^{n} c_k \phi_k \right\|^2 = \| f \|^2 - \sum_{k=1}^{n} |c_k|^2. \qquad (14.13)$$

From (14.12) and (14.13),

$$\left\| f - \sum_{k=1}^{n} c_k \phi_k \right\|^2 = \left\| f - \sum_{k=1}^{n} \alpha_k \phi_k \right\|^2 - \sum_{k=1}^{n} |c_k - \alpha_k|^2$$

$$\leqslant \left\| f - \sum_{k=1}^{n} \alpha_k \phi_k \right\|^2. \qquad (14.14)$$

Since $\displaystyle\sum_{k=1}^{n} |c_k - \alpha_k|^2 > 0$ unless $\alpha_k = c_k$ for all k, $1 \leqslant k \leqslant n$, the first part is proved. From (14.13) we have

$$\sum_{k=1}^{n} |c_k|^2 \leqslant \|f\|^2$$

for every n. This implies the convergence of the series $\displaystyle\sum_{k=1}^{\infty} |c_k|^2$ and (14.11) follows. ∎

From (14.12) and (14.13) we deduce the following ∎

Corollary 1. The equation

$$\sum_{n=1}^{\infty} |c_n|^2 = \int_a^b |f(x)|^2 \, dx \qquad \text{(Parseval's formula)}$$

holds, if and only if we have

$$\lim_{n \to \infty} \|f - s_n\| = 0$$

where $s_n(x) = \displaystyle\sum_{k=1}^{n} c_k \phi_k(x)$ is the nth partial sum of the series (14.9).

Corollary 2. If $s_n^*(x) = \displaystyle\sum_{k=1}^{n} \alpha_k \phi_k(x)$ then the integral $\displaystyle\int_a^b |f - s_n^*|^2 \, dx$ is a minimum when $s_n^*(x)$ is the nth partial sum of the Fourier series of f with respect to $\{\phi_k\}_1^{\infty}$.

EXAMPLE 14.2.3.   Let $f \in L^2 [-\pi, \pi]$. Since the system $\left\{\dfrac{e^{inx}}{\sqrt{2\pi}}\right\}_{-\infty}^{\infty}$ is orthonormal on $[-\pi, \pi]$, we have

$$f(x) \simeq \sum_{n=-\infty}^{\infty} c_n e^{inx} = \sum_{n=-\infty}^{\infty} (c_n \sqrt{2\pi}) \left(\frac{e^{inx}}{\sqrt{2\pi}}\right)$$

where

$$c_n = \frac{1}{2\pi} \int_{-\pi}^{\pi} f(x) e^{-int} \, dt.$$

The corresponding Bessel inequality is

$$\sum_{n=-\infty}^{\infty} |c_n|^2 \leqslant \frac{1}{2\pi} \int_{-\pi}^{\pi} |f(x)|^2 \, dx.$$

EXAMPLE 14.2.4. Let $f(x) \in L^2[-\pi, \pi]$ and be real-valued. Then f is certainly $L[-\pi, \pi]$. The Fourier series of f with respect to the trigonometric system (14.6) can be written as

$$f(x) \simeq \frac{1}{\sqrt{2\pi}} \left(\sqrt{2\pi} \, \frac{a_0}{2} \right) + \sum_{n=1}^{\infty} \left(\frac{\cos nx}{\sqrt{\pi}} \, a_n \sqrt{\pi} + \frac{\sin nx}{\sqrt{\pi}} \, b_n \sqrt{\pi} \right)$$

[See equations (14.3), (14.4).] Hence the corresponding Bessel inequality is

$$\left(\sqrt{2\pi} \, \frac{a_0}{2} \right)^2 + \sum_{n=1}^{\infty} (a_n^2 \pi + b_n^2 \pi) \leqslant \int_{-\pi}^{\pi} (f(x))^2 \, dx$$

that is,

$$\frac{a_0^2}{2} + \sum_{n=1}^{\infty} (a_n^2 + b_n^2) \leqslant \frac{1}{\pi} \int_{-\pi}^{\pi} (f(x))^2 \, dx. \tag{14.15}$$

Note that (14.14) reduces to

$$\|f - s_n\| = \left(\int_{-\pi}^{\pi} |f - s_n|^2 \, dx \right)^{1/2} \leqslant \|f - t_n\|. \tag{14.16}$$

Here $s_n(x) = \frac{1}{2} a_0 + \sum_{k=1}^{n} (a_k \cos kx + b_k \sin kx)$ and $t_n(x)$ is a trigonometric polynomial of the form

$$\frac{1}{2} \alpha_0 + \sum_{k=1}^{n} (\alpha_k \cos kx + \beta_k \sin kx)$$

where α_k and β_k are any real numbers.

Remarks. In order that an orthogonal system $\{\phi_n\}$ be useful for expanding a function in a series of the form (14.9), it must be complete.

Definition 14.2.3. *An orthonormal system* $\{\phi_k\}_1^\infty \in L^2[a, b]$ *is said to be complete if there is no strictly larger set of orthonormal elements. That is* $\psi \in L^2[a, b]$ *and* $(\psi, \phi_k) = 0$, $k = 1, 2, \ldots$ *implies* $\psi = 0$ *a.e.*

Thus if the system $\{\phi_k\}_1^\infty$ is not complete, then there will be a function $\phi \in$

$L^2[a,b]$ not zero a.e. and whose Fourier series with respect to the system $\{\phi_k\}_1^\infty$ would consist only of zeros.

Definition 14.2.4. *An orthonormal system* $\{\phi_k\}_1^\infty \in L^2[a,b]$ *is said to be closed if every f in* $L^2[a,b]$ *can be approximated closely by finite linear combinations of the* ϕ_k. *That is given* $f \in L^2[a,b]$ *and* $\epsilon > 0$ *we can find constants* $\alpha_1, \ldots, \alpha_n$ *such that*

$$\left\| f - \sum_1^n \alpha_k \phi_k \right\| < \epsilon.$$

It can be shown that the following three statements are equivalent.
    (i) $\{\phi_k\}_1^\infty$ is complete.
    (ii) $\{\phi_k\}_1^\infty$ is closed.
    (iii) Parseval's formula (Corollary 1 of Theorem 14.2.1) holds.
For the proof we require the following

Theorem 14.2.2 (Riesz–Fischer). Let $c_n(n=1,2,\cdots)$ be any sequence of numbers for which $\sum_{n=1}^\infty |c_n|^2 < \infty$, and let $\{\phi_n(x)\}_1^\infty$ be any orthonormal system. Then there is a function $F \epsilon L^2[a,b]$ such that $(F,\phi_k) = c_k$ for $k = 1,2,\cdots$.

Proof of Theorem 14.2.2. Let $s_n(x) = \sum_{k=1}^n c_k \phi_k(x)$. Then

$$\|s_{n+p} - s_n\|^2 = \sum_{k=n+1}^{n+p} |c_k|^2. \tag{14.17}$$

Our hypothesis shows that $\{s_n\}$ is a Cauchy sequence. Since $L^2[a,b]$ is a complete metric space (see 13.4.3; see Hewitt and Stromberg, *Real and Abstract Analysis*, pp. 192-4) there is a function $F \epsilon L^2[a,b]$ such that

$$\|s_n - F\| = o(1), n \longrightarrow \infty. \tag{14.18}$$

Now $(F,\phi_k) = (F - s_n,\phi_k) + (s_n,\phi_k)$ and by Schwarz inequality (Theorem 13.4.1) and (14.18),

$$|(F - s_n,\phi_k)|^2 \leqslant \|F - s_n\|^2 \|\phi_k\|^2 = o(1), n \longrightarrow \infty.$$

Further

$$\lim_{n\to\infty} (s_n,\phi_k) = \lim_{n\to\infty} \int_a^b \left(\sum_1^n c_j\phi_j \right) \phi_k \, dx = c_k.$$

Hence letting $n \longrightarrow \infty$, we have

$$(F,\phi_k) = c_k. \quad \blacksquare$$

Proof. 1 (i) $\implies$ (ii).

Let $f \epsilon L^2[a,b]$. Then Bessel's inequality shows that

$$\sum_1^\infty |c_n|^2 < \infty, c_n = (f, \phi_n).$$

Hence by Riesz-Fischer theorem, there exists a function $F \in L^2[a,b]$ such that

$$(F, \phi_n) = c_n \quad (n = 1, 2, \ldots).$$

Hence

$$c_n = (f, \phi_n) = (F, \phi_n) \text{ and so } (f - F, \phi_n) = 0 \quad n = 1, 2, \ldots.$$

Since $\{\phi_n\}_1^\infty$ is complete $f = F$ a.e. From (14.18) we see that

$$\| s_n - f \| = o(1), n \longrightarrow \infty.$$

This implies (ii).

II (ii) $\implies$ (iii).

For any $f \in L^2[a,b]$ we have by (14.14)

$$\left\| f - \sum_1^n c_k \phi_k \right\| \leqslant \left\| f - \sum_1^n \alpha_k \phi_k \right\| < \epsilon, c_k = (f, \phi_k).$$

Hence

$$\lim_{n \to \infty} \| f - s_n \| = 0.$$

Now see Corollary 1 of Theorem 14.2.1.

III (iii) $\implies$ (i).

If $\psi \in L^2[a,b]$ and is such that

$$c_n = (\psi, \phi_n) = 0, n = 1, 2, \ldots$$

then (see Corollary 1 of Theorem 14.2.1)

$$\sum_1^\infty |c_n|^2 = \| \psi \|^2 = 0.$$

Hence $\psi = 0$ a.e. That is $\{\phi_n\}_1^\infty$ is complete. $\blacksquare$

From Bessel's inequality (14.11) it follows that the Fourier coefficients c_n, of $f \in L^2[a,b]$, with respect to the orthonormal system $\{\phi_n\}_1^\infty$, tend to zero as $n \longrightarrow \infty$. If ϕ_n are uniformly bounded then this result holds also for the Fourier coefficients of any $f \in L[a,b]$. We shall prove this for the trigonometric system (14.6). *We will assume, in the rest of this chapter that $f(x)$ is real-valued, 2π-periodic and integrable $L[-\pi, \pi]$.*

Theorem 14.3.3 (Riemann-Lebesgue). Let $-\pi \leqslant a < b \leqslant \pi$. If $f \in L[-\pi, \pi]$ then

$$\lim_{n \to \infty} \int_a^b f(x) \cos nx\, dx = 0 = \lim_{n \to \infty} \int_a^b f(x) \sin nx\, dx.$$

To prove this we require the following

Lemma. Let $f \in L[a,b]$. Given $\epsilon > 0$, there exists $g \in L^2[a,b]$ such that

$$\int_a^b |f - g|\, dx < \epsilon.$$

Proof. Since $f \in L[a,b]$, there exists $\delta > 0$ such that if A is any measurable set in $[a,b]$ and $m(A) < \delta$ then $\int_A |f|\, dx < \dfrac{\epsilon}{2}$. For n sufficiently large $m\{x : |f| > n\} < \delta$. Call this set A. Then $m(A) < \delta$. There exists a simple function S such that $|S - f| < \dfrac{\epsilon}{2(b-a)}$ on $[a,b] - A$. Let $g = S\chi_{[a,b]-A}$ where χ is the characteristic function. Then g is real-valued measurable and bounded, and so $g \in L^2[a,b]$. Also,

$$\int_a^b |g - f|\, dx = \int_A |f|\, dx + \int_{[a,b]-A} |S - f|\, dx$$

$$< \frac{\epsilon}{2} + \frac{\epsilon}{2(b-a)} (b-a) = \epsilon \; \blacksquare$$

Proof of Theorem 14.2.3. Consider the cosine integral. Given $\epsilon > 0$, by the lemma there exists a real-valued function $g \in L^2[-\pi, \pi]$ such that $\int_{-\pi}^{\pi} |g - f|\, dx < \epsilon$. Further there exists a positive integer $n_0(\epsilon)$ such that for $n \geqslant n_0$

$$\left| \int_{-\pi}^{\pi} g(x) \cos nx\, dx \right| < \epsilon.$$

[See (14.15).] Hence for $n \geqslant n_0$,

$$\left| \int_{-\pi}^{\pi} f(x) \cos nx\, dx \right| \leqslant \left| \int_{-\pi}^{\pi} g(x) \cos nx\, dx \right| + \int_{-\pi}^{\pi} |f(x) - g(x)|\, dx < 2\epsilon$$

It follows that

$$\lim_{n \to \infty} \int_{-\pi}^{\pi} f(x) \cos nx\, dx = 0$$

Now replace f by $f\chi_{[a,b]}$ and the result for the cosine integral follows. The proof for the sine integral is similar and is left to the reader. ▮

14.3. Dirichlet's Integral and Convergence Tests.

Let $f(x) \in L[-\pi, \pi]$, be 2π-periodic, and real-valued. In this section we obtain an expression for the nth partial sum, $s_n(x)$, of the Fourier series (14.2), and tests for convergence of this series. We have

$$a_k = \frac{1}{\pi} \int_{-\pi}^{\pi} f(t) \cos kt \, dt$$

Now, by periodicity,

$$\int_{x-\pi}^{-\pi} f(t) \cos kt \, dt = -\int_{\pi}^{x+\pi} f(t) \cos kt \, dt \ (-\pi \leqslant x \leqslant \pi),$$

and so

$$a_k = \frac{1}{\pi} \int_{x-\pi}^{x+\pi} f(t) \cos kt \, dt, \qquad k = 0, 1, \ldots .$$

Similarly, we have

$$b_k = \frac{1}{\pi} \int_{-\pi}^{\pi} f(t) \sin kt \, dt = \frac{1}{\pi} \int_{x-\pi}^{x+\pi} f(t) \sin kt \, dt, \qquad k = 1, 2, \ldots .$$

Hence

$$s_n(x) = \frac{a_0}{2} + \sum_{k=1}^{n} (a_k \cos kx + b_k \sin kx)$$

$$= \frac{1}{\pi} \int_{x-\pi}^{x+\pi} f(t) \, dt + \sum_{k=1}^{n} \frac{1}{\pi} \int_{x-\pi}^{x+\pi} f(t) (\cos kt \cos kx + \sin kt \sin kx) \, dt$$

$$= \frac{1}{\pi} \int_{x-\pi}^{x+\pi} f(t) \left\{ \frac{1}{2} + \sum_{k=1}^{n} \cos k(t-x) \right\} dt. \tag{14.17}$$

Let

$$D_n(x) = \frac{1}{2} + \sum_{k=1}^{n} \cos kx. \tag{14.18}$$

Then

$$2D_n(x) \sin\frac{x}{2} = \sin\frac{x}{2} + \sum_{k=1}^{n} 2 \sin\frac{x}{2}\cos kx$$

$$= \sin\frac{x}{2} + \sum_{k=1}^{n}\left\{\sin\left(k+\frac{1}{2}\right)x - \sin\left(k-\frac{1}{2}\right)x\right\}$$

$$= \sin\left(n+\frac{1}{2}\right)x.$$

Hence

$$s_n(x) = \frac{1}{\pi}\int_{x-\pi}^{x+\pi} f(t)\, \frac{\sin\left(n+\frac{1}{2}\right)(t-x)}{2\sin\dfrac{t-x}{2}}\, dt$$

$$= \frac{1}{\pi}\int_{-\pi}^{\pi} f(u+x)\, \frac{\sin\left(n+\frac{1}{2}\right)u}{2\sin\dfrac{u}{2}}\, du$$

$$= \frac{1}{\pi}\left[\int_{-\pi}^{0} + \int_{0}^{\pi}\left(f(u+x)\, \frac{\sin\left(n+\frac{1}{2}\right)u}{2\sin\dfrac{u}{2}}\right)du\right]$$

$$= \frac{1}{\pi}\int_{0}^{\pi} (f(x+u)+f(x-u))\, \frac{\sin\left(n+\frac{1}{2}\right)u}{2\sin\dfrac{u}{2}}\, du. \qquad (14.19)$$

The function D_n is called the *Dirichlet's kernel* and the last integral in (14.19) is known as *Dirichlet's integral.* If we take $f(x) = 1$ then $\dfrac{a_0}{2} = 1$, $a_n = b_n = 0$, $s_n(x) = 1$ and consequently from (14.19) it follows that

$$1 = \frac{1}{\pi}\int_{0}^{\pi} 2\, \frac{\sin\left(n+\frac{1}{2}\right)u}{2\sin\dfrac{u}{2}}\, du.$$

Multiplying this by a real number s and subtracting from (14.19) we get

$$s_n(x) - s = \frac{1}{\pi} \int_0^\pi (f(x + u) + f(x - u) - 2s) \frac{\sin\left(n + \frac{1}{2}\right)u}{2 \sin \frac{u}{2}} \, du \quad (14.20)$$

A necessary and sufficient condition that the series (14.2) should converge at x to the sum $s = s(x)$ is therefore that this integral should tend to zero as $n \longrightarrow \infty$. We simplify the expression on the right. Fix x and s and let

$$\phi(u) = f(x + u) + f(x - u) - 2s.$$

Then $\phi(u) \in L[-\pi, \pi]$, and for $0 < \delta < \pi$, we have

$$s_n(x) - s = \frac{1}{\pi} \left[\int_0^\delta \phi(u) \frac{\sin\left(n + \frac{1}{2}\right)u}{2 \sin \frac{u}{2}} \, du + \int_\delta^\pi \phi(u) \frac{\sin\left(n + \frac{1}{2}\right)u}{2 \sin \frac{u}{2}} \, du \right] \quad (14.21)$$

Now $\dfrac{\phi(u)}{\sin \dfrac{u}{2}}$ is integrable L on $[\delta, \pi]$ and so by the Riemann-Lebesgue theorem, the second integral in (14.21) tends to zero as $n \longrightarrow \infty$. For the first integral we write

$$\frac{\sin\left(n + \frac{1}{2}\right)u}{2 \sin \frac{u}{2}} = \frac{\sin nu}{u} + \left(\frac{1}{2 \tan \frac{u}{2}} - \frac{1}{u}\right) \sin nu + \frac{1}{2} \cos nu$$

and note that by the Riemann-Lebesgue theorem, $\displaystyle\int_0^\delta \phi(u) \cos nu \, du = o(1)$, $n \longrightarrow \infty$. Further $u - 2 \tan \dfrac{u}{2} = O(u^3)$ as $u \longrightarrow 0$, and so the function

$$\theta(u) = \begin{cases} \dfrac{1}{2 \tan \dfrac{u}{2}} - \dfrac{1}{u}, & u > 0 \\ \\ 0, & u = 0 \end{cases}$$

is continuous on $[0, \delta]$. Hence

$$\int_0^\delta \phi(u) \left(\frac{1}{2 \tan \frac{u}{2}} - \frac{1}{u}\right) \sin nu \, du = o(1), \quad n \longrightarrow \infty$$

and the relation (14.21) simplifies to

$$s_n(x) - s = \frac{1}{\pi} \int_0^\delta \phi(u) \frac{\sin nu}{u} du + o(1). \qquad (14.22)$$

Take $s = 0$, and we have

$$s_n(x) = \frac{1}{\pi} \int_0^\delta (f(x+u) + f(x-u)) \frac{\sin nu}{u} du + o(1)$$

$$= \frac{1}{\pi} \int_{-\delta}^\delta f(x+u) \frac{\sin nu}{u} du + o(1)$$

Since the integral utilizes the values of f in $(x - \delta, x + \delta)$ we have proved

Theorem 14.3.1 (Riemann's Principle of Localization). The convergence or divergence of the Fourier series at a point x depends only upon the behavior of the function f in the immediate neighborhood of the point x.

Formula (14.22) gives also the following necessary and sufficient condition for convergence.

Theorem 14.3.2. The Fourier series of a function $f \in L[-\pi,\pi]$ converges at a point x to the value s if and only if

$$\lim_{n\to\infty} \int_0^\delta \frac{\phi(u)}{u} \sin nu \, du = 0$$

for some $\delta \in (0,\pi)$.

The following sufficient condition for convergence follows immediately from Theorem 14.3.2.

Theorem 14.3.3 (Dini's Test). If $\dfrac{\phi(u)}{u} \in L[0,\delta]$ for some $\delta > 0$, then $\lim_{n\to\infty} s_n(x) = s$.

In the following corollary we consider functions of satisfying the condition

$$|f(x+h) - f(x)| \leqslant c|h|^\alpha, \alpha > 0, c > 0;$$

at a point x. Then f is said to satisfy a Lipschitz condition of order α at the point x.

Corollary 1. If f satisfies a Lipschitz condition of order α, $0 < \alpha$, at the point x then the series converges at x to $f(x)$.

Proof. Let $s = f(x)$. Then $\dfrac{|\phi(u)|}{u} \leqslant \dfrac{|f(x+u) - f(x)| + |f(x-u) - f(x)|}{u} \leqslant$ $2cu^{\alpha-1}$ and Corollary 1 is proved by applying Dini's test with $s = f(x)$. $\blacksquare$

Corollary 2. If f is differentiable at x then the series converges at x to $f(x)$.

Exercise 14.3

1. Use the Fourier series for $|x|$ to find the sum S of the series

$$1 + \frac{1}{3^2} + \frac{1}{5^2} + \cdots .$$

$\left(\textit{Hint}: \text{ See Exercise 14.1(1) and use Corollary 1. } S = \dfrac{\pi^2}{8} \, . \right)$

2. Find the sum of the series

$$1 - \frac{1}{2^2} + \frac{1}{3^2} - \frac{1}{4^2} + \cdots .$$

$\left(\textit{Hint}: \text{ See Exercise 14.1(1) and use Corollary 1. } S = \dfrac{\pi^2}{12}. \right)$

3. Find the sum of the series

$$1 + \frac{1}{2^2} + \frac{1}{3^2} + \cdots .$$

$\Big(\textit{Hint}$: The series is absolutely convergent. Let its sum be S. Then

$$S = 1 + \frac{1}{3^2} + \frac{1}{5^2} + \cdots$$

$$+ \frac{1}{2^2}\left(1 + \frac{1}{2^2} + \frac{1}{3^2} + \cdots\right) = \frac{\pi^2}{8} + \frac{S}{4}$$

This gives $S = \dfrac{\pi^2}{6}$. $\Big)$

14.4. Summability of Fourier Series

In this section we will consider the $(C,1)$ summability of the Fourier series (14.2). Note that $f \in L\,[-\pi,\pi]$, is 2π-periodic, and real-valued. Write

$$K_n(t) = \frac{1}{n+1}\,(D_0(t) + \cdots + D_n(t))$$

where $D_n(t)$ is Dirichlet's kernel. $K_n(t)$ is called the *Fejér kernel.* Since

$$2 \sin \left(k + \frac{1}{2} \right) t \sin \frac{t}{2} = \cos kt - \cos (k + 1) t$$

we obtain

$$K_n(t) = \begin{cases} \dfrac{1 - \cos (n + 1)t}{4 (n + 1) \sin^2 \dfrac{t}{2}} & \text{if } t \neq 0 \,(\text{mod } 2\pi) \\[4mm] \dfrac{n + 1}{2} & \text{if } t = 0 \,(\text{mod } 2\pi). \end{cases} \qquad (14.23)$$

One can easily prove the following:

(i) $D_n(u) = D_n(-u) \leqslant n + \dfrac{1}{2}$

(ii) $0 \leqslant K_n(u) = K_n(-u) \leqslant \dfrac{1}{2} (n + 1)$ (14.24)

(iii) $\dfrac{1}{\pi} \displaystyle\int_{-\pi}^{\pi} D_n(u) \, du = 1 = \dfrac{1}{\pi} \int_{-\pi}^{\pi} K_n(u) \, du.$

Since $\sin \theta \geqslant \dfrac{2\theta}{\pi}$, for $0 \leqslant \theta \leqslant \pi/2$, one obtains

(iv) $K_n(u) \leqslant \dfrac{\pi^2}{2(n + 1)u^2}$ for $0 < |u| \leqslant \pi$

and

(v) $\displaystyle\lim_{n \to \infty} \int_{\delta}^{\pi} K_n(u) \, du = 0$ for $0 < \delta < \pi.$

We follow the notation of Sec. 11.6 and write

$$\sigma_n(x) = \sigma_n(x,f) = \frac{1}{n + 1} (s_0(x) + \cdots + s_n(x))$$

$$= \frac{1}{n + 1} \left\{ (n + 1) \frac{a_0}{2} + n(a_1 \cos x + b_1 \sin x) + \cdots \right.$$

$$\left. + (a_n \cos nx + b_n \sin nx) \right\}$$

$$= \frac{a_0}{2} + \sum_{k=1}^{n} \left(1 - \frac{k}{n + 1} \right) (a_k \cos kx + b_k \sin kx). \qquad (14.25)$$

From (14.19),

$$\sigma_n(x) = \frac{1}{\pi} \int_0^\pi (f(x+u) + f(x-u)) \left(\frac{D_0(u) + \cdots + D_n(u)}{n+1} \right) du$$

$$= \frac{1}{\pi} \int_0^\pi (f(x+u) + f(x-u)) K_n(u)\, du. \tag{14.26}$$

Since $K_n(u)$ is even it follows from (14.24) (iii) that

$$s = \frac{2}{\pi} \int_0^\pi s K_n(u)\, du. \tag{14.27}$$

Subtracting (14.27) from (14.26) we get

$$\sigma_n(x) - s = \frac{1}{\pi} \int_0^\pi (f(x+u) + f(x-u) - 2s) K_n(u)\, du. \tag{14.28}$$

We write $C[a,b]$ for the class of functions continuous on $[a,b]$ and prove

Theorem 14.4.1. (Fejér). (i) If for some x, $f(x+0)$ and $f(x-0)$ exist then the Fourier series of f is $(C,1)$ summable at the point x to the sum

$$\frac{f(x+0) + f(x-0)}{2}.$$

(ii) If $f \in C[a,b]$ and $[a',b'] \subseteq (a,b)$ then $\lim_{n \to \infty} \sigma_n(x) = f(x)$ uniformly on $[a',b']$.

(iii) If $f \in C[-\pi, \pi]$ then $\lim_{n \to \infty} \sigma_n(x) = f(x)$ uniformly on $[-\pi, \pi]$.

Proof. (i) Let $s = \frac{1}{2}(f(x+0) + f(x-0))$ in (14.28) and write $\phi(u) = f(x+u) + f(x-u) - f(x+0) - f(x-0)$. Then $\phi(u) \longrightarrow 0$ as $u \longrightarrow 0$. Hence given $\epsilon > 0$ there exists $\delta > 0$ such that $|\phi(u)| < \epsilon$ for $|u| \leqslant \delta$. Now

$$\sigma_n(x) - s = \frac{1}{\pi} \int_0^\delta \phi(u) K_n(u)\, du + \frac{1}{\pi} \int_\delta^\pi \phi(u) K_n(u)\, du$$

$= I_1 + I_2$, say. Then

$$|I_1| < \frac{\epsilon}{\pi} \int_0^\delta K_n(u)\, du < \frac{\epsilon}{\pi} \int_0^\pi K_n(u)\, du = \frac{\epsilon}{2}$$

$$|I_2| < \frac{1}{\pi} \int_\delta^\pi |\phi(u)| K_n(u)\, du$$

$$< \frac{1}{\pi} \frac{\pi^2}{2(n+1)\delta^2} \int_\delta^\pi |\phi(u)| \, du$$

by (14.24) (iv). Since $\phi(u) \in L[\delta, \pi]$ we have, for $n > n_0$,

$$|I_2| < \frac{\epsilon}{2}$$

and so, for $n > n_0$,

$$|\sigma_n(x) - s| < \epsilon.$$

This completes the proof of the first part.

(ii) Let $\lambda = \min (\pi, b - b', a' - a)$. Since f is continuous on $[a, b]$, given $\epsilon > 0$, we can choose $\delta = \delta(\epsilon) \leqslant \lambda$ such that for $u \leqslant \delta$ and $a' \leqslant x \leqslant b'$,

$$|f(x + u) - f(x)| < \frac{\epsilon}{2} \text{ and } |f(x - u) - f(x)| < \frac{\epsilon}{2}.$$

Take $s = f(x)$. Then for $0 \leqslant u \leqslant \delta$ and $a' \leqslant x \leqslant b'$, $|\phi(x)| < \epsilon$. As in part (i)

$$\sigma_n(x) - s = I_1 + I_2$$

where $|I_1| < \frac{\epsilon}{2}$. Let $M = \sup_{a \leqslant x \leqslant b} |f(x)|$. Then $|\phi(u)| \leqslant 4M$ for $u \leqslant \delta$ and $a' \leqslant x \leqslant b'$. Hence

$$|I_2| < \left\{ \frac{\pi}{2(n+1)\delta^2} \right\} 4M\pi < \frac{\epsilon}{2}.$$

for $n \geqslant n_1(\epsilon)$ and $a' \leqslant x \leqslant b'$. Thus $\sigma_n(x)$ converges uniformly to $f(x)$ on $[a', b']$.

(iii) By periodicity, $f(x)$ is continuous on every interval $[a, b]$, and consequently (iii) follows from (ii). ∎

Corollary 1. Suppose, for some x, $f(x + 0)$ and $f(x - 0)$ exist. If $s_n(x)$ converges then it must converge to $\frac{1}{2} \{f(x + 0) + f(x - 0)\}$.

Proof. The series is $(C, 1)$ summable to $\frac{1}{2} \{f(x + 0) + f(x - 0)\}$. Since $(C, 1)$ method is regular, that is a convergent series is summable $(C, 1)$ and to the same sum, Corollary 1 follows. ∎

Corollary 2. Let $f \in C[-\pi, \pi]$. Then

(i) $\displaystyle \lim_{n \to \infty} \int_a^b |f(x) - s_n(x)|^2 \, dx = 0 \quad$ for every interval $[a, b]$.

(ii) $\dfrac{1}{\pi} \displaystyle\int_{-\pi}^{\pi} \{f(x)\}^2 \, dx = \dfrac{a_0^2}{2} + \displaystyle\sum_{1}^{\infty} (a_n^2 + b_n^2)$ (Parseval's formula)

(iii) $\displaystyle\int_{0}^{x} f(t) \, dt = \dfrac{a_0 x}{2} + \displaystyle\sum_{n=1}^{\infty} \int_{0}^{x} (a_n \cos nt + b_n \sin nt) \, dt$

the last series being uniformly convergent on every interval $[a,b]$.

Proof.   (i)   We use (14.16) with $t_n = \sigma_n$ to obtain the inequality

$$\int_{-\pi}^{\pi} |f - s_n|^2 \, dx \leqslant \int_{-\pi}^{\pi} |f - \sigma_n|^2 \, dx$$

But since $\sigma_n(x) \longrightarrow f(x)$ uniformly on $[-\pi, \pi]$ we get

$$\lim_{n \to \infty} \int_{-\pi}^{\pi} |f - s_n|^2 \, dx = 0 \tag{14.29}$$

If $-\pi \leqslant a < b \leqslant \pi$, the result follows.   Otherwise let, p and k be integers such that $p\pi \leqslant a < (p + 2)\pi, (p + 2k - 2)\pi \leqslant b < (p + 2k)\pi$.   Then

$$\int_{a}^{b} |f - s_n|^2 \, dx \leqslant \int_{p\pi}^{(p+2)\pi} + \cdots + \int_{(p+2k-2)\pi}^{(p+2k)\pi} |f - s_n|^2 \, dx$$

$$= k \int_{p\pi}^{(p+2)\pi} |f - s_n|^2 \, dx$$

$$= k \int_{-\pi}^{\pi} |f - s_n|^2 \, dx$$

$$\to 0 \text{ as } n \longrightarrow \infty.$$

(ii)   Since [see (14.13) and Example 14.2.4]

$$\int_{-\pi}^{\pi} |f - s_n|^2 \, dx = \|f - s_n\|^2 = \|f\|^2 - \left(\pi \frac{a_0^2}{2} + \pi \sum_{1}^{n} (a_k^2 + b_k^2) \right)$$

part (ii) follows from (14.29).

(iii)   Let $a \leqslant x \leqslant b$.   Without loss of generality we may suppose that $a \leqslant 0$. Then, by Schwarz inequality and part (i).

$$0 \leqslant \left\{ \int_{0}^{x} |f(t) - s_n(t)| dt \right\}^2$$

$$\leqslant \left(\int_0^x |f - s_n|^2 \, dt \right) \left(\int_0^x 1 \, dt \right)$$

$$\leqslant \left(\int_a^b |f - s_n|^2 \, dt \right) (b - a)$$

$$< \epsilon$$

provided $n \geqslant n_0(\epsilon)$. Hence

$$\int_0^x f(t) \, dt = \lim_{n \to \infty} \int_0^x s_n(t) \, dt$$

$$= \lim_{n \to \infty} \left\{ \frac{a_0 x}{2} + \sum_{k=1}^n \int_0^x (a_k \cos kt + b_k \sin kt) \, dt \right\}$$

This proves that the series in (iii) converges for every x and uniformly on every interval $[a,b]$. █

Note that the series in (iii) converges for every x whether the Fourier series (14.2) converges or not. In fact Fejér has given the example of a continuous function with Fourier series divergent at a point. Note also that for the relation (iii) (of Corollary 2) to hold it is not necessary to assume that $f \in C[-\pi,\pi]$. The hypothesis on f stated in the beginning of this section is sufficient. For the proof of this and for Fejér's example we refer the reader to N. K. Bary, *A Treatise on Trigonometric Series*, Vol. 1, Chapter 1.

Theorem 14.4.2 (Weierstrass Approximation Theorem 1). Let $f(x)$ be 2π-periodic, real-valued and continuous on $[-\pi,\pi]$. Given $\epsilon > 0$ there exists a trigonometric polynomial $T(x)$ such that

$$|f(x) - T(x)| \leqslant \epsilon \qquad \text{for all } x$$

The proof follows from Theorem 14.4.1 (iii), for each $\sigma_n(x)$ is a trigonometric polynomial [see (14.25)].

Theorem 14.4.3 (Weierstrass Approximation Theorem 2). Let f be continuous and real valued on $[a,b]$. Given $\epsilon > 0$ there exists an algebraic polynomial $p(x)$ such that $|f(x) - p(x)| \leqslant \epsilon$ for every x in $[a,b]$.

Proof. Consider $F(y) = f\left(a + \frac{b-a}{\pi} y \right)$. Then $F \in C[0,\pi]$. We extend F to $(-\pi,0)$ by $F(y) = F(-y)$ for $-\pi < y < 0$ and then by periodicity so that F is continuous and 2π-periodic. By Theorem 14.4.1(iii) we can find $n_0(\epsilon)$ such that for $n > n_0(\epsilon), |F(y) - \sigma_n(y)| < \frac{\epsilon}{2}$ for every y in $[0,\pi]$. Each term in $\sigma_n(y)$ is a linear combination of $\cos ky$ (see (14.25); $b_k = 0$ since F is even). Further

$$\cos ky = \sum_{j=0}^{\infty} (-1)^j \frac{(ky)^{2j}}{(2j)!} \, ,$$

the convergence being uniform on every interval $[a,b]$, and so we can replace each $\cos ky$ in $\sigma_n(y)$ by a polynomial $r_k(y)$ and thus obtain a polynomial $r(y)$ such that $|\sigma_n(y) - r(y)| < \dfrac{\epsilon}{2}$ for every y in $[0,\pi]$. Hence $|F(y) - r(y)| < \epsilon$, $0 \leqslant y \leqslant \pi$. Set $p(x) = r\left(\dfrac{\pi(x - a)}{b - a}\right)$ which is an algebraic polynomial in x, and we get

$$|f(x) - p(x)| \leqslant \epsilon \text{ for every } x \text{ in } [a,b] . \blacksquare$$

Miscellaneous Exercises for Chapter 14.

1. If f is continuous at $x \in [-\pi,\pi]$ prove that

$$f(x) = \lim_{r \to 1} \left\{ \frac{a_0}{2} + \sum_{k=1}^{\infty} (a_k \cos kx + b_k \sin kx) r^k \right\} .$$

[*Hint*: See Theorem 14.4.1 (i) and Theorem 11.6.5.]

2. Prove that if $0 < r < 1$ then

$$\frac{a_0}{2} + \sum_{k=1}^{\infty} (a_k \cos kx + b_k \sin kx) r^k$$

$$= \frac{1}{2\pi} \int_{-\pi}^{\pi} f(t) \, \frac{1 - r^2}{1 - 2r \cos (t - x) + r^2} \, dt.$$

$\Bigg[$ *Hint*: The expression on the left is equal to

$$\frac{1}{2\pi} \int_{-\pi}^{\pi} f \, dt + \frac{1}{\pi} \sum_{n=1}^{\infty} r^n \int_{-\pi}^{\pi} f(t) \cos n(t - x) \, dt$$

The series $\displaystyle\sum_{n=1}^{\infty} r^n \cos n(t - x)$ converges uniformly with respect to t. Further,

$$\frac{1}{2} + \sum_{n=1}^{\infty} r^n \cos n\alpha = \frac{1 - r^2}{2(1 - 2r \cos \alpha + r^2)} \Bigg]$$

3. Prove that a necessary and sufficient condition for the Fourier series of f to be $(C,1)$ summable at a point x to the value s is that

$$\lim_{n\to\infty} \int_0^\delta \{f(x+u)+f(x-u)-2s\}\, K_n(u)\, du = 0$$

for some $\delta > 0$.
[*Hint*: Use (14.28) and (14.24) (iv).]

4. Prove that if $\displaystyle\sum_{n=1}^\infty (|a_n|+|b_n|) < \infty$, then the series (14.2) is uniformly and absolutely convergent on every interval $[a,b]$ and that it is the Fourier series of a continuous 2π-periodic function.

5. Prove the following:
   (a)

$$\int_{-\pi}^{\pi} |f(x)-\sigma_n(x)|^2\, dx = \int_{-\pi}^{\pi} |f(x)|^2\, dx - \frac{\pi}{2}\, a_0^2 - \pi \sum_{k=1}^n (a_k^2 + b_k^2)$$

$$+ \frac{\pi}{(n+1)^2} \sum_{k=1}^n k^2\, (a_k^2 + b_k^2)$$

   (b) If $f \in C[-\pi,\pi]$, then

$$\sum_{k=1}^n k^2\, (a_k^2 + b_k^2) = o(n^2).$$

   [*Hint*: Use (14.25) and Corollary 2(ii) of Theorem 14.4.1.]

6. Suppose that a_k and b_k are real in the trigonometric series (14.1). Prove that if the corresponding $\{\sigma_n(x)\}_{n=0}^\infty$ converges uniformly on $[-\pi,\pi]$, then (14.1) is the Fourier series of a continuous and 2π-periodic function f.
   [*Hint*: By periodicity $\{\sigma_n(x)\}_0^\infty$ converges uniformally on every interval $[a,b]$. Let $f(x) = \lim_{n\to\infty} \sigma_n(x)$. Then, for $0 \leqslant k \leqslant n$,

$$\frac{1}{\pi}\int_{-\pi}^{\pi} \sigma_n(x)\cos kx\, dx = \left(1 - \frac{k}{n+1}\right) a_k$$

   Fix k and let $n \longrightarrow \infty$. The case is similar for b_k.]

7. Suppose that a_k and b_k are real in the trigonometric series (14.1). If the partial sums $s_n(x)$ remain bounded, or bounded by an L-integrable function on $[-\pi,\pi]$ and if (14.1) converges to a function $f(x)$ almost everywhere, then show that $f \in L[-\pi,\pi]$ and (14.1) is the Fourier series of f.

8. Let $0 < \alpha \leqslant 1$. Prove that the series

$$\sum_{n=1}^{\infty} \frac{\sin nx}{n^{\alpha}}$$

is the Fourier series of a function $f \in L [-\pi,\pi]$.

Hint: The series is convergent for every x (Example 8.6.3). Let its sum be $f(x)$. Write $B_k = \sum_{n=1}^{k} \frac{1}{n^{\alpha}}$. Then

$$\sum_{k=1}^{\infty} \frac{B_k}{k(k+1)} = \sum_{k=1}^{\infty} \frac{1}{k(k+1)} \sum_{n=1}^{k} \frac{1}{n^{\alpha}}$$

$$= \sum_{n=1}^{\infty} \frac{1}{n^{\alpha}} \sum_{k=n}^{\infty} \frac{1}{k(k+1)} = \sum_{n=1}^{\infty} \frac{1}{n^{1+\alpha}} = B, \text{ say}.$$

Let $\dfrac{\pi}{k+1} \leqslant x < \dfrac{\pi}{k}$. Then

$$f(x) = \sum_{1}^{k-1} \frac{\sin nx}{n^{\alpha}} + \sum_{k}^{\infty} \frac{\sin nx}{n^{\alpha}}$$

and [see (8.8) and (8.9)]

$$|f(x)| \leqslant B_k + \frac{1}{k^{\alpha} \sin \dfrac{x}{2}} \leqslant B_k + \frac{\pi}{k^{\alpha}} \frac{1}{x} \leqslant B_k + \frac{(k+1)}{k^{\alpha}}.$$

Hence

$$\int_{0}^{\pi} |f(x)| \, dx = \sum_{1}^{\infty} \int_{\pi/(k+1)}^{\pi/k} |f| \, dx$$

$$\leqslant \sum_{1}^{\infty} \frac{\pi}{k(k+1)} \left(B_k + \frac{k+1}{k^{\alpha}} \right) = \pi B + \pi B$$

Hence $f(x) \in L [0,\pi]$. And since $f(x)$ is odd, $f(x) \in L [-\pi,\pi]$. Let m be any fixed integer. Then the series $\sum_{n=1}^{\infty} \dfrac{\sin nx}{n^{\alpha}} \sin mx$ is uniformly convergent on $[0,\pi]$, since [see (8.8) and (8.9)]

$$\left| \sin mx \sum_{n=p}^{q} \frac{\sin nx}{n^{\alpha}} \right| \leqslant mx \left| \sum_{n=p}^{q} \frac{\sin nx}{n^{\alpha}} \right|$$

$$\leqslant \frac{mx}{p^{\alpha}} \frac{1}{\sin \dfrac{x}{2}} \leqslant \frac{\pi m}{p^{\alpha}}$$

Hence integration term-by-term over $[0,\pi]$ gives

$$\frac{2}{\pi} \int_{0}^{\pi} f(x) \sin mx \, dx = \frac{1}{m^{\alpha}} \Bigg]$$

9. Is the series

$$\sum_{1}^{\infty} \frac{\sin nx}{\sqrt{n}}$$

the Fourier series of some function in $L^2 \; [-\pi,\pi]$?

Appendix: Property of Completeness

In this book no idea was stressed more than that of completeness in one form or another. We list below many properties which were discussed at various places in the text and are equivalent to each other.

1. Dedekind property (for R only):

Let R be decomposed into nonempty disjoint sets A and B such that $a \in A$ and $b \in B$ implies $a < b$. Then either A has the last element or B has the first element.

2. Greatest lower bound property (for R only):

Every nonempty subset of R which is bounded below has the greatest lower bound.

3. Least upper bound property (for R only);

Every nonempty set of R which is bounded above has the least upper bound.

4. Cauchy's criterion:

Every Cauchy sequence is convergent.

5. Principle of monotone convergence.

Every bounded monotone sequence is convergent.

6. Bounded sequence property:

Every bounded sequence has a convergent subsequence.

7. Nested spheres property:

A nest of nonempty closed bounded spheres has a nonempty intersection.

8. Bolzano Weierstrass property:

Every infinite bounded set has a limit point.

9. Heine-Borel property:

Every open covering of a closed and bounded set has a finite subcovering.

It may be added here the properties 4 and 7 are always implied by any one of the remaining property, and if the Archimedian property is assumed, then all the properties mentioned here are equivalent to each other.

Suggestions for Further Reading

Set Theory

Eves, H. and Newsom, C. V., *The Foundation and Fundamental Concepts of Mathematics*, Holt, Rinehart & Winston, New York, 1965.

Halmos, P. R., *Naive Set Theory*, D. Van Nostrand Company, Inc., Princeton, 1960.

Stoll, R. R., *Foundations of Mathematics*, W. H. Freeman and Company, San Francisco, 1963.

Wilder, R. L., *Foundations of Mathematics* 2d ed., John Wiley & Sons, Inc., New York, 1965.

Topology and Metric Spaces

Mendelson, B., *Introduction to Topology*, 2d ed, Allyn and Bacon, Boston, 1968.

Simmons, G., *Introduction to Topology and Analysis*, McGraw Hill, New York, 1963.

Yelbaum, B. and Olmsted, J. M. H., *Counterexamples in Analysis*, Holden-Day, San Francisco, 1964.

Infinite Series and Fourier Series

Bary, N. K., *A Treatise on Trigonometric Series*, Vol. I and II, English translation by M. F. Mullins, MacMillan and Co., New York, 1964.

Knopp, K., *Theory and Applications of Infinite Series* English translation by Miss R. C. Young, Blackie and Son, London, 1928.

Powell, R. E. and Shah, S. M., *An Introduction to Summability Theory and Applications*, Van Nostrand, Reinhold Co., London, 1972.

Zygmund A., *Trigonometric Series* Vol. I & II, Cambridge University Press, London, 1959.

Inequalities

Hardy, G., Littlewood, J. E. and Polya, G., *Inequalities*, Cambridge University Press, New York, 1964.
Mitrinovic, D. S., *Analytic Inequalities*, Springer-Verlag, New York, 1970.

Measure and Integration

Berberian, S. K., *Measure and Integration*, MacMillan Company, New York, 1968.
Hewitt, E. and Stromberg, K., *Real and Abstract Analysis*, Springer-Verlag, New York, 1965.
Kolmogorov, A. N. and Fomin, S. V., *Measure, Lebesgue Integrals, and Hilbert Space*, Academic Press, New York, 1961.
Lebesgue, H., *Measure and Integral*, Holden-Day, San Francisco, 1966.
Munroe, M. E., *Introduction to Measure and Integration* Addison-Wesley Publishing Co., Reading, Mass., 1953.
Phillips, E. R., *An Introduction to Analysis and Integration Theory*, Intext Educational Publishers, Scranton, Pa., 1971.
Royden, H. L., *Real Analysis*, 2nd ed., MacMillan Company, New York, 1968.

Complex Analysis

Copson, E. T., *An Introduction to the Theory of Functions of a Complex Variable*, Oxford University Press, London, 1948.
Hardy, G. H., *Pure Mathematics*, Cambridge University Press, New York, 1964.
Titchmarsh, E. C., *The Theory of Functions*, Oxford University Press, London, 1939.

List of Symbols and Notations

Symbols	Meaning	Page
$\in$	belongs to	1
$\notin$	does not belong to	1
$\mathfrak{N}$	the set of all natural numbers	1
$\mathfrak{g}$	the set of all integers	1
$\mathfrak{Q}$	the set of all rational numbers	1
R	the set of all real numbers	2
R^n	the Euclidean space of n-dimensions	2
$\subset$	is a subset of, is contained in	2
$\supset$	contains	2
$\cup$	union	2
$\cap$	intersection	2
$\Longrightarrow$	implies, implying	2
$\Longleftrightarrow$	if and only if, logical equivalence	2
$\exists$	there exists, there is	2
$\nexists$	there does not exist	99
$A - B$	the set of all points in A which are not in B	3
$A \times B$	Cartesian product of A and B	5
$x\not R y$	x is not R related to y	11
$\forall$	for every, for all	8
$\ni$	such that	13
cA	the complement of A	3
$\emptyset$	the null set (empty set)	3
$f : X \longrightarrow Y$	a mapping (or a function) from X into Y.	6
$f^{-1}(y)$	the set of points whose image is y.	8
xRy	x is R-related to y	11
glb or inf	the greatest lower bound, infimum	25
lub or sup	the least upper bound or supremum	25

Symbols	Meaning	Page
los	abbreviation for a "linearly ordered set"	12
wos	abbreviation for a "well ordered set"	13
f_D	the restriction a function on D	10
R_e	extended real number system	26
$Re\, z$	real part of the complex number z	33
$\overline{z}$	conjugate of the complex number z	35
$\lvert z \rvert$	magnitude (or absolute value) of the Complex number z	34
$(L \mid U)$	Dedekind cut with lower part L and upper part U.	17
$\Vert\ \Vert$	norm, magnitude	28
$\sim$	is equivalent to, asymptotic equivalence	37, 188
$\aleph_0$	The cardinal number of a denumerable set.	39
ω	The ordinal number of the set of all natural numbers	49
∞	Infinity	26
$-\infty$	Negative infinity	26
lim sup or $\overline{\lim}$	limit superior or upper limit	61
lim inf or $\underline{\lim}$	limit inferior or lower limit	61
S_i	the interior of a set S	75
$\overline{S}$	the closure of S	74
$sci(F)$	the smallest closed interval containing a closed set F	74
$\{M, d\}$	the metric space M with metric d	82
$d(x, y)$	the metric distance between x and y	83
$d(x, A)$	distance between the point x and the set A	86
$d(A, B)$	distance between the sets A and B	86
diam (S)	diameter of the set S	30
$\lim_{n \to \infty} a_n$	limit of the sequence $\{a_n\}$	55
$\lim_{x \to x_0} f(x)$	limit of the function as $x \longrightarrow x_0$	110
F_σ, G^δ, etc.	Borel sets	101
S_c	the set of points of condensation of S	105
$O_s(f, D)$	the oscillation of a function on a set D	119
$O_s(f, x_o)$	the oscillation of a function at a point x_o	119
$O_s(x)$	the oscillation function of f	119
WIVT	abbreviation for Weierstrass intermediate value theorem	115
$\{x : f(x) < k\}$	the set of all points in the domain of f such that the functional value at each of these points is less than k.	116
$\displaystyle\sum_{n=1}^{\infty} a_n$	series whose nth turn is a_n; if convergent the sum is $\displaystyle\sum_{n=1}^{\infty} a_n$	130
Σa_n	the series whose nth term is a_n.	130
S_n	the nth partial sum of a series	130

Symbols	Meaning	Page
$\downarrow$	monotone nonincreasing	139
$\uparrow$	monotone nondecreasing	139
$Df(x_0)$ or $f'(x_0)$ or $\left(\dfrac{df}{dx}\right)_{x=x_0}$	derivative of f at x_0	163
D^+f	the upper right derivative of f	164
D_+f	the lower right derivative of f	164
D^-f	the upper left derivative of f	164
D_-f	the lower left derivative of f	164
$D_R f$	the right derivative of f	164
$D_L f$	the left derivative of f	164
$\overline{Df}$	the upper derivative of f	164
$\underline{Df}$	the lower derivative of f	165
$\overline{s}(f,P)$	lower Darboux (Lebesgue) sum of f for a partition (generalized partition) P	171, 268
$S(f,P)$	upper Darboux (Lebesgue) sum of f for a partition (generalized partition) P	171, 268
S'	derived set of S	73
$R\displaystyle\underline{\int}_a^b f$	the lower Riemann integral of f	173
$R\displaystyle\overline{\int}_a^b f$	the upper Riemann integral of f	173
$R\displaystyle\int_a^b f$	the Riemann integral	173
O	the capital order	188
o	the small order	188
$[x]$	integer part of x	199
$\overline{z}$	the conjugate complex of z	181
$(C,1)$	Cesáro's method of first order	218
$l(G)$	the length of an open set G	234
$l(F)$	the length of a closed set F	235
$m_e(S)$	the external measure of a set S	240
$m_i(S)$	the internal measure of a set S	240
$m(S)$	the measure of S	240
$L\displaystyle\underline{\int}_D f$	the lower Lebesgue integral of f on D	269
$L\displaystyle\overline{\int}_D f$	the upper Lebesgue integral of f on D	270

Symbols	Meaning	Page
$L \int_D f$ or $\int_D f$	the Lebesgue integral	270
$L^2(D)$	the class of square Lebesgue integrable functions over D	289
$L^2[a,b]$	the class of square Lebesgue integrable functions on closed interval $[a,b]$	298
$L[-\pi,\pi]$	the class of Lebesgue integrable function on $[-\pi,\pi]$	294
$S(f)$	the Fourier series of f	295
(f,g)	inner product of f and g both belonging to $L^2[a,b]$	298
$\|f\|$	norm of f in $L^2[a,b]$.	298

Index